AF389099

Urban Ecosystem Services

Urban Ecosystem Services

Editors

Alessio Russo
Giuseppe T. Cirella

MDPI • Basel • Beijing • Wuhan • Barcelona • Belgrade • Manchester • Tokyo • Cluj • Tianjin

Editors
Alessio Russo
University of Gloucestershire
UK

Giuseppe T. Cirella
University of Gdansk
Poland

Editorial Office
MDPI
St. Alban-Anlage 66
4052 Basel, Switzerland

This is a reprint of articles from the Special Issue published online in the open access journal *Land* (ISSN 2073-445X) (available at: https://www.mdpi.com/journal/land/special_issues/Urban_Ecosystem_Services).

For citation purposes, cite each article independently as indicated on the article page online and as indicated below:

LastName, A.A.; LastName, B.B.; LastName, C.C. Article Title. *Journal Name* **Year**, *Volume Number*, Page Range.

ISBN 978-3-0365-0582-4 (Hbk)
ISBN 978-3-0365-0583-1 (PDF)

Cover image courtesy of Alessio Russo.

Contents

About the Editors

Alessio Russo is Senior Lecturer and Academic Course Leader in the Master of Landscape Architecture at the University of Gloucestershire, Cheltenham, United Kingdom. Before joining the University of Gloucestershire, he worked in Russia as an Associate Professor at RUDN University in Moscow and Professor and Head of Laboratory of Urban and Landscape Design at Far Eastern Federal University in Vladivostok. He holds a Bachelor in Science in Plant Production from the University of Naples, Post-Graduate Specialization in Healing Garden Design from the University of Milan, and Master in Science in Landscape Design and Planning from the University of Pisa. He received his Ph.D. in Urban Forestry from the University of Bologna. Outside of academia, Dr. Russo has worked as a Landscape Architect in the United Kingdom, Italy, and the United Arab Emirates, dealing with sustainable design and planning. He is a member of the International Scientific Committee on Cultural Landscapes ICOMOS-IFLA, International Federation of Landscape Architects (IFLA) Advisory Circle, and International Union for Conservation of Nature Commission on Ecosystem Management.

Giuseppe T. Cirella is Professor of Human Geography at the Faculty of Economics, University of Gdansk, Sopot, Poland. He received his Ph.D. in Environmental Engineering (specialization: Sustainability) from Griffith University, Australia. He is the founder of the Polo Centre of Sustainability and is Director and Head of Research. He has acted as a principal investigator and coordinator in a number of international projects and is a reviewer and member of the editorial board of several reputed international journals on sustainability and the environment. He has extensive interdisciplinary and cross-cultural experience in socioeconomics as well as expertise in landscape architecture, urban planning, and societal development.

Preface to "Urban Ecosystem Services"

This book is inspired by the two decades of exploratory urban research the two editors have ensued, specifically, sustainability and ecosystem services. In particular, the editors believe that ecosystem services should not be underestimated by those involved in city policy and the design and planning of urban environments. We live in an era of societal and environmental issues and challenges that have shaped the way we interact with nature. Homo sapiens have evolved jestingly into "Homo urbanus", where people live in urbanised areas and can regularly spend up to 90% of their time indoors with potential negative impacts on mental health and wellbeing. There are only nine years left to achieve the United Nations Sustainable Development Goals in which several studies point out that the benefits of incorporating ecosystem services into urban design and planning are fundamental.

To this end, this book contains 13 thoroughly refereed contributions published within the Special Issue "Urban Ecosystem Services". The contributions underscore key ecosystem service benefits that complement the nutrient cycle, soil formation, habitat provision and biodiversity—aiding in the provision, regulation and cultural services of the human-nature relationship. Best practices in urban greening interrelates the ecological, technological, societal and economic by combining stakeholder values and opinions into action. This book is intended for students, researchers and practitioners who wish to delve deeper and expand their knowledge with trending state-of-the-art research on urban ecosystem services. It is the first publication in the "Urban Ecosystem Services" series and exemplifies the developing readership and importance of the topic for contemporary urban designers, landscape architects and urban futurists.

Alessio Russo, Giuseppe T. Cirella
Editors

 *land*

Editorial

Urban Ecosystem Services: New Findings for Landscape Architects, Urban Planners, and Policymakers

Alessio Russo [1,*] **and Giuseppe T. Cirella** [2]

[1]   School of Arts, University of Gloucestershire, Cheltenham GL50 4AZ, UK
[2]   Faculty of Economics, University of Gdansk, 81-824 Sopot, Poland; gt.cirella@ug.edu.pl
*   Correspondence: arusso@glos.ac.uk; Tel.: +44-(0)1242-714557

Received: 11 January 2021; Accepted: 15 January 2021; Published: 19 January 2021

More than half of the world's population lives in urban ecosystems. The United Nations has projected that 28% of people worldwide will be concentrated in cities with at least 1 million inhabitants by 2030 [1]. Worldwide megacities are projected to rise from 33 in 2018 to 43 in 2030 [1]. Urbanisation has a profound impact on how we as human beings interact with the world around us [2]. Cities are often described as new ecosystems that did not have a natural analogue before the expansion of the urban population [3]. Cities themselves are microcosms of the kinds of modifications that are occurring, making them informative test cases for understanding the dynamics of the socioecological system and response to change [4]. The concept of urban ecosystems has been defined as "those in which people live at high densities, and where built structures and infrastructure cover much of the land surface" [5]. However, the ecological interpretation of urban systems must also include less densely populated areas due to reciprocal flows and influences between densely and sporadically populated areas [5]. Research of urban ecosystems is a relatively new topic in ecology that dates back to the mid-1970s [6,7]. This concept is discussed across a wide range of science fields, including urban ecology, landscape ecology, environmental science, ecosystem services science, and public health [8]. The concept is also increasingly addressed in sustainability science, landscape architecture, architecture, engineering, urban design, and urban planning [8].

Nature in cities plays a vital role in urbanised systems as the ecological basis for human–nature relations and the production of urban ecosystem services [9]. Several early landscape architects, particularly Fredrick Law Olmsted (1822–1903), attempted not only to improve the aesthetic of the city but also to improve health and provide the crowded urban population with areas for rest and recreation [9]. Olmsted became a leading landscape architect, park builder, and advocate of the 19th century based on an intuitive understanding of the connection between nature and human wellbeing that underpins what we know today as ecosystem services [10]. Researchers define "urban ecosystem services" as "the benefits urban residents derive from local and regional ecosystem functions" that "are co-produced by people and ecosystems" [8,11–13]. Recently, Tan et al. (2020) examined the terminological grouping of "urban ecosystem services" over the past two decades. Two different—but equally valid—interpretations were disseminated: (1) from analogues of natural and seminatural ecosystems within urban boundaries and (2) a much broader definition that incorporates the former as well as urban city-based services [14].

Landscape architects and urban planners recognise urban ecosystem service as a powerful concept that guides the development of urban landscapes towards greater sustainability, resilience, and liveability [14]. It is also evident that urban ecosystem services contribute to the quality of urban life, even though urban citizens still rely on global ecosystem services to survive [15]. The UN Agenda 2030 clearly considers the role of ecosystem services in urban settings [16]. SDG (Sustainable Development Goal) 15 targets the conservation and restoration of using terrestrial ecosystems, reducing the loss of natural habitats and biodiversity, which play an important part in our common future and heritage. Cities, in particular, need to become "more inclusive, safe, resilient and sustainable", as stated in the title of SDG 11 [17], if they are to become "greener" and supportive of current mass urbanisation. Partial solutions to this phenomenon will be the widespread use of urban green technologies via ecosystem service-based features [16].

A number of studies demonstrate linkages between urban ecosystems and public health through a range of benefits such as the mitigation of heat hazards, improvement of mental health and wellbeing through contact with nature, and stormwater management. [18,19]. In particular, green and blue infrastructures provide several ecosystem services, such as pollution removal, carbon storage and sequestration, food production, noise reduction, and recreational and cultural values [15,20,21] (Figure 1).

Figure 1. Ecosystem services provided by green and blue infrastructure: (**a**) regulation of microclimate, (**b**) noise reduction, (**c**) food production, (**d**) carbon storage and sequestration, (**e**) habitat provision, (**f**) run-off retention and water filtration, (**g**) recreational and cultural values, and (**h**) air purification (image modified from Macrovector/Freepik).

However, changes in ecological conditions resulting from human actions in urban environments ultimately impact human health and wellbeing [22]. Sustainability needs complete understanding at all levels of direct and indirect human interventions affecting ecological processes and ecosystem states [23]. According to Gómez-Baggethun and Barton [24], "urban ecosystems are still an open frontier in ecosystem service research", and the interface between economic costs and sociocultural impacts must be taken into account to "enhance resilience and quality of life in cities".

This Special Issue contains 12 peer-reviewed papers. The contributions are written by authors from several countries, including Australia, Chile, Ireland, Italy, Norway, Pakistan, Poland, South Africa, Sweden, the Netherlands, the United Kingdom, and the United States.

Brzoska and Spāǵe [25] reviewed urban ecosystem services of different types of green infrastructure. The review identified 40 different ecosystem service classes assessed in relation to different types of green infrastructure. The results show that the majority of the studies focused on assessing regulation ecosystem services classes such as filtration, sequestration, storage, and accumulation by "microorganisms, algae, plants, and animals" and "regulation of temperature and humidity, including ventilation and transpiration" in "urban green spaces" at the city dimension [25]. The results also show that the number of assessments of provisioning ecosystem services has been increasing over the last 6 years. Only a few studies have considered individual small structures, such as green roofs or single gardens; moreover, green spaces are often aggregated into one object of investigation, mostly city-wide, leading to oversimplifications [25]. Colding et al. [26] focused on the incremental change in green spaces—a fate that is largely undetectable for urban residents. The research illustrates the set of drivers resulting in the subtle loss of urban green space and elaborates on the consequences of this for the resilience planning of ecosystem services [26]. Semeraro et al. [27] analysed a case study in Lecce, Italy, applying a top-down and bottom-up approach to dispute resolution at the institutional level in the use of urban space. The research suggests that in the socioecological system, the bottom institutional level can introduce innovation or a new vision in the use of free urban space, and, as a result, bottom-up participation can stimulate or trigger the evolution of the urban ecosystem, while the top institutional level drives a change from top-down to bottom-up participation information in planning actions between decision-makers [27].

Giliani et al. [28] studied the dynamics of urban landscape ecology in the Islamabad Capital Territory in Pakistan during the period of 1976–2016 [28]. The outcomes of their study show a consistent increase in the settlement class, with the highest annual rate of 8.79% during the period of 2000–2010. Tree cover >40% and <40% canopies decreased at annual rates of 0.81% and 0.77%, respectively, between 1976 and 2016. Forest fragmentation analysis indicates that core forests of >500 acres decreased from 392 (i.e., 65.41%) to 241 km^2 (i.e., 55%) and patch forest increased from 15 (i.e., 2.46%) to 20 km^2 (i.e., 4.54%) from 1976 to 2016. SDG indicator 11.3.1 land consumption rate to the population growth rate ratio was 0.62 from 1976 to 2000, increasing to 1.36 from 2000 to 2016 [28].

Holloway and Field [29] composed richness and abundance data for 771 quadrats across three counties, finding a total of 81 species, with 48 species on the groynes and 71 species on the natural rocky shores. Their research found similar degrees across structures for algae, higher diversity and abundance for lichens and mobile animals on natural shores, and higher numbers of sessile animals on groynes. The study points out that groynes host similar ecological communities to those found on natural shores, in which differences do exist, particularly with respect to rock-pool habitats [29].

Combrinck et al. [30] investigated the property values and distance to urban green space in Potchefstroom, South Africa. Potchefstroom residents recognise the social, environmental, and economic value of green spaces; however, fewer residents recognise the economic value of green spaces. Over half of the respondents agreed that green spaces are perceived as crime hotspots and thus as contributing to unsafe neighbourhoods and indicating a related ecosystem disservice. Approximately 60% of those surveyed agreed that they would pay more for a property that is located next to a green space.

Professional planners that had been surveyed agreed that unattractive green spaces are due to a lack of maintenance by local authorities and a lack of community engagement. Half of the planners involved in the survey stated that environmental considerations are not prioritised in the planning process, even though environmental management is considered a critical component of local urban planning approaches, policy, and legislative frameworks [30]. Parker and Simpson [31] developed a theoretical framework to support human–nature connections and urban resilience via green infrastructure. In particular, they explored how urban resilience theory and human–nature connection theory can inform urban development. The urban resilience theory advocates the improvement of policy and planning frameworks, risk reduction techniques, adaptation strategies, disaster recovery mechanisms, environmentally sustainable alternatives to fossil fuel energy, the building of social capital, and integration of ecologically sustainable urban green infrastructure [31].

In Sweden, to identify key factors for fostering the incorporation of ecosystem services into municipal planning practise, Khoshkar et al. [32] examined and evaluated the views and experiences of practitioners of local spatial planning practises in municipalities in Stockholm County [32]. The practitioners stressed the need to establish legal support and ecosystem services regulation at Swedish and EU policy levels. Moreover, focus was placed on the need for local capacity building and awareness of ecosystem services as well as increased regional support for enhancing local information sharing and learning. In order to fully integrate ecosystem services in urban planning for sustainable development in a decentralised local government structure, such as in Sweden, locally adapted practical tools and monitoring procedures were considered important [32].

In Poland, Rędzińska and Piotrkowska [33] developed a procedure of building neighbourhood resilience to climate threats, embedded in the planning and design process, and focused on the use of natural adaptive potential. This procedure was applied at the strategic level to the city of Warsaw by drawing up the ranking of districts with a view to prioritising adaptation measures based on climate threats, demographic vulnerability, and the assessment of the potential of Warsaw's green infrastructure [33]. In Oregon, Elderbrock et al. [34] developed a clear method for deciding between alternatives to green infrastructure based on their quantitative and qualitative potential to provide high-priority urban ecosystem services to different stakeholders in a specific area. They used and evaluated this method in Eugene by investigating the potential for increased urban ecosystem services resulting from the conversion of public grass to alternative planting regimes that align with expressed stakeholder priorities [34]. In Norway, Venvik and Boogaard [35] tested the hydraulic infiltration capacity of a rain garden. This research leads to an understanding of the dynamics of infiltration systems, including how a rain garden interacts with the hydrological and hydrogeological aspects of the local urban water cycle [35]. Finally, the contribution from Restemeyer and Boogaard [36] explored how online citizen science platforms, demonstrated by the case of ClimateScan, can stimulate stakeholder participation and encourage nature-based solutions. This has culminated in an illustrated map of over 5000 nature-based solutions initiatives around the world and an average of more than 100 visitors a day within 6 years [36].

As cities are expected to grow rapidly in the coming decades, it is important that urban ecosystem services are understood and valued by city planners and policymakers [15]. We do hope that the results of this Special Issue can be used by landscape architects, urban planners, and policymakers to make cities sustainable, safer, resilient, and adaptable to climate change and other future risks. The survival and wellbeing of the human species in urban environments depend on how we manage to provide ecosystem services for future generations.

Author Contributions: Writing—original draft preparation, A.R.; writing—review and editing, A.R. and G.T.C.; visualization, A.R. All authors have read and agreed to the published version of the manuscript.

Funding: This research received no external funding.

Acknowledgments: We are grateful to the MDPI *Land* team of academic editors and reviewers for assisting with the Special Issue's academic excellence. Vector imaging was designed by Macrovector/Freepik.

Conflicts of Interest: The authors declare no conflict of interest.

References

1. United Nations. *The World's Cities in 2018: Data booklet*; United Nations: New York, NY, USA, 2018; p. 29.
2. Russo, A.; Cirella, G.T. Urban Sustainability: Integrating Ecology in City Design and Planning. In *Sustainable Human–Nature Relations: Environmental Scholarship, Economic Evaluation, Urban Strategies*; Cirella, G.T., Ed.; Springer Singapore: Singapore, 2020; pp. 187–204. ISBN 978-981-15-3049-4.
3. Pataki, D.E. Grand challenges in urban ecology. *Front. Ecol. Evol.* **2015**, *3*. [CrossRef]
4. Grimm, N.B.; Faeth, S.H.; Golubiewski, N.E.; Redman, C.L.; Wu, J.; Bai, X.; Briggs, J.M. Global change and the ecology of cities. *Science* **2008**, *319*, 756–760. [CrossRef] [PubMed]
5. Pickett, S.T.A.; Cadenasso, M.L.; Grove, J.M.; Nilon, C.H.; Pouyat, R.V.; Zipperer, W.C.; Costanza, R. Urban Ecological Systems: Linking Terrestrial Ecological, Physical, and Socioeconomic Components of Metropolitan Areas. *Annu. Rev. Ecol. Syst.* **2001**, *32*, 127–157. [CrossRef]
6. Pickett, S.T.A.; Grove, J.M. Urban ecosystems: What would Tansley do? *Urban Ecosyst.* **2009**, *12*, 1–8. [CrossRef]
7. Stearns, F. Urban Ecology Today. *Science* **1970**, *170*, 1006–1007. [CrossRef] [PubMed]

8. Isaac, B.T. Cities and the Environment (CATE) Managing Cities as Urban Ecosystems: Fundamentals and a Framework for Los Angeles, California Managing Cities as Urban Ecosystems: Fundamentals and a Framework for Los Angeles, California. *Cities Environ.* **2017**, *10*, 1–30.

9. McPhearson, T.; Hamstead, Z.A.; Kremer, P. Urban Ecosystem Services for Resilience Planning and Management in New York City. *Ambio* **2014**, *43*, 502–515. [CrossRef]

10. Eisenman, T.S. Frederick Law Olmsted, Green Infrastructure, and the Evolving City. *J. Plan. Hist.* **2013**, *12*, 287–311. [CrossRef]

11. McPhearson, T.; Pickett, S.T.A.; Grimm, N.B.; Niemelä, J.; Alberti, M.; Elmqvist, T.; Weber, C.; Haase, D.; Breuste, J.; Qureshi, S. Advancing Urban Ecology toward a Science of Cities. *Bioscience* **2016**, *66*, 198–212. [CrossRef]

12. Larondelle, N.; Hamstead, Z.A.; Kremer, P.; Haase, D.; McPhearson, T. Applying a novel urban structure classification to compare the relationships of urban structure and surface temperature in Berlin and New York City. *Appl. Geogr.* **2014**, *53*, 427–437. [CrossRef]

13. Andersson, E.; McPhearson, T.; Kremer, P.; Gomez-Baggethun, E.; Haase, D.; Tuvendal, M.; Wurster, D. Scale and context dependence of ecosystem service providing units. *Ecosyst. Serv.* **2015**, *12*, 157–164. [CrossRef]

14. Tan, P.Y.; Zhang, J.; Masoudi, M.; Alemu, J.B.; Edwards, P.J.; Grêt-Regamey, A.; Richards, D.R.; Saunders, J.; Song, X.P.; Wong, L.W. A conceptual framework to untangle the concept of urban ecosystem services. *Landsc. Urban Plan.* **2020**, *200*, 103837. [CrossRef] [PubMed]

15. Bolund, P.; Hunhammar, S. Ecosystem services in urban areas. *Ecol. Econ.* **1999**, *29*, 293–301. [CrossRef]

16. Filho, W.L.; Barbir, J.; Sima, M.; Kalbus, A.; Nagy, G.J.; Paletta, A.; Villamizar, A.; Martinez, R.; Azeiteiro, U.M.; Pereira, M.J.; et al. Reviewing the role of ecosystems services in the sustainability of the urban environment: A multi-country analysis. *J. Clean. Prod.* **2020**, *262*, 121338. [CrossRef]

17. United Nations. *The Sustainable Development Goals Report 2017*; United Nations: New York, NY, USA, 2017.

18. Jennings, V.; Floyd, M.F.; Shanahan, D.; Coutts, C.; Sinykin, A. Emerging issues in urban ecology: Implications for research, social justice, human health, and well-being. *Popul. Environ.* **2017**, *39*, 69–86. [CrossRef]

19. Andreucci, M.B.; Russo, A.; Olszewska-Guizzo, A. Designing Urban Green Blue Infrastructure for Mental Health and Elderly Wellbeing. *Sustainability* **2019**, *11*, 6425. [CrossRef]

20. Russo, A.; Escobedo, F.J.; Timilsina, N.; Zerbe, S. Transportation carbon dioxide emission offsets by public urban trees: A case study in Bolzano, Italy. *Urban For. Urban Green.* **2015**, *14*, 398–403. [CrossRef]

21. Russo, A.; Cirella, G.T. Edible urbanism 5.0. *Palgrave Commun.* **2019**, *5*, 163. [CrossRef]

22. Alberti, M. The Effects of Urban Patterns on Ecosystem Function. *Int. Reg. Sci. Rev.* **2005**, *28*, 168–192. [CrossRef]

23. Douglas, I. Urban ecology and urban ecosystems: Understanding the links to human health and well-being. *Curr. Opin. Environ. Sustain.* **2012**, *4*, 385–392. [CrossRef]

24. Gómez-Baggethun, E.; Barton, D.N. Classifying and valuing ecosystem services for urban planning. *Ecol. Econ.* **2013**, *86*, 235–245. [CrossRef]

25. Brzoska, P.; Spāǵe, A. From City-to Site-Dimension: Assessing the Urban Ecosystem Services of Different Types of Green Infrastructure. *Land* **2020**, *9*, 150. [CrossRef]

26. Colding, J.; Gren, Å.; Barthel, S. The Incremental Demise of Urban Green Spaces. *Land* **2020**, *9*, 162. [CrossRef]

27. Semeraro, T.; Nicola, Z.; Lara, A.; Sergi Cucinelli, F.; Aretano, R. A Bottom-Up and Top-Down Participatory Approach to Planning and Designing Local Urban Development: Evidence from an Urban University Center. *Land* **2020**, *9*, 98. [CrossRef]

28. Gilani, H.; Ahmad, S.; Qazi, W.A.; Abubakar, S.M.; Khalid, M. Monitoring of Urban Landscape Ecology Dynamics of Islamabad Capital Territory (ICT), Pakistan, Over Four Decades (1976–2016). *Land* **2020**, *9*, 123. [CrossRef]

29. Holloway, P.; Field, R. Can Rock-Rubble Groynes Support Similar Intertidal Ecological Communities to Natural Rocky Shores? *Land* **2020**, *9*, 131. [CrossRef]

30. Combrinck, Z.; Cilliers, E.J.; Lategan, L.; Cilliers, S. Revisiting the Proximity Principle with Stakeholder Input: Investigating Property Values and Distance to Urban Green Space in Potchefstroom. *Land* **2020**, *9*, 235. [CrossRef]

31. Parker, J.; Simpson, G.D. A Theoretical Framework for Bolstering Human-Nature Connections and Urban Resilience via Green Infrastructure. *Land* **2020**, *9*, 252. [CrossRef]

32. Khoshkar, S.; Hammer, M.; Borgström, S.; Balfors, B. Ways Forward for Advancing Ecosystem Services in Municipal Planning—Experiences from Stockholm County. *Land* **2020**, *9*, 296. [CrossRef]
33. Rędzińska, K.; Piotrkowska, M. Urban Planning and Design for Building Neighborhood Resilience to Climate Change. *Land* **2020**, *9*, 387. [CrossRef]
34. Elderbrock, E.; Enright, C.; Lynch, K.A.; Rempel, A.R. A Guide to Public Green Space Planning for Urban Ecosystem Services. *Land* **2020**, *9*, 391. [CrossRef]
35. Venvik, G.; Boogaard, F. Infiltration Capacity of Rain Gardens Using Full-Scale Test Method: Effect of Infiltration System on Groundwater Levels in Bergen, Norway. *Land* **2020**, *9*, 520. [CrossRef]
36. Restemeyer, B.; Boogaard, F.C. Potentials and Pitfalls of Mapping Nature-Based Solutions with the Online Citizen Science Platform ClimateScan. *Land* **2020**, *10*, 5. [CrossRef]

Publisher's Note: MDPI stays neutral with regard to jurisdictional claims in published maps and institutional affiliations.

 *land*

Review

From City- to Site-Dimension: Assessing the Urban Ecosystem Services of Different Types of Green Infrastructure

Patrycia Brzoska [1,*] **and Aiga Spāģe** [2]

[1] Leibniz Institute of Ecological Urban and Regional Development, 01217 Dresden, Germany
[2] Department of Landscape Architecture and Planning, Faculty of Environment and Civil Engineering, Latvia University of Life Sciences and Technologies, LV-3004 Jelgava, Latvia; aiga.spage@llu.lv
* Correspondence: p.brzoska@ioer.de; Tel.: +49-351-4679-258

Received: 22 April 2020; Accepted: 13 May 2020; Published: 14 May 2020

Abstract: Cities have a wide variety of green infrastructure types, such as parks and gardens. These structures can provide important ecosystem services (ES) with a major impact on human well-being. With respect to urban planning, special consideration must be given to such green infrastructure types when implementing measures to maintain and enhance the quality of life. Therefore, generating knowledge on the urban ES of differently scaled green infrastructure types is important. This systematic literature review provides an overview of existing studies which have explicitly investigated the urban ES of differently spatial-scaled green infrastructure types. By reviewing 76 publications, we confirm rising academic interest in this topic. The most frequently assessed urban ES belong to the category Regulating and Maintenance. Only a few have considered individual small structures such as green roofs or single gardens; green spaces are often aggregated into one, mostly city-wide, object of investigation, with resulting oversimplifications. Moreover, generalizing methods are mostly applied. Simultaneously, many studies have applied methods to evaluate location-specific primary data. More research is needed on small-scale structures, in particular to consider site-, and thus location-specific, parameters in order to successfully implement the ES concept into urban planning and to obtain realistic results for ES assessments.

Keywords: ecosystem services; assessment; urban ecosystem services; site; green infrastructure; cities; systematic literature review; urban planning

---

1. Introduction

Today, more than half the world's population is living in cities [1]. The continuous growth in urban populations combined with a more extreme urban climate due to global warming are having a detrimental impact on urban ecosystems [2]. In order to maintain the quality of life for the burgeoning numbers of urban residents, it is becoming increasingly important to protect and promote urban ecosystems and their services [3–6]. Concepts such as green and blue infrastructure have been developed in recent years to help tackle the environmental challenges of cities. The strategic planning of urban green structures improves the well-being of inhabitants while simultaneously boosting the resilience of cities to climatic changes [7]. Yet, such strategic planning requires comprehensive insights and information on the multiple functions and services of green infrastructure on different spatial scales. In particular, knowledge and expertise are needed on ecosystem services (ES) on the small spatial scales where planned measures are realized [8–12]. More research into urban ES on small spatial scales will improve our understanding of this planning factor, thereby aiding the integration of the urban ES concept into urban planning as an important factor for sustainable urban development. The correct application of this concept has the potential to better exploit the multiple benefits of urban ecosystems, so that urban planning can be more closely oriented to natural conditions and resources [13–15]. Furthermore, the development of standards and indicators to assess and describe ES in urban contexts can help politicians, urban planners as well as practitioners create ecological and sustainable cities [16]. While we can already point to a few practical examples of the successful integration of ES-related subjects into diverse planning documents and tools, there are still several unresolved problems limiting a more general implementation [17].

One limiting factor is the lack of information on urban ecosystems on different spatial scales and their services. The poor quality of available spatial or other relevant data on small, local scales often complicates the integration of ES into planning frameworks or decision-making processes [13–20]. A further limitation is the lack of suitable methods to assess ES at such spatial scales; hitherto, most assessment methods have referred to the global, national or regional scale. Clearly, if we wish to promote the inclusion of ES in decision making at the urban level, it is necessary to improve our knowledge of this subject at city-wide but most importantly also on local, and thus site-, scales [9,19]. Over the last few decades, urban ES has become a widely investigated topic in different research fields, with scholars recognizing its importance in mitigating climatic extremes and contributing in diverse ways to human well-being [19,21]. In this study, urban ecosystems are defined as areas largely dominated by the built environment and which comprise gray and green infrastructures [4,22,23]. Of course, urban ecosystems only provide a fraction of the ES used by city dwellers – the larger part of these services are provided by widely distributed ecosystems in the city surroundings. Yet in relation to the size of urban ecosystems, they benefit a large number of citizens [24]. Thus, urban ES have a high anthropogenic impact, representing an explicit type of ES that needs to be considered more closely.

Locally provided ES generally play an important role in promoting the quality of life of urban residents. Yet, the issue of the ES of urban small-scale structures is an underrepresented research field [14]. The small spaces within cities are designed by urban planners in great detail, and it is exactly this spatial scale and structure that is directly perceived by residents and thus strongly influences the quality of life in the city. Previous studies on the assessment of urban ES have stressed the importance of the spatial scale of investigations [14,25,26]. Hitherto, many assessments have been conducted on larger spatial scales (city, region, nation-wide) with results often presented in a generalized way. To obtain more realistic results, it is necessary to conduct empirical ES studies of smaller urban structures. In order to ensure the practical implementation of the ES concept, we have to focus on spatial structures and scales that are recognized by existing planning tools, e.g., neighborhoods, small single parks, etc. [26]. Furthermore, previous reviews have revealed a large number of different methods used to assess urban ES [14,26,27]. Most of these involve spatial proxy methods, for example, utilizing land use and land cover data to estimate ES supply capacities. Primary data is rarely collected in urban ES assessments [27]. Another approach to the assessment of ES is to consider the complexity of urban structures [14,26]. In this case, it is important to take account not only of built structures but also urban open spaces, for example, the various types of green open space [14,26]. In particular, Haase et al. [14] found that most previous studies assessed regulative ES in cities, with only a few looking at cultural and provisioning ES.

These aspects determine the scope of this review and shape the key questions. We aim to review the current state of knowledge on methods to assess the urban ES of different types of green infrastructures from city to site scales. To this end, we have only considered studies that examine individual spatial structures or forms of land use in cities such as parks, gardens or trees. The review will answer the following three questions: (1) Which urban ES are assessed in relation to green infrastructure types? (2) Which specific spatial structures are the subjects of investigation? (3) Which methods are used to assess ES on larger (city) and smaller (site) spatial scales? Furthermore, we will look at the motivations of studies in assessing the urban ES of different types of green infrastructure types, as well as check which data type (i.e., primary or secondary) has been used by the reviewed studies.

2. Materials and Methods

2.1. Review Approach

The first step was to carry out a systematic quantitative literature review after Pickering and Byrne [28]. In comparison to classical meta-analytical reviews, the methodology after Pickering and Byrne [28] aims to determine general aspects of studies (e.g., numbers, types, and geographical aspects), research trends and gaps as well as methodological patterns. To this end, the literature databases "Web of Science" and "Scopus" were searched for relevant peer-reviewed articles published in international scientific journals. This search was conducted from March to April 2019. Several filter criteria were applied to specify the review but still to identify as many relevant articles as possible:

- The article should be written in English and published sometime between 2000 (chosen to reflect Bolund and Hunhammer's article of 1999 [3], which was the first to write about "urban ecosystem services") and April 2019;
- The article should explicitly address "urban ecosystem services";
- The article should deal directly with the assessment or valuation of urban ES (the studies should not merely map urban ES);
- The study should examine larger and smaller scale green infrastructure types or land use types in cities such as parks, gardens or single trees.

Systematic searches were conducted of the database "Web of Science" for each possible urban green infrastructure type (see Table 1), adding the search terms "urban ecosystem services" and "assessment" or "valuation" (for example, "urban ecosystem services" AND assessment AND park). After completing individual searches for various spatial scales (see examples in the "Scopus" search terms), all results were cross-checked to exclude repeated articles. The search procedure of the database "Web of Science" identified a total of 35 studies.

Table 1. Range and definitions of investigated spatial objects.

Structural type	Studies that investigate a spatial pattern consisting of different types of land use typical of a city. In contrast to Neighborhood, this is not an administratively confined spatial unit, but in general reflects typical urban structures.
Neighborhood	Studies that focus on a part of a city as a community within the urban context, which is administratively defined and localized.
Urban green spaces	Studies that examine in aggregation many different green spaces (defined as any vegetated areas found in the urban landscape, e.g., parks, urban forests, lawns, home gardens, street trees).
Forest	Studies that investigate urban forests as ecosystems.
Water bodies and wetland	Studies that investigate wetlands and flowing or still waters.
Park	Studies that focus on (usually larger) green areas designed for recreation and landscaping.
Allotment/community gardens	Studies that investigate allotment or community gardens as plots of land for individual and own use to grow food.
Brownfield	Studies that examine areas within the residential area which were formerly used for different purposes and which are temporarily or permanently no longer used as they were originally used.
Trees	Studies that examine single or several trees in a city.
Green roof/wall	Studies that focus on different forms of building greenery.

In the database "Scopus", a general search was conducted for all possible green infrastructure types using the following keyword combination:

TS = (neighborhood* OR district* OR estate* OR meadow* OR brownfield* OR allotment* OR "community garden*" OR park* OR woodland* OR "green space*" OR "green infrastructure" OR residential OR cemeter* OR wetland* OR "urban tree*" OR "urban forest*" OR lake* OR waterbod* OR river* OR stream*) AND ("urban ecosystem service*" AND assessment OR valuation).

This search identified a total of 31 papers. After cross-checking and combining the results from both databases, two articles were excluded, resulting in a selection of 29 articles. Further articles could be added to this list by screening the bibliographies. This procedure led to a final total of 63 scientific articles.

The second step of the review approach was to search the relevant gray literature, such as reports and documents that were compiled by organizations and institutions that do not belong to the "traditional" academic instances (e.g., government departments, non-governmental organizations or civil society). For this purpose, an internet search was conducted to identify some initial potential international funding bodies of projects on urban ecosystems across Europe. Their webpages were then screened for relevant projects, after which the project webpages were studied. This search procedure led to a snowball effect, resulting in the identification of 13 relevant project documents, of which six were additional scientific articles drawn from the bibliographies of the project documents.

From these two review steps, we were able to identify a total of 76 articles. An overview of the assessed studies is presented in Supplementary Materials (S1). Each publication was analyzed and added in an Excel databank, where specific information was extracted and combined in one table.

2.2. Analysis Approach of Included Articles

The authors of the various articles made use of a range of different terms and expressions for ES. For the purposes of our study, it was first necessary to consolidate these terms to allow us to summarize and compare the investigated ES (cf. key question 1). For this reason, all identified ES from the articles were classified into the corresponding ES sections, groups and classes of the latest version of the Common International Classification of Ecosystem Services (CICES V 5.1).

To provide a comprehensive overview of the investigated differently scaled green infrastructure types (cf. key question 2), the spatial objects in the papers were first assigned to one of two dimensions, i.e,. city- or site-dimension (see Figure 1). This classification was intended to reflect the scope of the investigated objects in each study. Thus, whenever an assessment method was applied to several spatially distributed (yet urban) objects, the dimension of this study was classified as "city" (e.g., ES assessment of various urban parks in a city). Alternatively, if only one object was assessed, e.g., a city park, the dimension was defined as "site". In this case, only situation and location-specific conditions can be said to apply.

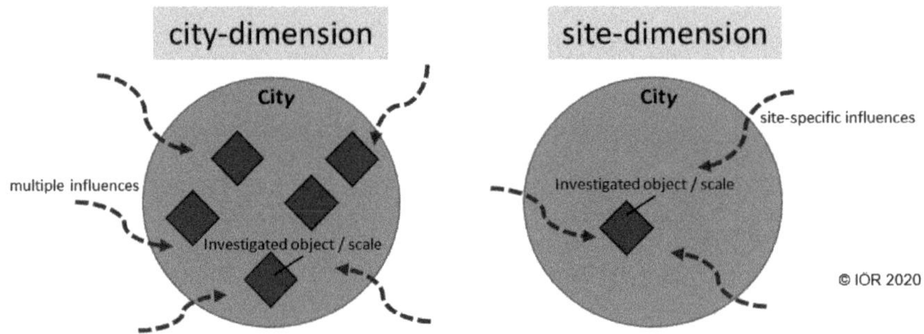

Figure 1. Schematic illustration of the assignment of studies to city- or site-dimension.

Following this initial classification into city- or site-dimension, the urban green infrastructure types in each dimension were summarized and classified into more precise types, e.g., park, garden, forest, etc. An overview of this classificatory system is given in Table 1.

Furthermore, we identified seven different categories for the method classification of all reviewed articles (cf. key question 3). These categories are as follows: "spatial proxy methods", "samplings/field mapping and observations", "surveys and questionnaires", "economic valuation methods", "model-based methods", "social media-based methods" and "remote sensing and earth observations" (see Supplementary Materials (S2) for a detailed description of the categories).

On the basis of the described classifications, the extracted information from the 76 reviewed articles was then evaluated to answer the key questions.

3. Results

3.1. General Overview of Articles

In Figure 2, we can see, firstly, that most of the articles focus on European cities, and, secondly, that research on the assessment of the urban ES of differently spatial-scaled green infrastructure types is growing. Although the databases were screened for published articles from the year 2000, published articles corresponding to the described search criteria were only identified from 2008.

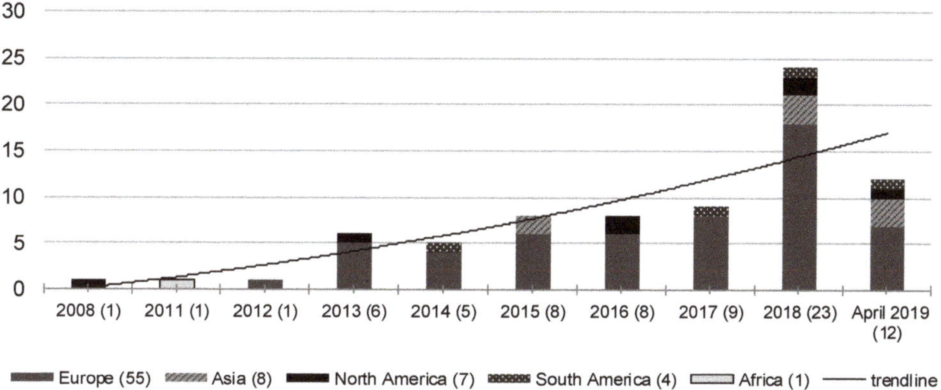

Figure 2. Numbers and geographical distribution of case studies in the articles published between 2008 and April 2019. The figures in brackets beside each year are the total number of articles published in that year; the total numbers of articles from each continent are also indicated. A polynomial trend line illustrates a predicted increase of the number of articles. (Note: One study is excluded due to its analysis of a fictional urban model).

There is a clear rise in the number of studies assessing the urban ES of green infrastructure types on different spatial scales, especially over the last two years. The first paper to offer a smaller-scale assessment of explicitly urban ES was published in 2008. While the highest number of relevant publications was identified in the year 2018, by April 2019 more than 50% of the total number of articles from 2018 had already been published. The polynomial trend line also draws a predicted increase in the number of articles in this field (see Figure 2). Most of the reviewed publications focus on green infrastructures within European cities (55); a much smaller number deal with cities in Asia (8) or North (7) and South America (4). Only one publication dealt with the assessment of urban ES in an African city, specifically Cape Town (found in the gray literature document TEEB (The Economics of Ecosystems and Biodiversity) [29], referring to [30,31]). The scientific articles were sourced from 32 different journals in diverse fields, such as ecosystem services, environmental science, ecology and urban planning.

3.2. Assessed Es of Urban Green Infrastructure Types

In total, we identified 40 different ES classes assessed in regard to different green infrastructure types (Table 2, for detailed ES class overview see Supplementary Materials (S3)). The ES section Regulation & Maintenance according to CICES V5.1 specifies 16 different ES classes. The ES section Provisioning examines 12 different ES classes. A total of nine different ES classes were studied within the Cultural ES section.

Table 2. Overview of assessed ecosystem services (ES) sections and classes (numbers in brackets) according to CICES V5.1 in regard to urban green infrastructures.

ES Section	Number
Provisioning (12)	37
Regulation & Maintenance (16)	177
Cultural (9)	115

With a closer look at the analyzed ES classes (see Supplementary Materials (S3)) we can see the varying frequency of the examined individual ES classes. The most frequently assessed ES classes were "Filtration/sequestration/storage/accumulation by microorganisms, algae, plants, and animals" (2.1.1.2) and "Regulation of temperature and humidity, including ventilation and transpiration" (2.2.6.2). In general, most studies focused on an assessment of ES classes within the ES section Regulation & Maintenance (a total of 177 instances, see Table 2). The ES classes within the Cultural ES section were the second most frequently examined. In total, ES classes were assessed 115 times in this section (see Table 2). Specifically, the most frequently assessed classes were "Characteristics of living systems that enable activities promoting health, recuperation or enjoyment through active or immersive interactions" (3.1.1.1) and "Characteristics of living systems that enable activities promoting health, recuperation or enjoyment through passive or observational interactions" (3.1.1.2). ES were assessed 37 times in the Provisioning ES section (see Table 2). This represents the most rarely assessed ES for urban green infrastructure types.

3.3. Investigated Green Infrastructure Types and Dimensions

Figure 3 shows the number of investigated green infrastructure types at city- and site-dimensions in the reviewed publications. In general, the articles investigated a wide range of green infrastructure types. It can be seen that the majority of the studies analyzed ES at city-dimension (Figure 3, left side), especially in regard to "urban green spaces" and "structural types" with a much smaller number of studies analyzing single structures in cities (Figure 3, right side). The two dimensions show a similar number of assessments of "parks".

Regarding the assessment of ES for green infrastructure types (Table 3), our results show that most of the studies concentrated on assessing the Regulation ES classes "Filtration/sequestration/ storage/accumulation by microorganisms, algae, plants, and animals" (2.1.1.2) and "Regulation of temperature and humidity, including ventilation and transpiration" (2.2.6.2) in "urban green spaces" at city-dimension. This spatial structure type was also the focus of a large number of assessments of the cultural ES classes "Characteristics of living systems that enable activities promoting health, recuperation or enjoyment through active or immersive interactions" (3.1.1.1) and "Characteristics of living systems that enable activities promoting health, recuperation or enjoyment through passive or observational interactions" (3.1.1.2). Within the site-dimension, the ES class "Filtration/sequestration/storage/accumulation by microorganisms, algae, plants, and animals" (2.1.1.2) was most frequently assessed in "neighborhoods" and "parks".

Table 3. Assessed ES classes according to CICES V5.1 in relation to the considered green infrastructure types. The figures in brackets indicate how often an ES was assessed for the corresponding spatial structure. Since multiple ES were analyzed in some articles, the sum in brackets is higher and does not represent the total number of articles reviewed. (Note: The specification of the considered ES class is missing in some articles; these were assigned to the ES section "in general, specifications missing" in each ES section.)

| | | Assessed Green Infrastructure Types | | | | | | | | | | | | | | |
| | | City-Dimension | | | | | | | | | | Site-Dimension | | | | |
ES Section	ES Class	Structural Types	Neighborhood	Urban Green Spaces	Forest	Water Bodies and Wetland	Park	Allotment/Community Garden	Brownfield	Trees	Green Roof/Wall	Neighborhood	Urban Green Spaces	Water Bodies and Wetland	Park	Trees
Provisioning	Cultivated terrestrial plants (incl. fungi, algae) grown for nutritional purposes	(3)	(1)	(5)			(1)	(3)		(1)						
	Fibres and other materials from cultivated plants, fungi, algae and bacteria for direct use or processing (excluding genetic materials)			(4)												
	Cultivated plants (including fungi, algae) grown as a source of energy			(2)												
	Animals reared for nutritional purposes			(1)												
	Fibres and other materials from reared animals for direct use or processing (excluding genetic materials)			(2)												
	Wild plants (terrestrial and aquatic, including fungi, algae) used for nutrition			(1)								(1)				
	Fibres and other materials from wild plants for direct use or processing (excluding genetic materials)			(1)				(1)								
	Wild animals (terrestrial and aquatic) used for nutritional purposes			(1)												
	Fibres and other materials from wild animals for direct use or processing (excluding genetic materials)			(1)												
	Surface water for drinking			(1)								(1)				
	Surface water used as a material (non-drinking purposes)			(1)												
	Ground (and subsurface) water for drinking	(1)		(1)								(1)				
	Provisioning services (in general, specifications missing)	(1)													(1)	

Table 3. *Cont.*

ES Section	ES Class	Assessed Green Infrastructure Types														
		City-Dimension										Site-Dimension				
		Structural Types	Neighborhood	Urban Green Spaces	Forest	Water Bodies and Wetland	Park	Allotment/Community Garden	Brownfield	Trees	Green Roof/Wall	Neighborhood	Urban Green Spaces	Water Bodies and Wetland	Park	Trees
Regulation & Maintenance	Filtration/sequestration/storage/accumulation by micro-organisms, algae, plants, and animals	(6)	(2)	(14)	(1)	(1)	(2)	(1)		(8)	(1)	(3)	(2)		(3)	(1)
	Noise attenuation	(3)	(1)	(2)			(1)			(1)		(1)				
	Control of erosion rates	(1)		(2)											(1)	
	Hydrological cycle and water flow regulation (Including flood control, and coastal protection)	(4)		(7)			(3)	(1)		(3)	(1)	(2)	(1)		(1)	
	Wind protection			(1)												
	Fire protection			(1)												
	Pollination (or 'gamete' dispersal in a marine context)	(4)		(3)			(1)	(2)								
	Seed dispersal			(2)											(1)	
	Maintaining nursery populations and habitats (Including gene pool protection)	(4)		(6)	(1)	(1)	(1)	(4)			(1)	(1)	(1)		(1)	
	Pest control (including invasive species)			(1)												
	Disease control			(1)												
	Weathering processes and their effect on soil quality			(1)												
	Decomposition and fixing processes and their effect on soil quality	(1)		(3)				(3)								
	Regulation of the chemical condition of freshwaters by living processes		(1)	(1)								(1)			(1)	
	Regulation of chemical composition of atmosphere and oceans			(4)				(1)								
	Regulation of temperature and humidity, including ventilation and transpiration	(7)	(2)	(11)	(1)	(1)	(2)	(2)	(1)	(5)		(2)	(1)	(1)	(2)	(1)
	Regulation & maintenance services (in general, specifications missing)	(1)													(1)	

Table 3. *Cont.*

ES Section	ES Class	Assessed Green Infrastructure Types														
		City-Dimension										Site-Dimension				
		Structural Types	Neighborhood	Urban Green Spaces	Forest	Water Bodies and Wetland	Park	Allotment/Community Garden	Brownfield	Trees	Green Roof/Wall	Neighborhood	Urban Green Spaces	Water Bodies and Wetland	Park	Trees
Cultural	Characteristics of living systems that that enable activities promoting health, recuperation or enjoyment through active or immersive interactions	(5)	(2)	(10)			(3)	(2)	(1)	(1)		(1)	(2)		(2)	(1)
	Characteristics of living systems that enable activities promoting health, recuperation or enjoyment through passive or observational interactions	(3)	(2)	(10)			(3)	(2)	(1)	(1)		(1)	(1)		(2)	(1)
	Characteristics of living systems that enable scientific investigation or the creation of traditional ecological knowledge			(3)			(1)	(1)				(1)	(1)			
	Characteristics of living systems that enable education and training			(2)			(1)			(1)		(1)				
	Characteristics of living systems that are resonant in terms of culture or heritage			(4)			(2)	(1)								
	Characteristics of living systems that enable aesthetic experiences	(3)		(7)			(2)	(1)				(1)				
	Elements of living systems that have sacred or religious meaning	(1)		(3)				(1)								
	Elements of living systems used for entertainment or representation	(1)		(3)			(1)	(1)								
	Other (Communication and interaction with other people)			(2)			(1)	(1)								
	Cultural services (in general, specifications missing)	(1)	(1)	(1)		(2)	(2)		(1)					(1)	(1)	

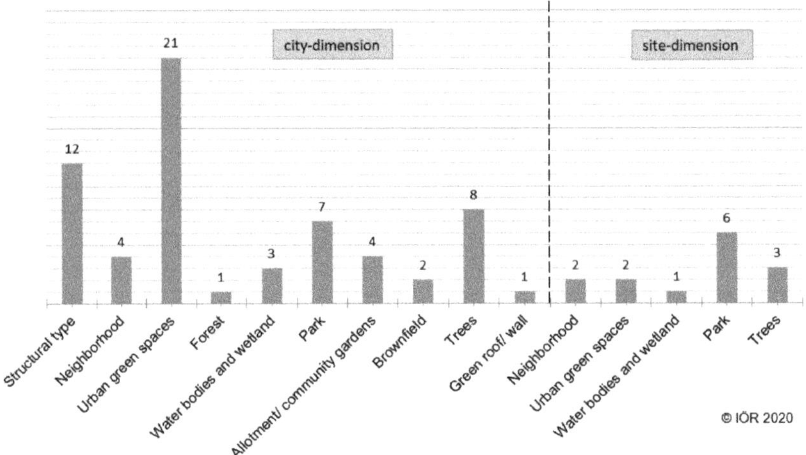

Figure 3. Urban green infrastructure types investigated by the articles, subdivided into city- (**left**) and site-dimension (**right**).

3.4. Methods Used to Assess the Urban Es of Green Infrastructure Types

The majority of the reviewed publications applied "spatial proxy methods" followed by "surveys and questionnaires" (see Figure 4). "Social media-based methods" were the least commonly used method for ES assessment for different green infrastructure types in cities. Our analysis showed that over a quarter of the studies used more than one method to assess urban ES. "Surveys and questionnaires" and "model-based methods" were most frequently combined, followed by "spatial proxy methods" and "model-based methods".

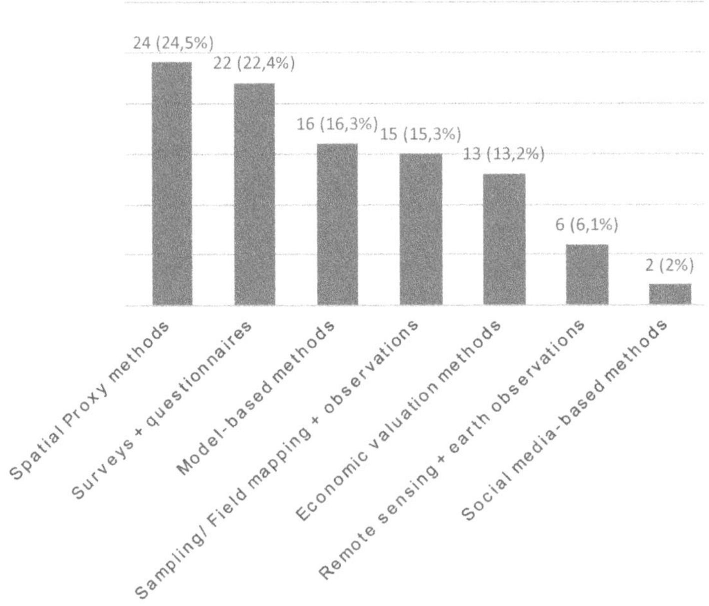

Figure 4. Numbers (and % of total share) of methods used in the reviewed articles.

Table 4 categorizes the different methods used to assess urban green infrastructure types. At city-dimension, the reviewed studies most frequently applied "spatial proxy methods" and "surveys and questionnaires". The ES of "urban green spaces" were the most studied in this dimension by applying methods that belonged to the categories "spatial proxy methods", "surveys and questionnaires" and "model-based methods". "Structural types" and "trees" were also the focus of many assessments of urban ES. Here, the most common methods were "spatial proxy methods" and "model-based methods". At site-dimension, the most frequently used method was "surveys and questionnaires", largely to assess the urban ES of "parks".

Table 4. Numbers of methods identified in the reviewed articles to assess green infrastructure types. The figures in brackets show how often a method was applied to the corresponding structure type. Since multiple methods were used in some articles, the sum of methods is higher than the actual number of articles.

Method Categories	Assessed Small-Scale Structures														
	City-Dimension										Site-Dimension				
	Structural Types	Neighborhood	Urban Green Spaces	Forest	Water Bodies and Wetland	Park	Allotment/Community Garden	Brownfield	Trees	Green Roof/Wall	Neighborhood	Urban Green Spaces	Water Bodies and Wetland	Park	Trees
Spatial Proxy methods	(5)		(7)	(1)		(3)	(1)		(2)	(1)		(2)	(1)		(1)
Sampling/Field mapping + observations	(2)		(3)				(2)		(3)		(1)	(2)		(2)	(1)
Surveys and questionnaires		(2)	(7)		(2)	(3)	(3)	(1)			(1)			(3)	(1)
Economic valuation methods	(2)		(4)	(1)		(2)			(1)		(1)			(2)	
Model-based methods	(2)	(1)	(7)					(1)	(5)		(2)			(1)	
Social media-based methods			(1)				(1)								
Remote sensing and earth observations	(1)	(1)	(3)					(1)							

4. Discussion

The results show a rise in research interest in this area. In particular, the growing numbers of articles, especially since the year 2018, as well as the trend line confirm the increasing significance of this topic (see Figure 2). Already in the first four months of 2019, more than 50% of the previous year's total number of publications on the investigated topic had been published. The review still confirms the finding of Haase et al. [14] that most studies are carried out in European cities. Due to the higher numbers of publications in the last two years, it can be assumed that the topic is still evolving as a research field and that more studies will be published in the future.

Referring to key question (1), we found that Provisioning ES are the least assessed ES section for urban green infrastructure types (see Table 1). This also tallies with the finding of Haase et al. [14]. Unlike the ES sections Regulation & Maintenance and Cultural, Provisioning ES can generally be physically transported (e.g., deliveries of foodstuffs to stores). For this reason, cities mainly import provisioning ES from the surrounding countryside and even further afield. The demand for provisioning ES, however, is very high in densely populated areas. Our review indicates that the number of assessments of provisioning ES has been increasing over the last six years. This trend can perhaps be attributed to new initiatives fostering urban agriculture such as urban gardening, Edible Cities, etc. In today's society, we can identify changes that are serving to highlight the issue of Edible Cities or the role of urban gardening [32]. In the foreseeable future, provisioning ES could become an important element of the urban environment. Moreover, with regard to promoting the sustainable development of cities and the reduction in urban ecological footprints, Gómez Baggethun and Barton [13] have pointed out that ES should not be imported from distant locations. Concurring with the review by Haase et al. in 2014 [14], Regulation & Maintenance were the most frequent forms of ES investigated in the publications (see Table 1). This ES section contains a number of ES classes that play a special role in securing the well-being of inhabitants, although they are generally only perceived when lacking. Such ES are mostly used indirectly and are in great demand [33]. One reason for the high frequency of these ES assessments may be the current lively debates on climate change and its impact on cities. Cities are facing increasing environmental pressure as well as concerns about human health and the well-being of local residents [34]. Regulation and Maintenance ES can have a considerable impact on human well-being by regulating the microclimate, air pollution or water flows [35]. There is an increasing demand on cities to plan and develop important green structures to counteract the negative impacts of heat islands, heavy rainfall, air pollution, etc. As we also found in our review, more and more research has been conducted over the last few years into these ES classes in urban areas [14]. Our results show that the Cultural ES section is the second most investigated in the reviewed publications (see Table 1). This fact is unsurprising if we remember that the Cultural section contains ES classes, which are commonly referred to as "recreation services". Other studies have shown that ES classes related to recreational aspects (such as 3.1.1.1 and 3.1.1.2) are the most frequently investigated cultural ES within urban areas and especially within urban green spaces or parks, where this ES is mostly directly used [14,36]. Other publications have considered some less obvious small green infrastructure types; Mathey et al. [37] and Pueffel et al. [38], for instance, have conducted surveys on the recreational ES provided by urban brownfields (Table 2). Contrary to the findings of Haase et al. [14], our results show that many studies assess multiple ES. Those studies that evaluate only one class of ES, such as Takács et al. [39], Marando et al. [40] or Lehmann et al. [41], generally deal with Regulating & Maintenance ES. A closer look at these studies shows that they are largely based on primary data drawn from on-site inspections and include biotope mapping or climatic field measurements.

In regard to key question (2), our findings show that previous ES assessments have looked at a wide variety of urban green infrastructure types at larger and smaller spatial scales. The majority of examined urban structures considered the city-dimension (Figure 3). Taking a closer look at the studies dealing with site-dimension, we found that many applied time- and cost-intensive methods, in particular self-collected (primary) datasets (e.g., [42–44]). Consequently, the small number of studies at the site-dimension can be attributed to the requirement for more precise and site-dependent data, usually primary data collected in the field. The high number of publications that assessed ES in "urban green spaces" and "structural types" (Figure 3) can be explained by the aggregation of different spatial structures, in particular treating diverse green spaces as one undifferentiated study object (e.g., [42,45,46]). On the other hand, this frequency can also be attributed to the data used in the individual articles. Many of the articles took land-use data as a basis for their assessments, resulting in an investigation of different land-use types that usually exist in aggregated classes (e.g., green spaces, streets, buildings, etc.). In addition to this, the frequently examined green infrastructure type "parks" (Figure 3) is one of those structures mostly associated with the urban context; we can thus expect a large number of investigations in such spaces. Many diverse analyses can be carried out in parks, which can also be investigated in regard to direct usage by local residents, especially within the section of Cultural ES. In contrast to parks, for example, which are present in almost all cities, the smaller number of ES assessments of green infrastructure types, such as "forest" or "green roof/wall" (Figure 3), can be attributed to their rarity in many cities. Consequently, ES assessments of such green infrastructure types are less frequent.

Based on this review, we identified four main motivations for assessing the ES of urban green infrastructure types on different scales. Most studies based their investigations on the need to successfully implement ES in urban planning and for the development of appropriate measures to preserve and protect these ES (e.g., [46–48]). Others, such as Czembrowski et al. [49], justified their investigations by pointing out the services and value of urban green areas as well as the need to facilitate communication between environmental experts and decision makers. Another stated intention in assessing ES is to ensure the inclusion of user demands in the planning and design of cities and green spaces [33]. Studies such as Mathey et al. [37] aimed to increase the acceptance of less popular green structures such as brownfields by the public as well as by urban planners. This is achieved by highlighting the values of associated ES and striving to prevent the automatic redevelopment of urban brownfields, instead maintaining these as areas of green space.

In a large number of publications, green spaces were generally considered in an aggregated form and thus examined in a larger scale study, i.e., defined as "urban green structures" according to our classification (Figure 2). For such green spaces, the studies particularly investigated the Regulative ES. On the other hand, studies examining single and smaller sites (site-dimension) most often focused on spatial structures that have a stronger impact at small-scale levels, such as the usually high perception and significance of parks within a district.

While previous studies have particularly highlighted economic valuation methods to assess ecosystem services (e.g., [50,51]), we only identified a small number of articles applying "economic valuation methods" (Figure 4). Regarding key question (3), the most frequently used approach is the "spatial proxy method". This method is particularly applied to the assessment of ES in regard to "structural types", "urban green spaces" as well as "trees". As in the previous section, this observation can be explained by the adopted base data. In most cases, the land-use types of a study area provide the basis for the application of different proxies, as performed in the studies of Speak et al. [52] and Kremer et al. [53], for example. Such studies used proxy variables such as land cover maps and other geoinformation system (GIS) datasets that depict special ecosystem processes as a basis for ES assessments. One reason for this high number of studies applying spatial proxy methods can be limited time and financial resources available to researchers as well as a lack of primary data. However, spatial proxy methods have been criticized for the overgeneralization of obtained results that represent only a fraction of reality [54]. On the other hand, they allow for the investigation of ES in study regions that lack primary data while offering the advantage of good comparability of results and generally low costs [54]. Only a few of the reviewed studies applied "social media-based methods" (Figure 4). The inclusion of "modern" (in particular social media-based) datasets such as photographs in Richards and Tunçer [55] is still an under-researched and rather new field in regard to ES assessments. In the near future, this method could potentially become more popular due to the continuous expansion of social media and uptake of mobile technology by the public as well as for research purposes (e.g., citizen science). However, there still exist several limitations in the use of such social-media-based datasets and methods. For example, it can be argued that the real world is not being accurately represented as data generation is currently mainly carried out by young people [55]. The finding that many studies use a mix of methods (Figure 4) agrees with previous work by Seppelt et al. [27]. "Model-based methods" such as "I-tree Eco", a free peer-reviewed software to assess the ES of trees provided by the USDA Forest Service, were applied by some studies, especially for the assessment of Regulating and Maintenance ES. In these cases, the particular issue of interest was indeed "trees", such as in the studies of Parsa et al. [56] and Baró et al. [57]. These models use primary data (in these cases, tree inventories) and apply benefit transfers under location-dependent variables such as climatic parameters. This can also limit the accuracy of results. Therefore, while such investigations can be compared in their underlying methodology, at the same time, they are based on generalizations, especially with regard to ecological parameters, rendering them difficult to transfer to other geographic patterns. Although the popular "I-tree Eco" model also provides results expressed in monetary values and thus integrates economic valuation methods, it is surprising, that most of the reviewed studies using this model are limited to the presentation of biophysical results [35,52,58–60]. This results in a lower number for "economic valuation methods" in Figure 4. Only a few studies also discuss the economic results provided by "I-tree Eco" [56,57,61].

Regarding our research design, it should be noted that the criteria used in this study placed restrictions on the literature review and thus led to a smaller sample size. In an initial search, in which we searched for the term "ecosystem services" instead of "urban ecosystem services", we received a high amount of results. After reviewing some of the results, we found that a large part did not match our search scope. Therefore, we decided to narrow the search terms to explicitly "urban ecosystem services" in order to reduce the large number of non-relevant articles identified after the initial scan of the databases. We are aware that our narrow selection meant that some potentially relevant studies, which assessed ES in cities but did not explicitly use the term "urban ecosystem services" (in article titles, keywords or abstracts), could not be included. Nevertheless, all studies that refer to "urban ecosystem services" are included in this review. A further restriction results from only searching for the keyword "green infrastructure" (apart from the different green and blue elements) and with this not explicitly integrating other planning concept terms that are sometimes synonymously used especially in the context of regulative ES, e.g., Sustainable Drainage Systems (SuDS), Low Impact Development (LID), or the term Nature-Based Solutions (NBS). An investigation of the underrepresented blue elements and the inclusion of the mentioned concepts represent an interesting field for further research. In this context, it can also be expected that the results are increasingly reflecting studies from European cities, as the term "green infrastructure" is commonly used within the EU member states.

5. Conclusions

While the importance of ecosystem services is today widely acknowledged, our results also show that there is still a gap in assessment methods on urban local scales [14,19,27]. The aim of our review was to raise awareness of this issue and provide a framework for further research. Previous reviews on the assessment of urban ES have mainly focused on larger spatial scales within cities. Our findings follow on from here, giving an update as well as checking whether any additional studies have examined even smaller spatial structures in the context of urban ES assessments (such as "green roofs", small garden patches, individual trees, etc.). Our results show that a research gap still exists at this point.

The significance of urban ES is readily acknowledged by scientists, and first steps have already been undertaken to adapt this concept for policymaking [17,62]. The high numbers of studies in the last years indicate that this topic is still evolving, especially in Europe. Several published reports have acknowledged that Europe, North America and northeast Asia are the main centers of research into the interrelationship between human well-being and green spaces [34,63,64].

In recent years, multiple ES classes have been investigated in cities as well as in their local spatial structures. Since the review by Haase et al. [14], more studies have been published on ES on the level of small urban structures; yet, the majority of work still focuses on larger spatial structures, mostly applying generalizing methods that provide results with a poor fit to reality [54]. Our results have shown a distinction between the numbers of ES assessments at city- and site-dimension. For a more rigorous implementation of the ES concept in urban planning as well as to develop adapted measures and design clearly additional research is needed, especially in evaluation methods on local urban scales. Improving our understanding of the ecosystem services provided by site-scaled green infrastructure types will constitute an important step towards setting policy objectives and creating suitable measures for sustainable urban development [9,19].

Approaches are needed that require data gathering [19], as these lead to credible and more realistic assessments of urban ES [54]. Furthermore, Beichler et al. [25] have stressed the importance of spatial scale in ES assessments, arguing that the exclusion of settlements and built-up areas from investigations can cause us to overlook ES provided by small ecosystems within such structures. In the urban context, results from spatially comprehensive (natural and built-up areas) approaches ultimately form more convincing arguments for the ecological and sustainable design of future cities with their small green structures and, in this way, could usefully influence the decision-making processes.

Supplementary Materials: The following are available online at http://www.mdpi.com/2073-445X/9/5/150/s1, **S1.** Overview of the studies assessed in this review sorted according to reading order; **S2.** Short descriptions of method categories; **S3.** Overview of assessed ES sections and classes according to CICES V5.1 in regard to urban green infrastructures.

Author Contributions: P.B. designed the study and conceptualized the methodology. Both P.B. and A.S. conducted the literature search and reviewed the articles. P.B. performed the analysis, visualized the data and wrote the manuscript in consultation with A.S. All authors have read and agreed to the published version of the manuscript.

Funding: This research was mainly funded by the Federal Ministry of Education and Research (BMBF) in the joint project "HeatResilientCity" (subproject grant number: 01LR1724A). The promoter of this project is the DLR project management agency (DLR-PT).

Acknowledgments: The authors would like to thank Derek Henderson for language editing and proofreading. We also thank the three anonymous reviewers for their very constructive comments and valuable hints.

Conflicts of Interest: The authors declare no conflict of interest.

References

1. United Nations. *World Urbanization Prospects: The 2018 Revision (ST/ESA/SER.A/420)*; United Nations, Department of Economic and Social Affairs, Population Division: New York, NY, USA, 2019.

2. IPCC SRCCL. *Climate Change and Land. An IPCC Special Report on Climate Change, Desertification, Land Degradation, Sustainable Land Management, Food Security, and Greenhouse Gas Fluxes in Terrestrial Ecosystems.* Summary for Policymakers: 2019; Approved Draft, Genf, Switzerland. Available online: https://www.ipcc.ch/srccl/ (accessed on 13 May 2020).

3. Bolund, P.; Hunhammer, S. Ecosystem Services in Urban Areas. *Ecol. Econ.* **1999**, *29*, 293–301. [CrossRef]

4. Gómez-Baggethun, E.; Barton, D.N. Classifying and Valuing Ecosystem Services for Urban Planning. *Ecol. Econ.* **2013**, *86*, 235–245. [CrossRef]

5. Haase, D.; Fantzeskaki, N.; Elmqvist, T. Ecosystem Services in Urban Landscapes: Practical Applications and Governance Implications. *AMBIO* **2014**, *43*, 407–412. [CrossRef] [PubMed]

6. United Nations. *World Urbanization Prospects: The 2014 Revision, Highlights*; United Nations, Department of Economic and Social Affairs, Population Division: New York, NY, USA, 2014.

7. Communication from the Commission to the European Parliament, the Council, the European Economic and Social Committee and the Committee of the Regions. In *Green Infrastructure (GI)—Enhancing Europe's Natural Capital—COM (2013) 249*; European Commission: Brussels, Belgium, 2013; p. 149.

8. Burkhard, B.; Kroll, F.; Nedkov, S.; Müller, F. Mapping Ecosystem Service Supply, Demand and Budgets. *Ecol. Indic.* **2012**, *21*, 17–29. [CrossRef]

9. Demuzere, M.; Orru, K.; Heidrich, O.; Olazabal, E.; Geneletti, D.; Orru, H.; Bhave, A.G.; Mittal, N.; Feliu, E.; Faehnle, M. Mitigating and Adapting to Climate Change: Multi-functional and Mulit-Scale Assessment of Green Urban Infrastructure. *J. Environ. Manag.* **2014**, *146*, 107–115. [CrossRef]

10. Niemelä, J.; Saarela, S.R.; Soderman, T.; Kopperoinen, L.; Yli-Pelkonen, V.; Vare, S.; Kotze, D.J. Using the Ecosystem Services Approach for Better Planning and Conservation of Urban Green Spaces. A Finland case study. *Biodivers. Conserv.* **2010**, *19*, 3225–3243. [CrossRef]

11. Pauleit, S.; Liu, L.; Ahern, J.; Kazmierczak, A. Multifunctional Green Infrastructure Planning to Promote Ecological Services in the City. In *Urban Ecology. Patterns, Processes, and Applications*; Niemelä, J., Ed.; Oxford Univnversity Press: Oxford, UK, 2011; pp. 272–285. [CrossRef]

12. Radford, K.G.; James, P. Changes in the Value of Ecosystem Services along a Rural-Urban Gradient: A case study of Greater Manchester, UK. *Landsc. Urban Plan.* **2013**, *109*, 117–127. [CrossRef]

13. Artmann, M.; Bastian, O.; Grunewald, K. Using the Concepts of Green Infrastructure and Ecosystem Services to Specify Leitbilder for Compact and Green Cities—The Example of the Landscape Plan of Dresden (Germany). *Sustainability* **2017**, *9*, 198. [CrossRef]

14. Haase, D.; Larondelle, N.; Andersson, E.; Artmann, M.; Borgström, S.; Breuste, J.; Gomez-Baggethun, E.; Gren, A.; Hamstead, Z.; Hansen, R.; et al. A Quantitative Review of Urban Ecosystem Service Assessments: Concepts, Models, and Implementation. *AMBIO* **2014**, *43*, 413–433. [CrossRef]

15. Wende, W. Ecosystem Services and Landscape Planning. How to Integrate Two Different Worlds in a High-Density Urban Setting. In *Urban Landscapes in High—Density Cities*; Rinaldi, B.M., Tan, P.J., Eds.; Birkhäuser: Berlin, Germany; Birkhäuser: Basel, Switzerland, 2019; pp. 154–164. [CrossRef]

16. Grunewald, K.; Richter, B.; Behnisch, M. Multi-Indicator Approach for Characterising Urban Green Space Provision at City and City-District Level in Germany. *Int. J. Environ. Res. Public Health* **2019**, *16*, 2300. [CrossRef]

17. Cortinovis, C.; Geneletti, D. Ecosystem Services in Urban Plans: What is there, and what is still needed for better decisions. *Land Use Policy* **2018**, *70*, 298–312. [CrossRef]

18. Kroll, F.; Müller, F.; Haase, D.; Fohrer, N. Rural-Urban Gradient Analysis of Ecosystem Services Supply and Demand Dynamics. *Land Use Policy* **2012**, *29*, 521–535. [CrossRef]

19. Pandeya, B.; Buytaert, W.; Zulkafli, Z.; Karpouzoglou, T.; Mao, F.; Hannah, D.M. A Comparative Analysis of Ecosystem Services Valuation Approaches for Application at the Local Scale and in Data Scarce Regions. *Ecosyst. Serv.* **2016**, *22*, 250–259. [CrossRef]

20. Wurster, D.; Artmann, M. Development of a Concept for Non-monetary Assessment of Urban Ecosystem Services at the Site Level. *AMBIO* **2014**, *43*, 454–465. [CrossRef]

21. Tammi, I.; Mustajärvi, K.; Rasinmäki, J. Integrating Spatial Valuation of Ecosystem Services into Regional Planning and Development. *Ecosyst. Serv.* **2017**, *26*, 329–344. [CrossRef]

22. Pickett, S.T.A.; Cadenasso, M.L.; Grove, J.M.; Nilon, C.H.; Pouyat, R.V.; Zipperer, W.C.; Costanza, R. Urban Ecological Systems: Linking Terrestrial Ecological, Physical, and Socioeconomic Components of Metropolitan Areas. *Annu. Rev. Ecol. Syst.* **2001**, *32*, 127–157. [CrossRef]

23. Gómez-Baggethun, E.; Gren, A.; Barton, D.N.; Langemeyer, J.; McPhearson, T.; O'Farrell, P.; Andersson, E.; Hamstead, Z.; Kremer, P. Chapter 11: Urban Ecosystem Services. In *Urbanization, Biodiversity and Ecosystem Services: Challenges and Opportunities. A Global Assessment*; Elmqvist, T., Fragkias, M., Goodness, J., Güneralp, B., Marcotullio, P.J., McDonald, R.I., Parnell, S., Schewenius, M., Sendstad, M., Seto, K.C., et al., Eds.; Springer: Dordrecht, The Netherlands, 2013; pp. 175–251. [CrossRef]

24. Grunewald, K.; Bastian, O. Special Issue: "Maintaining Ecosystem Services to Support Urban Needs". *Sustainabilty* **2017**, *9*, 1647. [CrossRef]

25. Beichler, S.A.; Bastian, O.; Haase, D.; Heiland, S.; Kabisch, N.; Müller, F. Does the Ecosystem Service Concept Reach its Limits in Urban Environments? *Landsc. Online* **2017**, *51*, 1–21. [CrossRef]

26. Kremer, P.; Hamstead, Z.; Haase, D.; McPhearson, T.; Frantzeskaki, N.; Andersson, E.; Kabisch, N.; Larondelle, N.; Lorance Rall, E.; Voigt, A.; et al. Key Insights for the Future of Urban Ecosystem Services Research. *Ecol. Soc.* **2016**, *21*, 29. [CrossRef]

27. Seppelt, R.; Dormann, C.F.; Eppink, F.V.; Lautenbach, S.; Schmidt, S. A Quantitative Review of Ecosystem Service Studies: Approaches, Shortcomings and the Road Ahead. *J. Appl. Ecol.* **2011**, *48*, 630–636. [CrossRef]

28. Pickering, C.; Byrne, J. The Benefits of Publishing Systematic Quantitative Literature Reviews for PhD Candidates and other Early-Career Researchers. *High. Educ. Res. Dev.* **2014**, *33*, 534–548. [CrossRef]

29. TEEB—The Economics of Ecosystems and Biodiversity. TEEB Manual for Cities: Ecosystem Services in Urban Management. 2011. Available online: www.teebweb.org (accessed on 13 May 2020).

30. De Wit, M.; van Zyl, H.; Crookes, D.; Blignaut, J.; Jayiya, T.; Goiset, V.; Mahumani, B. Including the economic value of well-functioning urban ecosystems in financial decisions: Evidence from a process in Cape Town. *Ecosyst. Serv.* **2012**, *2*, 38–44. [CrossRef]

31. De Wit, M.; Van Zyl, H.; Crookes, D.; Blignaut, J.; Jayiya, T.; Goiset, V.; Mahumani, B. Investing in Natural assets. A Business Case for the Environment in the City of Cape Town. *Cape Town* **2009**. [CrossRef]

32. Säumel, I.; Reddy, E.S.; Wachtel, T. Edible City Solutions—One Step Further to Foster Social Resilience through Enhanced Socio-Cultural Ecosystem Services in Cities. *Sustainability* **2019**, *11*, 972. [CrossRef]

33. Buchel, S.; Frantzeskaki, N. Citizens' Voice. A Case Study about Perceived Ecosystem Services by Urban Park Users in Rotterdam, the Netherlands. *Ecosyst. Serv.* **2015**, *12*, 169–177. [CrossRef]

34. Hegetschweiler, T.; de Vries, S.; Arnberger, A.; Bell, S.; Brennan, M.; Siter, N.; Olafsson, A.; Voigt, A.; Hunziker, M. Linking Demand and Supply Factors in Identifying Cultural Ecosystem Services of Urban Green Infrastructures: A review of European studies. *Urban For. Urban Green.* **2017**, *21*, 48–59. [CrossRef]

35. Graça, M.; Alves, P.; Gonçalves, J.; Nowak, J.D.; Hoehn, R.; Farinha-Marques, P.; Cunha, M. Assessing How Green Space Types Affect Ecosystem Services Delivery in Porto, Portugal. *Landsc. Urban Plan.* **2018**, *170*, 195–208. [CrossRef]

36. Hernández-Morcillo, M.; Plieninger, T.; Bieling, C. An Empirical Review of Cultural Ecosystem Service Indicators. *Ecol. Indic.* **2013**, *29*, 434–444. [CrossRef]
37. Mathey, J.; Rößler, S.; Banse, J.; Lehmann, I.; Bräuer, A. Brownfields as an Element of Green Infrastructure for Implementing Ecosystem Services into Urban Areas. *J. Urban Plan. Dev.* **2015**, *141*, A4015001. [CrossRef]
38. Pueffel, C.; Haase, D.; Priess, J.A. Mapping Ecosystem Services on Brownfields in Leipzig, Germany. *Ecosyst. Serv.* **2018**, *30*, 73–85. [CrossRef]
39. Takács, A.; Kiss, M.; Hof, A.; Tanács, E.; Gulyás, A.; Kántor, N. Microclimate Modification by Urban Shade Trees—An Integrated Approach to Aid Ecosystem Service Based Decision-Making. *Procedia Environ. Sci.* **2016**, *32*, 97–109. [CrossRef]
40. Marando, F.; Salvatori, E.; Sebastiani, A.; Fusaro, L.; Manes, F. Regulating Ecosystem Services and Green Infrastructure: Assessment of Urban Heat Island Effect Mitigation in the Municipality of Rome, Italy. *Ecol. Model.* **2019**, *392*, 92–102. [CrossRef]
41. Lehmann, I.; Mathey, J.; Rößler, S.; Bräuer, A.; Goldberg, V. Urban Vegetation Structure Types as a Methodological Approach for Identifying Ecosystem Services—Application to the Analysis of Micro-Climatic Effects. *Ecol. Indic.* **2014**, *42*, 58–72. [CrossRef]
42. De Valck, J.; Beames, A.; Liekens, I.; Bettens, M.; Seuntjens, P.; Broekx, S. Valuing Urban Ecosystem Services in Sustainable Brownfield Redevelopment. *Ecosyst. Serv.* **2019**, *35*, 139–149. [CrossRef]
43. Sun, F.; Xiang, J.; Tao, Y.; Tong, C.; Che, Y. Mapping the Social Values for Ecosystem Services in Urban Green Spaces: Integrating a Visitor-Employed Photography Method into SolVES. *Urban For. Urban Green.* **2019**, *38*, 105–113. [CrossRef]
44. Vieira, J.; Matos, P.; Mexia, T.; Silva, P.; Lopes, N.; Freitas, C.; Correia, O.; Santos-Reis, M.; Branquinho, C.; Pinho, P. Green Spaces are not all the same for the Provision of Air Purification and Climate Regulation Services: The Case of Urban Parks. *Environ. Res.* **2018**, *160*, 306–313. [CrossRef]
45. Ko, H.; Son, Y. Perceptions of cultural ecosystem services in urban green spaces: A case study in Gwacheon, Republic of Korea. *Ecol. Indic.* **2018**, *91*, 299–306. [CrossRef]
46. Pappalardo, V.; La Rosa, D.; Campisano, A.; La Greca, P. The Potential of Green Infrastructure Application in Urban Runoff Control for Land Use Planning: A Preliminary Evaluation from a Southern Italy Case Study. *Ecosyst. Serv.* **2017**, *26*, 345–354. [CrossRef]
47. Larondelle, N.; Lauf, S. Balancing Demand and Supply of Multiple Urban ecosystem Services on Different Spatial Scales. *Ecosyst. Serv.* **2016**, *22*, 18–31. [CrossRef]
48. Strohbach, M.W.; Haase, D. Above-Ground Carbon Storage by Urban Trees in Leipzig, Germany: Analysis of Patterns in a European City. *Landsc. Urban Plan.* **2012**, *104*, 95–104. [CrossRef]
49. Czembrowski, P.; Kronenberg, J.; Czepkiewicz, M. Integrating Non-Monetary and Monetary Valuation Methods—SoftGIS and Hedonic Pricing. *Ecol. Econ.* **2016**, *130*, 166–175. [CrossRef]
50. Costanza, R.; de Groot, R.; Sutton, P.; van der Ploeg, S.; Anderson, S.J.; Kubiszewski, I.; Farber, S.; Turner, R.K. Changes in the Global Value of Ecosystem Services. *Glob. Environ. Chang.* **2014**, *26*, 152–158. [CrossRef]
51. Sutton, P.; Anderson, S. Holistic Valuation of Urban Ecosystem Services in New York City's Central Park. *Ecosyst. Serv.* **2016**, *19*, 87–91. [CrossRef]
52. Speak, A.; Mizgajski, A.; Borysiak, J. Allotment Gardens and Parks: Provision of Ecosystem Services with an Emphasis on Biodiversity. *Urban For. Urban Green.* **2015**, *14*, 778–781. [CrossRef]
53. Kremer, P.; Hamstead, Z.A.; McPhearson, T. The Value of Urban Ecosysten Services in New York City: A Spatially Explicit Multicriteria Analysis of Landscape Scale Valuation Scenarios. *Environ. Sci. Policy* **2016**, *62*, 57–68. [CrossRef]
54. Eigenbrod, F.; Armsworth, P.R.; Anderson, B.J.; Heinemeyer, A.; Gillings, S.; Roy, D.B.; Thomas, C.D.; Gaston, K.J. The impact of proxy-based methods on mapping the distribution of ecosystem services. *J. Appl. Ecol.* **2010**, *47*, 377–385. [CrossRef]
55. Richards, D.R.; Tunçer, B. Using Image Recognition to Automate Assessment of Cultural Ecosystem Services from Social Media Photographs. *Ecosyst. Serv.* **2018**, *31*, 318–325. [CrossRef]
56. Parsa, V.A.; Salehi, E.; Yavari, A.R.; Van Bodegom, P.M. Analyzing Temporal Changes in Urban Forest Structure and the Effect on Air Quality Improvement. *Sustain. Cities Soc.* **2019**, *48*, 101548. [CrossRef]
57. Baró, F.; Chaparro, L.; Gómez-Baggethun, E.; Langemeyer, J.; Nowak, D.J.; Terradas, J. Contribution of Ecosystem Services to Air Quality and Climate Change Mitigation Policies. The Case of Urban Forests in Barcelona, Spain. *AMBIO* **2014**, *43*, 466–479. [CrossRef]

58. Graça, M.S.; Gonçalves, J.F.; Alves, P.J.M.; Nowak, D.J.; Hoehn, R.; Ellis, A.; Farinha-Marques, P.; Cunha, M. Assessing mismatches in ecosystem services proficiency across the urban fabric of Porto (Portugal): The influence of structural and socioeconomic variables. *Ecosyst. Serv.* **2017**, *23*, 82–93. [CrossRef]

59. Selmi, W.; Weber, C.; Rivière, E.; Blond, N.; Mehdi, L.; Nowak, D. Air pollution removal by trees in public green spaces in Strasbourg city, France. *Urban For. Urban Green.* **2016**, *17*, 192–201. [CrossRef]

60. Baró, F.; Haase, D.; Gómez-Baggethun, E.; Frantzeskaki, N. Mismatches between ecosystem services supply and demand in urban areas: A quantitative assessment in five European cities. *Ecol. Indic.* **2015**, *55*, 146–158. [CrossRef]

61. Kiss, M.; Takács, Á.; Pogácsás, R.; Gulyás, A. The role of ecosystem services in climate and air quality in urban areas: Evaluating carbon sequestration and air pollution removal by street and park trees in Szeged (Hungary). *Morav. Geogr. Rep.* **2015**, *23*, 36–46. [CrossRef]

62. Zepp, H.; Mizgajski, A.; Mess, C.; Zwierzchowska, I. A Preliminary Assessment of Urban Ecosystem Services in Central European Urban Areas. A Methodological Outline with Examples from Bochum (Germany) and Poznań (Poland). *Berichte. Geogr. Landeskd.* **2016**, *90*, 67–84.

63. Botzat, A.; Fischer, L.K.; Kowarik, I. Unexploited Opportunities in Understanding Liveable and Biodiverse Cities. A Review on Urban Biodiversity Perception and Valuation. *Glob. Environ. Chang.* **2016**, *39*, 220–233. [CrossRef]

64. Kabisch, N.; Qureshi, S.; Haase, D. Human-Environment Interactions in Urban Green Spaces—A Systematic Review of Contemporary Issues and Prospects for Future Research. *Environ. Impact Assess. Rev.* **2015**, *50*, 25–34. [CrossRef]

 *land*

Communication

The Incremental Demise of Urban Green Spaces

Johan Colding [1,2,*], **Åsa Gren** [2] **and Stephan Barthel** [1,3]

1   Department of Building Engineering, Energy Systems and Sustainability Science, University of Gävle, Kungsbäcksvägen 47, 80176 Gävle, Sweden; stephan.barthel@hig.se
2   The Beijer Institute of Ecological Economics, Royal Swedish Academy of Sciences, 104 05 Stockholm, Sweden; asa.gren@beijer.kva.se
3   Stockholm Resilience Centre, Stockholm University, Kräftriket 2B, 114 19 Stockholm, Sweden
*   Correspondence: johan.colding@hig.se

Received: 27 April 2020; Accepted: 19 May 2020; Published: 20 May 2020

Abstract: More precise explanations are needed to better understand why public green spaces are diminishing in cities, leading to the loss of ecosystem services that humans receive from natural systems. This paper is devoted to the incremental change of green spaces—a fate that is largely undetectable by urban residents. The paper elucidates a set of drivers resulting in the subtle loss of urban green spaces and elaborates on the consequences of this for resilience planning of ecosystem services. Incremental changes of greenspace trigger baseline shifts, where each generation of humans tends to take the current condition of an ecosystem as the normal state, disregarding its previous states. Even well-intended political land-use decisions, such as current privatization schemes, can cumulatively result in undesirable societal outcomes, leading to a gradual loss of opportunities for nature experience. Alfred E. Kahn referred to such decision making as 'the tyranny of small decisions.' This is mirrored in urban planning as problems that are dealt with in an ad hoc manner with no officially formulated vision for long-term spatial planning. Urban common property systems could provide interim solutions for local governments to survive periods of fiscal shortfalls. Transfer of proprietor rights to civil society groups can enhance the resilience of ecosystem services in cities.

Keywords: urban greenspace; privatization; property rights; incremental greenspace loss; ecosystem services; the tyranny of small decisions; resilience planning; urban densification; baseline shifts; urban nature connection

---

1. Introduction

Preserving greenspace quantity and quality in the face of increasing urbanization is a pressing global challenge [1]. Greenspaces provide invaluable ecosystem services to humans that are important to plan for in cities [2]. Economic motives and urban neoliberal policies are liable explanations behind the loss of public space in many cities [3,4]. Public greenspace is an important component of public space [5] and could be defined as "any vegetation found in the urban environment, including parks, open spaces, residential gardens, or street trees" [6] (p. 113). Here, by public space is meant spaces in cities that are "owned by the government, accessible to everyone without restriction, and/or fosters communication and interaction" [7] (p. 9). This definition encapsulates Louis Wirth's notion of urbanism [8], taking into account how individuals interact with one another and with spaces. Public spaces facilitate human exchange and interaction, as in the form of urban squares and market places that traditionally have served as arenas for public communication and social interaction [9]. However, there is a massive shift towards the privatization of public land and resources in many cities today [3,10], affecting green spaces in a multitude of ways with repercussion for long-term management of ecosystem services. While urbanization causes the direct loss of urban greenspace, comprising habitat fragmentation that involves both loss and/or the breaking apart of habitats [11], there exist more subtle forms of greenspace loss that ultimately are linked to property-rights arrangements. We here refer to such loss as incremental, occurring over a series of gradual declines or small steps but experienced at the cognitive level of urban space as 'baseline shifts' among urban residents.

Based on the literature related to institutions, economic geography, urban ecology, and social theory, we present a set of subtle drivers for why public green spaces gradually erode in cities. This gradual erosion, we argue, is often "invisible" in that it can almost exclusively be revealed by high spatial resolution remote sensing data [12] and that their effect can be translated into high social-ecological[1] costs that impinge negatively on human wellbeing [13,14]. Based on a set of examples of incremental greenspace change, this paper briefly discusses how urban planning authorities should avoid 'day-to-day planning' [15] and be more long-term oriented to meet an ever-increasing unpredictable future. Property rights theory in relation to natural resource management emphasizes institutions at the interface between social and natural systems [16,17], where the term 'institutions' signifies the rules and conventions of society that facilitate coordination among people regarding their behavior [18]. At a more general level, institutions are made up of formal constraints (rules, laws, constitutions), informal constraints (norms of behavior, conventions, and self-imposed codes of conduct), and their enforcement characteristics; thus, they shape incentives in human exchange, whether political, social, or economic [18]. Ecologically oriented scholars define institutions simply as working-rules or rules-in-use, meaning "the set of rules actually used by a set of individuals to organize repetitive activities" [19] (p. 19). Thus, property rights link people to nature, and have the potential to coordinate the social and natural systems in a complementary way for both ecological and human long-term objectives [20].

Property-rights could also be viewed as slow variables in urban transformation; however, their monoculturalization in favor of urban privatization schemes may gradually erode urban resilience (i.e., 'buffering capacity' to deal with disturbance and novel events) and make planning bodies and local authorities less pertinent to propel urban growth along more sustainable trajectories that value ecosystem services as risk insurance and as adaptive capacity for responding to known and unknown disturbances. The paper concludes by proposing common property systems as a viable alternative for local governments to survive economic disruptions and in turning public spaces into places that urban residents themselves can manage for improving and protecting greenspace and associated ecosystem services.

2. Urban Green Space Dynamics and Reasons for Their Incremental Demise

The availability of public green spaces is foremost linked to the geographical location of a city [21]. However, urban expansion in many cities takes place almost exclusively at the expense of farmland [22], with changes in greenspace predominantly occurring in the urban-rural periphery [12]. Urban greenspace consists mainly of semi-natural areas, such as different gardens, road and rail networks and their associated land, airfields, golf courses, parks, allotment areas, urban agriculture, etc. that together with formally protected nature reserves and Natura 2000-sites contribute to the generation of urban ecosystem services [23].

Quantifying spatiotemporal patterns of urban greenspace at more precise levels is reliant upon modern remote sensing techniques. Hence, more steadfast comprehensive assessments of detailed greenspace change are scarce in the ecological literature related to urban systems [24]. Previous studies have also come to different results. For example, Kabisch and Haase [6] could not find a significant change in Western and Southern European cities between 1990 and 2000; but found a significant increase of greenspace in the period from 2000 to 2006. In a study of 386 European cities, Fuller and Gaston [1] found a dramatic drop in per capita green space provision in cities with high population

---

[1] By the term 'social-ecological' is here meant a set of critical natural, socioeconomic, and cultural resources (or, capitals) whose flow and use is regulated by a combination of ecological and social systems.

density, likely due to more people being packed into the urban matrix rather than buildings replacing existing green spaces. They also found that access to greenspace rapidly declines as cities grow, decreasing opportunities for people to experience nature. Following post-socialist changes many East-European cities have experienced a decline in greenspace [25,26]. Similarly, McDonald et al. [24] found an open space loss between 1990 and 2000 for all the examined 274 metropolitan areas in the contiguous United States. While many Chinese cities show mixed results, with both increases and decreases, cities in many developing countries are losing green spaces at a rapid pace [12].

Despite the massive shift towards privatization of public land and resources [3,10], comprehensive studies that have examined the relationship between loss of green spaces and ownership regimes are greatly lacking. However, it may not be far-fetched to assume that much of the privatization of public space involves greenspace. While this loss comprises direct habitat loss and the breaking apart of habitats [11], few scholars have addressed more subtle causes behind urban greenspace loss, but see Mensah [27]. In the following we present a set of examples of incremental demise of public urban greenspace.

2.1. Lack of Financial Support

One of the major reasons behind the privatization of public space is financially strained local government budgets, that strain results in the outsourcing and the alienation of land to private interests and the privatization of services that previously were publicly delivered. There exist plenty of examples of how tightening budgets are leading to declines in the quality of green spaces and loss of ecosystem services, due to the lack of staff and maintenance resources [27,28]. The Heritage Lottery Fund—a large funder of heritage in the UK—reported in 2013 that almost half of the local park authorities were considering selling parks and green spaces or transferring their management to private entrepreneurs [29]. Considering the current recession due to the Covid-19 pandemic, one can only imagine how this situation will worsen.

There has also been an increase in long-term leaseholds to allow the transfer of public land, such as parks and other green spaces, to not-for-profit trusts and to resident-led management bodies [30]. While public land may be alienated from the ownership of local governments, privatization predominantly takes place through a mixture of transfers of governance and management responsibility from the public sector to a number of other actors in the private, voluntary, and community sectors [7,30], and with the degree of privatization ranging from full to partial outsourcing of responsibilities [4].

Public–private partnerships (PPP) constitute a well-known example of 'contractual governance', which increasingly is used to re-develop and manage public spaces, especially as capital investments [30]. Through a PPP, a local authority or a central-government agency sign a long-term contractual arrangement with a private supplier for the delivery of services and taking of responsibility for building infrastructure, financing the investment, and managing and upholding the facility [31]. PPPs are increasing across Europe, Canada, and the United States, as well in some developing countries [31]. Business improvement districts (BIDs) are an example of a PPP in which a local authority and a business community together develop schemes to benefit a local district area [32]. The services provided through this type of contractual arrangement could, for example, include improvements and attractiveness of physical areas [7] and management of public parks [33,34].

2.2. The Separation of Attributes

A critical fate that can affect underfinanced public spaces is the separation of attributes, which in economic theory can be expected when it is cost-effective and if sufficient demands for this exist [3]. In an overcrowded public domain, markets and governments will strive towards a separation of rights to land according to different attributes. This may be done in order to reduce potential conflicts and to lower transaction costs related to the governance of public space. In this way, the rights to different attributes of a park can be separated and granted to various user groups, such as devising demarcated land for leisure, habitats for wildlife, sporting areas, etc. [35]. Separation of attributes can also involve alienation or leaseholds of bits and pieces of public greenspace. There exist several cases where underfinanced public parks have been opened up to private interests, such as to restaurants, cafés and other social spaces. While local governments can reinvest the revenues from rents and/or property taxes for restoring degraded greenspace, profits are often instead used for other purposes [28,35].

2.3. Increased Private Control

Another subtle form of driver linked to urban greenspace change is the increased control and surveillance of public space [7,36]. Increased surveillance and policing over public space will likely intensify with an increased number of terror attacks as witnessed in many parts of the world. Today, much control of public space is outsourced to private corporations, making the boundary between public and private policing complex [37]. New York and Tokyo are telling examples of cities that lose much public space [36]. With increasing fear of terrorism after the Tokyo sarin gas attack in 1995 and the 9/11 attack in New York, many places where people formerly could relax from stress and annoyance have been eradicated [38]. Not only has an increased fear of terrorism acted as a vindication for imposing restrictions on the use of public sidewalks and plazas, but also in the use of natural habitats [38]. During the Covid-19 pandemic, the lack of public space in New York City prompted the banning of cars on certain streets in order to provide more space for pedestrians to upkeep social distancing. In contrast, green spaces in Sweden served as vital areas for social distancing [14].

Control of public space could be in the form of private police and/or surveillance equipment. While it may not directly lead to the loss of public space it can affect public space in more indirect ways, such as making people feel monitored and subsequently avoiding such spaces. The integrity of peoples' personal lives is increasingly also becoming jeopardized as digitalization is increasing in many cities [39]. While control can affect public green spaces the same way as it affects any other type of monitored public space, it especially can affect people's accessibility and use of urban green space. Many dwellers in urban areas display a fear of nature due to cultural reasons [40]. Lush green area habitats may be frightening for people due to lack of safety; hence, such habitats are sometimes replaced for safety reasons, e.g., increasing the width of sidewalks or increasing the occurrence and brightness of street lights, [41] thereby affecting greenspace negatively.

2.4. Under-Utilization

Under-utilization of public space represents yet another subtle driver behind greenspace loss. One example is the London Green Belt where green spaces are avoided by people due to poor management that have made them less accessible and attractive to be in. Residents may feel insecure and fearful of crime in unmanaged green spaces with short view distances [6]. So called 'boundary parks', located between vastly differing neighborhoods, are particularly susceptible of being underutilized which in turn can contribute to the decline of parks [42]. Accessibility of the public space itself can be argued as being one of the most effective factors and deterrents to utilization of a public space [9]. For example, public space will be less used by people if a user population does not live nearby [43].

2.5. Congestion

Direct over-use (congestion) of public greenspace is another indirect reason behind the demise of public urban space [3,35]. Congestion refers to the degree of competition within a public domain, or to "the numbers of individuals who jointly consume it and the range of tastes amongst those individuals (or groups)" [3] (p. 34). When congestion generates excessive transaction costs, such as the costs of queuing or resolving conflicts between different users, economic theory predicts that there is a high probability for pressures to reform the property rights and subdivide public space either into private domains or into smaller public domains (e.g., club goods). When public domains become congested, they must be governed in such a way that use rights to public space are clear and enforceable [3,44]. However, transaction costs for designing, creating, and administering such a system becomes increasingly high the more over-used a space becomes, and if costs for governance become too high in political or financial terms, then public space could become subject to land alienation where the local government seek to dispose of the property [3].

2.6. Activity Intensification

Activity intensification can also act as a subtle driver behind the transformation of public space [45]. To cut costs and for energy conservation, planners can, for example, resort to multifunctional land use. Haccou et al. [46] distinguish four types of multifunctional land use: interweaving, intensifying, layering, and timing. Interweaving combines different functions on the same piece of land; intensifying increases the effectiveness and efficiency of a certain land use on the same piece of land; layering mixes functions in the vertical dimension if possible; and timing uses the same building or space for different functions at different moments in time [47]. The first two types are also applicable to green spaces. In Stockholm—one of the fastest growing regions in Europe—planners have been taken by surprise by the sudden population boom, with over 300,000 new homes in need of being built in the coming decade. To cope with increased urban densification and population growth, multi-functionality in the use of public parklands has become a quick fix for dealing with overcrowding effects [48]. However, such a development may run contrary to planners' aspiration of making cities more resilient to various effects of climate-change and for securing biodiversity that depend on green spaces of a certain size and with certain ecological qualities that do not always rhyme well with social aspirations.

3. Baseline Shifts and Benefits of Nature

As cities develop over space and time, the different attributes of green spaces also change. Urban residents experience nature most of the time at "the cognitive level of urban space," that is, at the level where "people in the street" experience the city [49].

Human beings also constantly use memories of previous experiences to interpret current experiences [50]. When it comes to experience of urban nature, psychologists talk about baseline shifts, where each generation of humans tends to take the current condition of an ecosystem as the nondegraded state: the 'normal' experience, disregarding the fact that the ecosystem might have changed considerably over time [51]. Baseline shifts can lead to environmental generational amnesia [52], referring to the psychological process whereby each generation perceives the environment into which it is born as the norm, no matter how developed, urbanized, or polluted the environment is [51].

According to Hartig and Kahn [51], baseline shifts can help explain inaction on environmental problems in that people do not feel the urgency of the problems because the experiential baseline has shifted (Figure 1). Experiencing nature during childhood is crucial for shaping sustainable decision-making processes in adulthood [53].

The incremental changes of urban greenspace, previously dealt with herein, likely contribute to baseline shifts. To what extent is hard to determine. There might also be overlaps in the drivers previously dealt with, e.g., under-utilization of greenspace can at many times be explained by a lack of financial support. However, environmental generational amnesia may spread as more and more people are devoid of direct contact with nature in cities [54]. This in turn may make it more difficult to reach public acceptance of policies to deal with environmental problems, like climate change. Environmental illiteracy may further lead to the erosion of public green spaces since people will not value the ecosystem services they derive from nature (Figure 2). This is especially the case in countries where destruction of ecological sensitive areas and open spaces are rapidly urbanized due to weak urban planning institutions [55].

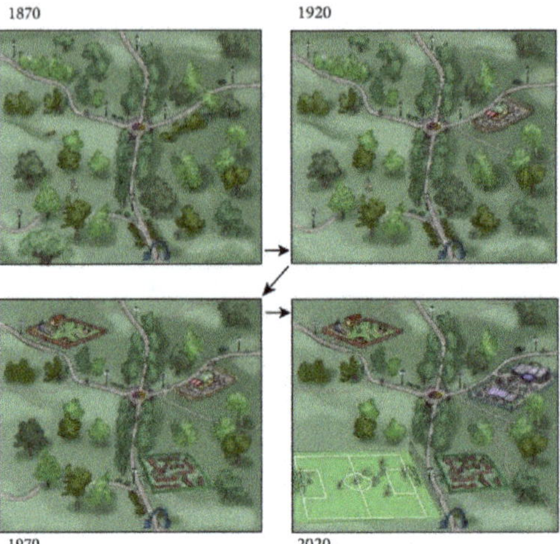

Figure 1. Schematic illustration of a baseline shift of a hypothetical public park over a 150-year period. Each development stage is perceived as being the nondegraded (normal) state by each new generation of humans—a situation that can contribute to environmental generational amnesia.

PROVISIONING SERVICES	REGULATING SERVICES	CULTURAL SERVICES
The "products" obtained from ecosystems	*Benefits obtained from the regulation of ecosystem processes*	*Nonmaterial benefits obtained from ecosystems*
Foods	Climate regulation	Educational
Fibers	Flood prevention	Recreational
Ornamentals	Erosion control	Sense of place
Medicines	Pest control	Spiritual
Biofuels	Pollination	Cognitive development
Fresh water	Seed dispersal	Stress relief
Genetic resources	Disease regulation	Gardening
SUPPORTING SERVICES		
Services necessary for the production of all other ecosystem services		
Biodiversity		
Nutrient recycling		
Primary productivity		

Figure 2. Ecosystem services from urban greenspaces. These include provisioning services, regulating services, cultural services, and supporting services. Source: Millennium Ecosystem Assessment. Ecosystems and Human Well-Being. Reprinted from Colding and Barthel [56].

4. Long-Term Resilience Planning

The incremental changes and loss of urban greenspace elaborated on in this paper fit the pattern of decision making that the economist Alfred E. Kahn once referred to as 'the tyranny of small decisions', representing a situation in which a number of decisions, individually small in size and time, result in non-optimized and socially undesirable outcomes [57]. According to Kahn the 'small decision effects' are common in market economics. In an urban planning context, many small-scale, independent decisions taken over time by a planning unit could culminate in outcomes that are neither intended nor preferred [58]; hence, even well-intended planning decisions in countries with strong state control can result in undesirable social-ecological outcomes that run contrary to the common good of a city and its inhabitants.

Nilsson [15] has described how leading politicians often abstain from restricting options in order to handle future planning issues—a phenomenon that often leads to an ad hoc form of planning, termed "day-to-day planning". It is mainly designed to manage acute problems and to resolve problems as soon as possible and is related to a discourse of economic development "in the way in which it tries to satisfy the short-term requirements of industry and commerce, and tends to emphasize the economic dimension of sustainable development" [15] (p. 441).

Accessibility to green spaces in cities is ultimately determined by property-rights arrangements that regulate their omission or entry. Institutional scholars have long recognized the role of property rights for linking people to nature, but as institutional research suggests, no single type of ownership regimes (i.e., state, private, or common property rights) can be prescribed as a remedy for resource overuse or environmental degradation [59]. Instead, policy should focus on establishing a multitude of property rights regimes that are designed to fit the cultural, economic, and geographic context in which they are to function [20].

How space and property rights are arranged to ensure access to urban nature will be of direct importance for building general urban resilience to human wellbeing in the long-term [14]. The gradual shift towards privatization of public property in cities, and even of common property systems [60,61] is a worrisome sign since it may negatively affect resilience planning, which needs to be long-term oriented. Property-rights could be viewed as slow variables in urban transformation. In resilience science, slow variables—like evolution, coral regrowth, or nutrient transportation through soils—play a substantial role for a system's resilience to change [62]. C.S. Holling's work [63] demonstrated that an ecosystem can absorb an entire chain of disturbances without being adversely affected until it suddenly changes to a completely different state where past functions and services no longer can be provided. A gradual loss of the system's slow variables often causes such abrupt change. Critical, slow social variables are characterized by normally being stable or changing slowly over long periods of time, providing continuity of functions. However, even property rights that have existed for millennia may change abruptly, causing changes in land use, social disruption, and even declines in human well-being [60,64].

Lindholm [10] describes how sightlines in cities commonly become blocked by new private developments, with less daylight reaching the ground and with more land areas constantly in shadow, resulting in poorer conditions for plants and biodiversity, and also how people at the eye level of cities experience their local socio-physical environment. New sightlines are often aggravated due to visible landmarks of private interest. Hence, indirect effects of private property may negatively impinge on ecosystem services that are nurtured in the public domain. While private building consortia often launch their development projects as "sustainable", they often constitute examples of "green washing" [10]. An often-time neglected aspect of green roofs and green walls that commonly replace natural areas on the ground, is that they only are accessible to a limited set of urban residents.

The slow variables, such as the amount and quality of public green spaces, determine how a fast variable, like daily visits to green spaces, is possible when external drivers are in operation, such as excessive heat waves or pandemics. The gradual transfer of public to private property rights and services will in the long run weaken the power of local planning units and local authorities. If left unchecked, the increasing privatization of public space could ultimately become a huge democratic problem.

5. Concluding Remarks

The ratio of public/private land is diminishing, affecting cities in a multitude of ways. The drivers of incremental change of urban greenspace presented herein are ultimately a result of urban densification schemes (spatial and/or population), which seemingly propel a monoculturalization of property-rights regimes towards economic efficiency and private and semi-private solutions. As elaborated on herein, congestion in the use of public space is but one outcome of urban densification, leading to such phenomena as the separation of attributes, activity intensification, and even to alienation of public space to private interests. Underutilization of greenspace coupled to inadequate funding for greenspace management make these spaces more vulnerable to urban encroachment, especially when cities grow inwardly, posing an overall negative effect on the maintenance and generation of ecosystem services.

To counter shifting baselines that may result from the incremental demise of urban green spaces, Hartig and Kahn [51] propose that cities should provide more opportunities for people to experience more stable, healthy, and even wilder forms of nature. Property rights, however, ultimately determine such opportunities, as no single space should be expected to meet the demands of all users at all times, nor will any single type of property-rights arrangements fulfill the multitude of functions that a vibrant city depends on. Designing urban commons as an alternative to land privatization is no guarantee, however, for halting the demise of urban greenspaces, as the effects of privatization and commodification of commons is also high in some countries, such as in Europe [65]. However, maintaining a well-balanced diversity and mixture of property-rights regimes could be a wise policy for planners to adhere to in order to increase preparedness to meet an increasingly disturbance-prone future.

Although common property resource systems have been known to collapse due to overuse, there are promising signs that these systems are experiencing a revival in urban settings and boosting collective environmental action [44,66]. There are important linkages among urban common property systems, social–ecological learning, and management of biodiversity and ecosystem services. Urban gardening projects provide not only cultural and provisioning ecosystem services but also regulating ecosystem services to city neighborhoods [44]. There are also examples of whole public parks being managed as common property systems [67]. By granting management rights of greenspace to residential collectives and social networks, local governments can still maintain control and ownership of public space. Long-term leaseholds and the establishment of 'user rights contracts' that transfer management rights of public space to civil society groups can be a viable interim solution for local governments in order to survive periods of fiscal shortcuts. Studies have shown that urban common property systems can serve a multitude of social-ecological purposes [11] and turn spaces into meaningful places that urban residents themselves can manage to improve and protect urban nature. The transferring of proprietor rights down to local levels can also lower transaction costs for greenspace management and governance [44]. To safeguard ecosystem services policy makers and planners should take advantage of these benefits and more meticulously address the incremental demise of public greenspace.

Author Contributions: Conceptualization, J.C. Writing—Original Draft Preparation, J.C., S.B., Å.G.; Leading of the writing process, J.C.; Corresponding author, J.C. All authors have read and agreed to the published version of the manuscript.

Funding: Johan Colding's and Stephan Barthel's work have partly been funded by University of Gävle. Barthel's work has also been funded by FORMAS/The Swedish Research Council for Environment, Agricultural Sciences and Spatial Planning. The project is called Spatial and Experiential Analyses for Urban Social Sustainability (ZEUS) (reference number: 2016-01193). Barthel's work is also funded by the Stockholm Resilience Centre. Johan Colding's and Åsa Gren's work has also been partly funded through a research grant (reference number: 2017-00937) received from the Swedish Research Council for Environment, Agricultural Sciences and Spatial Planning (FORMAS), and through means provided by the Stockholm County Council and the Stockholm University (SU-SLL Grant 2017: no. 20160884), and from core funding provided by the Beijer Institute of Ecological Economics, Stockholm, Sweden.

Acknowledgments: The authors wish to thank Jonas Adner for the nice Figure illustration.

Conflicts of Interest: The authors declare no conflict of interest.

References

1. Fuller, R.A.; Gaston, K.J. The scaling of green space coverage in European cities. *Biol. Lett.* **2009**, *5*, 352–355. [CrossRef]
2. Colding, J. The Role of Ecosystem Services in Contemporary Urban Planning. In *Urban Ecology*; Oxford University Press: Oxford, UK, 2011; pp. 228–237.
3. Lee, S.; Webster, C. Enclosure of the urban commons. *GeoJournal* **2006**, *66*, 27–42. [CrossRef]
4. Leclercq, E.; Pojani, D.; Van Bueren, E. Is public space privatization always bad for the public? Mixed evidence from the United Kingdom. *Cities* **2020**, *100*, 102649. [CrossRef]
5. World Health Organization. Urban green spaces: A brief for action. *Reg. Off. Eur.* **2017**. Available online: http://www.euro.who.int/__data/assets/pdf_file/0010/342289/Urban-Green-Spaces_EN_WHO_web3.pdf?ua=1 (accessed on 6 March 2020).

6. Kabisch, N.; Haase, D. Green spaces of European cities revisited for 1990–2006. *Landsc. Urban Plan.* **2013**, *110*, 113–122. [CrossRef]

7. Kohn, M. *Brave New Neighborhoods*; Routledge: Abingdon-on-Thames, UK, 2004; ISBN 9780203495117.

8. Wirth, L. Urbanism Way of Life. *Am. J. Sociol.* **1938**, *44*, 1–24. [CrossRef]

9. Pasaogullari, N.; Doratli, N. Measuring accessibility and utilization of public spaces in Famagusta. *Cities* **2004**, *21*, 225–232. [CrossRef]

10. Lindholm, G. Land and Landscape; Linking Use, Experience and Property Development in Urban Areas. *Land* **2019**, *8*, 137. [CrossRef]

11. Fahrig, L. Effects of Habitat Fragmentation on Biodiversity. *Annu. Rev. Ecol. Evol. Syst.* **2003**, *34*, 487–515. [CrossRef]

12. Zhou, W.; Wang, J.; Qian, Y.; Pickett, S.T.A.; Li, W.; Han, L. The rapid but "invisible" changes in urban greenspace: A comparative study of nine Chinese cities. *Sci. Total Environ.* **2018**, *627*, 1572–1584. [CrossRef]

13. Ogen, Y. Assessing nitrogen dioxide (NO2) levels as a contributing factor to coronavirus (COVID-19) fatality. *Sci. Total Environ.* **2020**, *726*, 138605. [CrossRef]

14. Samuelsson, K.; Barthel, S.; Colding, J.; Macassa, G.; Giusti, M. *Urban Nature as a Source of Resilience during Social Distancing Amidst the Coronavirus Pandemic*. OSF Preprint. 2020. Available online: https://osf.io/3wx5a/ (accessed on 6 March 2020). [CrossRef]

15. Nilsson, K.L. Managing Complex Spatial Planning Processes. *Plan. Theory Pract.* **2007**, *8*, 431–447. [CrossRef]

16. Ostrom, E. *Governing the Commons*; Cambridge University Press: Cambridge, UK, 2015; ISBN 9781316423936.

17. Berkes, F.; Colding, J.; Folke, C. *Navigating Social-Ecological Systems*; Berkes, F., Colding, J., Folke, C., Eds.; Cambridge University Press: Cambridge, UK, 2001; ISBN 9780521815925.

18. North, D.C. *Institutions, Institutional Change and Economic Performance*; Cambridge University Press: Cambridge, UK, 1990; ISBN 9780521397346.

19. Ostrom, E. *Crafting Institutions for Self-Governing Irrigation Systems*; ICS Press: San Francisco, CA, USA, 1992; ISBN 9781558151680.

20. Hanna, S.; Folke, C.; Mäler, K.-G. Property Rights and the Natural Environment. In *Rights to Nature: Ecological, Economic, Cultural, and Political Principles of Institutions for the Environment*; Island Press: Washington, DC, USA, 1996.

21. De Sousa Silva, C.; Viegas, I.; Panagopoulos, T.; Bell, S. Environmental Justice in Accessibility to Green Infrastructure in Two European Cities. *Land* **2018**, *7*, 134. [CrossRef]

22. Gren, Å.; Andersson, E. Being efficient and green by rethinking the urban-rural divide—Combining urban expansion and food production by integrating an ecosystem service perspective into urban planning. *Sustain. Cities Soc.* **2018**, *40*, 75–82. [CrossRef]

23. Colding, J.; Lundberg, J.; Folke, C. Incorporating Green-area User Groups in Urban Ecosystem Management. *AMBIO A J. Hum. Environ.* **2006**, *35*, 237–244. [CrossRef] [PubMed]

24. McDonald, R.I.; Forman, R.T.T.; Kareiva, P. Open Space Loss and Land Inequality in United States' Cities, 1990–2000. *PLoS ONE* **2010**, *5*, e9509. [CrossRef] [PubMed]

25. Hirt, S.; Kovachev, A. The changing spatial structure of post-socialist Sofia. In *The Urban Mosaic of Post-Socialist Europe. Contributions to Economics*; Tsenkova, S., Nedović-Budić, Z., Eds.; Physica-Verlag HD: Heidelber, Germany, 2006; pp. 113–130.

26. Haase, D.; Kabisch, S.; Haase, A.; Andersson, E.; Banzhaf, E.; Baró, F.; Brenck, M.; Fischer, L.K.; Frantzeskaki, N.; Kabisch, N.; et al. Greening cities—To be socially inclusive? About the alleged paradox of society and ecology in cities. *Habitat Int.* **2017**, *64*, 41–48. [CrossRef]

27. Mensah, C.A. Destruction of Urban Green Spaces: A Problem beyond Urbanization in Kumasi City (Ghana). *Am. J. Environ. Prot.* **2014**, *3*, 1–9. [CrossRef]

28. Jones, R. Managing the green spaces: Problems of maintaining quality in a local government service department. *Manag. Serv. Qual. Int. J.* **2000**, *10*, 19–31. [CrossRef]

29. Heritage Lottery Fund. *State of UK Public Parks*. 2016. Available online: https://www.heritagefund.org.uk/sites/default/files/media/attachments/state_of_uk_public_parks_2016_final_for_web%281%29.pdf (accessed on 10 March 2020).

30. De Magalhães, C.; Freire Trigo, S. Contracting out publicness: The private management of the urban public realm and its implications. *Prog. Plan.* **2017**, *115*, 1–28. [CrossRef]

31. Iossa, E.; Martimort, D. The Simple Microeconomics of Public-Private Partnerships. *J. Public Econ. Theory* **2015**, *17*, 4–48. [CrossRef]

32. Gross, J.S. Business improvement districts in New York: The private sector in public service or the public sector privatized? *Urban Res. Pract.* **2013**, *6*, 346–364. [CrossRef]

33. Anderson, L. A business improvement district for a park? Hey, it just might work. *Villager* **2008**, *78*, 14.

34. Kennedy, D.J. Restraining the Power of Business Improvement Districts: The Case of the Grand Central Partnership. *Yale Law Policy Rev.* **1996**, *15*, 283–330.

35. Colding, J. Creating Incentives for Increased Public Engagement in Ecosystem Management through Urban Commons. In *Adapting Institutions*; Boyd, E., Folke, C., Eds.; Cambridge University Press: Cambridge, UK, 2011; pp. 101–124, ISBN 9781139017237.

36. Cybriwsky, R. Changing patterns of urban public space. *Cities* **1999**, *16*, 223–231. [CrossRef]

37. Sparrow, M.K. Managing the Boundary between Public and Private Policing. In *Modern Perspectives on Policing: Selected Papers*; National Institute of Justice: Washington, DC, USA, 2015; ISBN 9781634639705.

38. Low, S.M. The Erosion of Public Space and the Public Realm: Paranoia, surveillance and privatization in New York City. *City* **2006**, *18*, 43–49. [CrossRef]

39. Colding, J.; Barthel, S.; Sörqvist, P. Wicked Problems of Smart Cities. *Smart Cities* **2019**, *2*, 31. [CrossRef]

40. Skår, M. Forest dear and forest fear: Dwellers' relationships to their neighbourhood forest. *Landsc. Urban Plan.* **2010**, *98*, 110–116. [CrossRef]

41. Cervero, R.; Duncan, M. Walking, Bicycling, and Urban Landscapes: Evidence from the San Francisco Bay Area. *Am. J. Public Health* **2003**, *93*, 1478–1483. [CrossRef]

42. Vischer, J.C.; Cranz, G. The Politics of Park Design: A History of Urban Parks in America. *Soc. Forces* **1984**, *63*, 287. [CrossRef]

43. Miller, H.C.; Gratz, R.B.; Mintz, N. *Cities Back from the Edge—New Life for Downtown*; John Wiley & Sons: Hoboken, NJ, USA, 2000.

44. Colding, J.; Barthel, S.; Bendt, P.; Snep, R.; van der Knaap, W.; Ernstson, H. Urban green commons: Insights on urban common property systems. *Glob. Environ. Chang.* **2013**, *23*, 1039–1051. [CrossRef]

45. Williams, K.; Burton, E.; Jenks, M. Achieving the Compact City through Intensification: An Acceptable Option? In *The Compact City*; Burton, E., Jenks, M., Williams, K., Eds.; Spon Press: London, UK, 2010; pp. 71–83.

46. Haccoû, H.A.; Deelstra, T.; Krośnicka, K.; Dol, M.; Kramer, M. *MILU Guide Practitioners' Handbook for Multifunctional and Intensive Land Use*; Habiforum Foundation: Gouda, The Netherlands, 2007; ISBN 978-90-806647-5-3.

47. Westerink, J.; Haase, D.; Bauer, A.; Ravetz, J.; Jarrige, F.; Aalbers, C.B.E.M. Dealing with Sustainability Trade-Offs of the Compact City in Peri-Urban Planning Across European City Regions. *Eur. Plan. Stud.* **2013**, *21*, 473–497. [CrossRef]

48. Stadsledningskontoret. *Grönare Stockholm: Riktlinjer för Planering, Genomförande och Förvaltning av Stadens Parker och Naturområden*; Stadsledningskontoret: Stockholm, Sweden, 2016.

49. Marcus, L.; Colding, J. Toward an integrated theory of spatial morphology and resilient urban systems. *Ecol. Soc.* **2014**, *19*, 55. [CrossRef]

50. Samuelsson, K.; Colding, J.; Barthel, S. Urban resilience at eye level: Spatial analysis of empirically defined experiential landscapes. *Landsc. Urban Plan.* **2019**, *187*, 70–80. [CrossRef]

51. Hartig, T.; Kahn, P.H. Living in cities, naturally. *Science* **2016**, *352*, 938–940. [CrossRef] [PubMed]

52. Miller, J.R. Biodiversity conservation and the extinction of experience. *Trends Ecol. Evol.* **2005**, *20*, 430–434. [CrossRef]

53. Giusti, M.; Barthel, S.; Marcus, L. Nature Routines and Affinity with the Biosphere: A Case Study of Preschool Children in Stockholm. *Child. Youth Environ.* **2014**, *24*, 16. [CrossRef]

54. Barthel, S.; Belton, S.; Raymond, C.M.; Giusti, M. Fostering Children's Connection to Nature Through Authentic Situations: The Case of Saving Salamanders at School. *Front. Psychol.* **2018**, *9*, 928. [CrossRef]

55. Korah, P.I.; Cobbinah, P.B.; Nunbogu, A.M.; Gyogluu, S. Spatial plans and urban development trajectory in Kumasi, Ghana. *GeoJournal* **2017**, *82*, 1113–1134. [CrossRef]

56. Colding, J.; Barthel, S. The Role of University Campuses in Reconnecting Humans to the Biosphere. *Sustainability* **2017**, *9*, 2349. [CrossRef]

57. Kahn, A.E. The Tyranny of Small Decisions: Market Failures, Imperfections, and the Limits of Economics. *Kyklos* **1966**, *19*, 23–47. [CrossRef]

58. Cooper, C.B.; Dickinson, J.; Phillips, T.; Bonney, R. Citizen Science as a Tool for Conservation in Residential Ecosystems. *Ecol. Soc.* **2007**, *12*, 11. [CrossRef]

59. Ostrom, E. Collective Action and the Evolution of Social Norms. *J. Econ. Perspect.* **2000**, *14*, 137–158. [CrossRef]

60. Ensminger, J. Changing Property Rights: Reconciling Formal and Informal Rights to Land in Africa. In *The Frontiers of The New Institutional Economics*; Drobak, J.N., Nye, V.C., Eds.; Academic Press: San Diego, CA, USA, 1997; pp. 165–196.

61. Schumacher, M.; Durán-Díaz, P.; Kurjenoja, A.K.; Gutiérrez-Juárez, E.; González-Rivas, D.A. Evolution and Collapse of Ejidos in Mexico—To What Extent Is Communal Land Used for Urban Development? *Land* **2019**, *8*, 146. [CrossRef]

62. Crépin, A.-S. Complexity, resilience and economics. In *Global Challenges, Governance, and Complexity*; Galaz, V., Ed.; Edward Elgar Publishing: Cheltenham, UK, 2019; pp. 166–187.

63. Holling, C.S. Resilience and Stability of Ecological Systems. *Annu. Rev. Ecol. Syst.* **1973**, *4*, 1–23. [CrossRef]

64. Folke, C.; Kofinas, G.P.; Chapin, F.S. *Principles of Ecosystem Stewardship*; Springer: New York, NY, USA, 2009; ISBN 978-0-387-73032-5.

65. Schicklinski, J. *The Governance of Urban Green Spaces in the EU*; Routledge: Abingdon-on-Thames, UK, 2017; ISBN 9781315403823.

66. Colding, J.; Barthel, S. The potential of 'Urban Green Commons' in the resilience building of cities. *Ecol. Econ.* **2013**, *86*, 156–166. [CrossRef]

67. Bendt, P.; Barthel, S.; Colding, J. Civic greening and environmental learning in public-access community gardens in Berlin. *Landsc. Urban Plan.* **2013**, *109*, 18–30. [CrossRef]

 *land*

Article

A Bottom-Up and Top-Down Participatory Approach to Planning and Designing Local Urban Development: Evidence from An Urban University Center

Teodoro Semeraro [1,*], Nicola Zaccarelli [2], Alejandro Lara [3], Francesco Sergi Cucinelli [4] and Roberta Aretano [5]

[1] University of Salento, Department of Biological and Environmental Sciences and Technologies, Ecotekne, Prov. le Lecce Monteroni, 73100 Lecce, Italy

[2] European Commission-DG-Joint Research Centre, Directorate for Energy, Transport and Climate, Westerduinweg, 3, NL-1755 LE Petten, The Netherlands; nicola.zaccarelli@ec.europa.eu

[3] University of Concepción Department of architecture Victor Lamas, 1290 Concepción, Chile; alejandrolara@udec.cl

[4] Francesco Sergi Cucinelli; Intercultural linguistic mediator: via cavalielli dell'ordine di Vittorio Veneto, 43. 73100 Lecce, Italy; fraseku@gmail.com

[5] Environmental consultant, Via San Leucio 15, 72100 Brindisi, Italy; roberta.aretano@gmail.com

* Correspondence: teodoro.semeraro@unisalento.it; Tel.: +39-320-8778174

Received: 28 February 2020; Accepted: 25 March 2020; Published: 27 March 2020

Abstract: The urban area is characterized by different urban ecosystems that interact with different institutional levels, including different stakeholders and decision-makers, such as public administrations and governments. This can create many institutional conflicts in planning and designing the urban space. It would arguably be ideal for an urban area to be planned like a socio-ecological system where the urban ecosystem and institutional levels interact with each other in a multi-scale analysis. This work embraces a planning process that aims at being applied to a multi-institutional level approach that is able to match different visions and stakeholders' needs, combining bottom-up and top-down participation approaches. At the urban scale, the use of this approach is sometimes criticized because it appears to increase conflicts between the different stakeholders. Starting from a case study in the Municipality of Lecce, South Italy, we apply a top-down and bottom-up participation approach to overcome conflicts at the institutional levels in the use of the urban space in the Plan of the Urban University Center. The bottom-up participation action analyzes the vision of people that frequent the urban context. After that, we share this vision in direct comparison with decision-makers to develop the planning and design solutions. The final result is a draft of the hypothetical Plan of the Urban University Center. In this way, the bottom-up and top-down approaches are useful to match the need of the community that uses the area with the vision of urban space development of decision-makers, reducing the conflicts that can arise between different institutional levels. In this study, it also emerges that the urban question is not green areas vs. new buildings, but it is important to focus on the social use of the space to develop human well-being. With the right transition of information and knowledge between different institutional levels, the bottom-up and top-down approaches help develop an operative effective transdisciplinary urban plan and design. Therefore, public participation with bottom-up and top-down approaches is not a tool to obtain maximum consensus, but mainly a moment of confrontation to better address social issues in urban planning and design.

Keywords: urban planning; urban space; urban regeneration; planning process; public participation

---

1. Introduction

Urban areas are ecosystems characterized by natural and artificial elements such as buildings, roofs, underground pipes, and green areas that are mainly related to human well-being. The urban ecosystem is not self-regulating but is "regulated" by humans [1–3]. The urban area is, therefore, a complex socio-ecological system where various communities can overlap and interact to a greater or lesser extent and co-evolve with their environment through change, instability, and mutual adaptation [2,4]. Therefore, the evolution of an urban ecosystem is influenced by social decisions or needs and by stakeholders' heterogeneity (for instance, culture, education, religion, vision, interest) [5–8]. These institutional levels include decision-makers like administrative and public institutions that plan the socio-ecological system at different scales: "ecosystems in urban areas", "urban areas within ecosystems", and "urban areas within regional/global ecosystems". At different scales, the boundaries of the urban ecosystem are not always well defined or clear, and therefore, boundaries of a survey area are defined considering the topic and interactions to be analyzed and on practical considerations. [5,9]. Institutional levels are hierarchical and consist of vertical relations between actors of the top and bottom of the levels. Therefore, the urban area has to be planned like a socio-ecological system where the urban ecosystems and institutional levels interact with each other in a multi-scale analysis. The urban planning and design have to create a synergy between different institutional levels (Figure 1) [10–13].

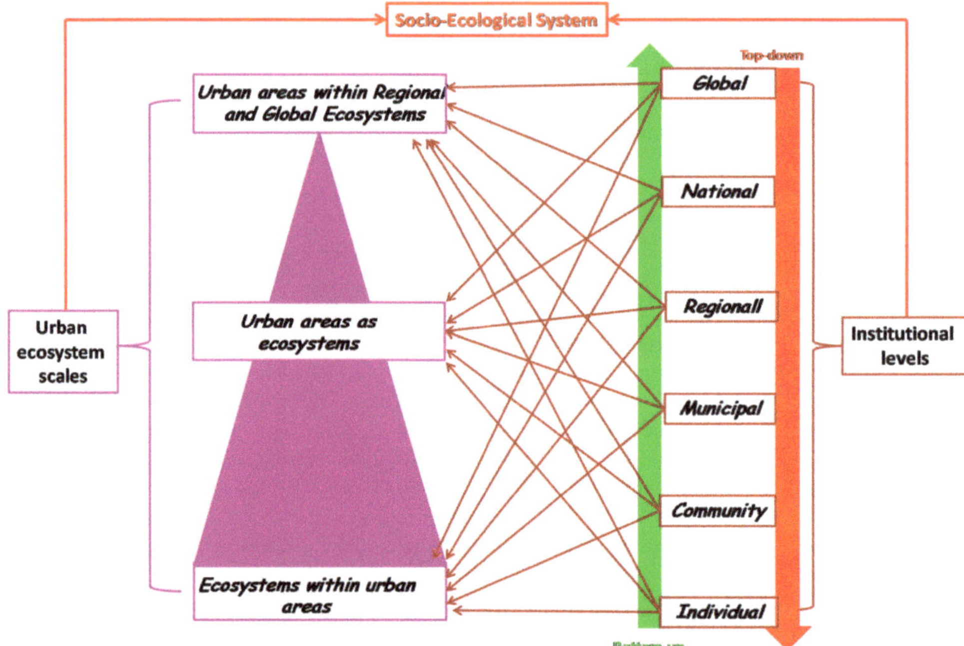

Figure 1. Schematic representation of the relationship between urban ecosystem scales and institutional levels in the socio-ecological system [5,6,9].

In many industrialized cities, urban planning must address the phenomenon of "shrinking cities" [14]. These cities have experienced a significant de-urbanization linked to the loss of functionality of some urban areas or buildings due to the decline of the manufacturing industry, migration, and depopulation [15,16]. Consequently, urban areas are characterized by free or temporarily not used urban spaces as a result of technological, economic, and social evolution. In many cases, these urban areas could be brownfield sites: "streets with vacant storefronts, underutilized social and technical infrastructures, and neglected parks and squares" [16,17].

Identifying new functions in urban spaces—either built or otherwise—in a transitioning economy and society is the main focus of resilient thinking, which has to recognize the complex and non-linear dynamic of economic and socio-ecological interactions [18–20].

Currently, the main urban planning and design use the top-down approach, where planners are considered "the experts" who put forward the proposal and then share it with others, mainly the decision-makers that can approve or reject the urban plan [13,21–24]. This generates stakeholders' conflicts in the type of use of the urban space, environmental protection, the interest of residents, labor conditions, economic development, and the identities of urban areas [8,25,26].

The planning of the urban space has to be considered a "public affair", aiming to envisage the right use of urban spaces considering the socio-ecological and cultural context of reference and solving conflicts in the choices or preferences in the use of destination of the urban space between stakeholder groups. Being able to evaluate the "awareness, value judgments, behavior, and attitudes" of the citizen in relation to urban space is an important task for a successful plan of urban transformations [27–30]. To create a social and shared vision of possible scenarios that can transform the territory, a prominent role must be given to stakeholders' needs, opinions, and interests, but also fears and doubts, in order to include their vision in the development of the urban space that they use [31–33].

Therefore, urban planning needs to combine bottom-up and top-down approaches, including stakeholder's participation with strategic spatial planning at different urban levels [34]. Public participation helps to understand the aspirations of stakeholders on possible urban development. Moreover, perception stimulates different stakeholders to develop ideas and proposals based on their knowledge, attitudes, and habits, providing greater awareness of their role in urban development. It is an action in urban design useful for increasing the ability to make effective planning choices [30]. For this reason, scholars consider stakeholder participation as one of the main aspects to take into consideration in order to guarantee the quality of urban planning [35,36].

On the one hand, there are many examples of bottom-up and top-down approaches in policy activities that are mainly focused on the management of natural resources or services (e.g., energy policy, climate change, watershed management, mobility, agricultural, environmental) on municipal, regional, national, and international programs. On the other hand, these approaches are less frequently used on small urban land use planning and design [37–39]. Although these approaches have attempted to include community stakeholders, this has often proved problematic, and planning guidelines do not yet consider design principles that foster social learning, knowledge exchange, and power-sharing [7,40–42]. Mainly, public participation may not always yield a mutually acceptable solution, especially when the interests of stakeholders are diverse and conflicting [38]. Often, top-down and bottom-up urban planning approaches are sometimes considered incompatible because they can produce conflict and fragmentation in the built new environment vision between different urban levels and stakeholders [13,43,44].

This work wants to develop a planning-process of the urban space transformation able to create feedback between different stakeholders at different institutional scales to reduce the mismatch between governance levels and the scales at which people benefit from urban space: from the need of the single individual to the development vision of the decision-makers [5,6,45]. Starting from a practical case study, we propose a combination of a bottom-up and top-down methodology capable of developing a participated urban plan, harmonizing the various stakeholders' interests that act at the different administrative levels and integrating ecological and socio-economic components in the context in which it is inserted [31,45].

Mainly, the study is focused on the Plan of the Urban University Center (PUUC), involving the creation of new university lecture halls in a university urban space that was the research site of tobacco production. Considering that the University represents the main stakeholder, as it is the owner of the urban space, the planners tried to satisfy university needs with the urban transformation vision of the decision-makers (top-down), also taking into account the aspirations of the citizens that act in the context area of the PUUC (bottom-up).

We hope to identify the best solutions for the use of urban spaces to integrate the citizens' visions with those of the planners and of the different public institutions that have an administrative role in choosing the final destination of the area.

2. Materials and Methods

2.1. Study Area

The study area is a free space of the university center in the Municipality of Lecce, Apulian region, South of Italy (Figure 2). The presence of the university has greatly influenced the economy of the district, favoring the opening of numerous commercial activities, such as bars, take-aways, restaurants, bookshops, and pubs. Furthermore, the real estate business linked to the rental or sale of student apartments has benefited from the situation. From the cultural and social point of view, the free urban space is located within the former Agricultural Research Center (ex CRA), which was used in the past for tobacco research activities. Currently, the ex CRA is employed for university lectures. The urban free space of interest is characterized by herbaceous vegetation with no ecological value (Figure 2B).

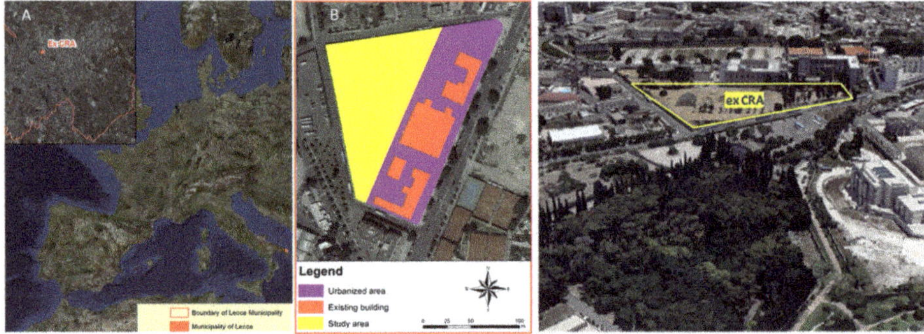

Figure 2. (**A**) Municipality of Lecce and location of the study area; (**B**) study area with reference to the context of the former Agricultural Research Center (ex CRA).

Near the ex CRA, there is a large green urban park, the cemetery of Lecce (classified as a historical asset for its architecture), and the Monastery of the Olivetans (founded at the end of the XII century, and currently used by the Department of Historical Studies of the University). To the west of the ex CRA, there is a main road, the Castle Charles V, and the ancient city walls. The north and south parts of the ex CRA have no significant neighboring elements.

2.2. Focus of the Planning Question

The Italian Inter-Ministerial Committee for the Economic Planning (CIPE) has identified and allocated resources in favor of interventions of strategic national and regional importance for the implementation of the national plan for the South's strategic priority: "Innovation, research, and competitiveness". One of the projects included in the plan for the South is located within the "Urban University Center of the ex CRA". Mainly, in an area of about 11,186 square meters, the university developed an urban plan involving the construction of a new building of about 3100 square meters for educational activities, a parking area of 1734 square meters, and a recreational green area of 6322 square meters. The new area will be realized near other university buildings, and together they will form a widespread urban university campus. The budget for the execution of the plan is EUR 8,000,000.

During the first phase of the planning activity, some regional authorities expressed a favorable opinion of the new PUUC because the plan did not cause negative environmental impact. However, the Ministry of Cultural Heritage, through the Superintendence, was the main institution that opposed the university urban plan. The Superintendence considers that the PUUC will alter the harmony between existing buildings and the identity of the area, and hence it only promotes the development of a green lung.

Considering the institutional conflict, this research wanted to develop bottom-up and top-down participation processes approach able to orient future use of free urban spaces.

2.3. Design Approach

This study was developed considering bottom-up and top-down approaches in the socio-ecological system [46,47]. Mainly, the stakeholders' participation process was designed considering different roles in the transformation of the socio-ecological system. The designers of the University and Superintendence are considered decision-makers that can directly choose the typology of the transformations (top-down). The citizens are considered as users of the urban space on which the choices of the decision-makers are reflected. However, at the same time, the citizens can revolt against the choice of decision-makers and condition the final result (bottom-up).

The work is organized as follows (Figure 3):

- Stage 1 develops a method to get information on citizens' perceptions, considering social issues linked with urban space;
- Stages 2 and 3 are dedicated to obtaining data and analyzing the results;
- Stage 4 develops the discussion of results of the citizens' participation;
- Stage 5 discusses the results of the questionnaires with the decision-makers.

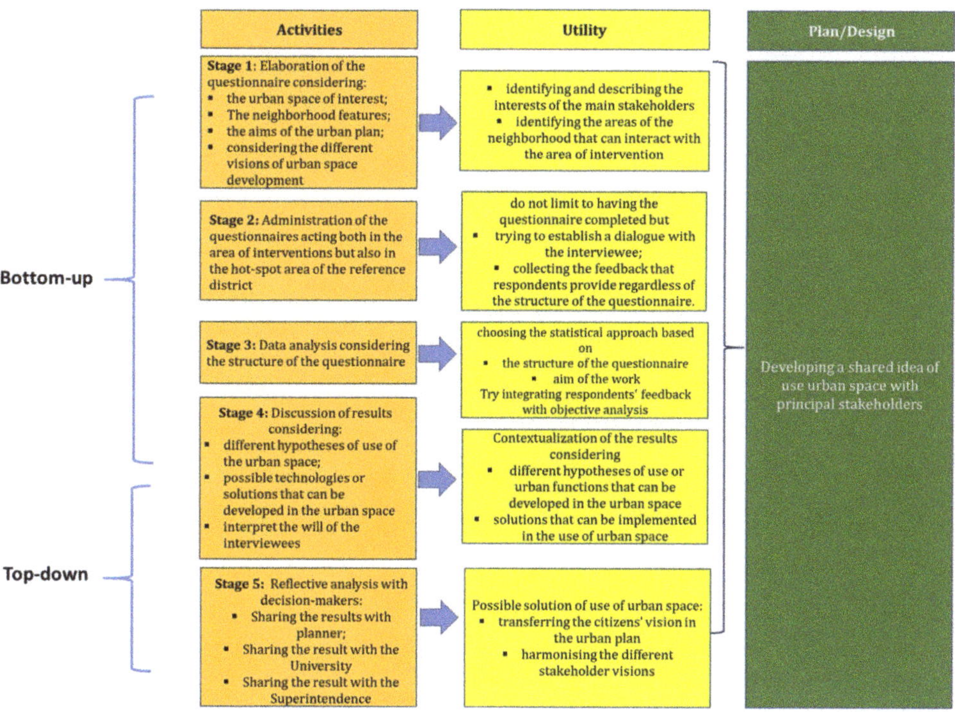

Figure 3. The conceptual work models that we have developed for the top-down and bottom-up approaches.

2.4. Bottom-Up Activity: Questionnaire Survey

The study was planned by taking into consideration the microscale, which encompasses the urban space, the structure of buildings, the relationship between them, and their interaction with other elements of the neighborhoods [48].

The work used face to face questionnaires to gather information about what citizens "feel, hope, wish, approve, or disapprove" for the future use or transformation of the identified urban space [49,50].

The survey was developed so as to include different types of citizens and users, such as students and people who live or work near the area where the project should be developed. This is important in order to explore their opinions and preferences on the possible uses of the free space of the ex CRA. Questionnaires were administered from 15 April 15 2018 to 30 May 2018, both in the morning and in the afternoon during working days and holidays. This was necessary to better characterize the typology of individuals who frequent the area of interest.

The questionnaires were delivered in three different places: the "ex CRA" area, the parking lot in front of the area of interest, and in the urban park "Belloluogo" (Figure 4), as these are the principal hotspot areas of the neighborhood in the context of the urban space of interest.

Figure 4. Localization of the points of administration of questionnaires, historical elements, and urban university sites.

We used Sierra's formula to detect the sample size of the number of individuals to interview [50]:

$$\text{Sample Size} = \frac{4 * N * p * (1-p)}{E^2(N-1) + 4 * p * (1-p)}$$

where N is the number of inhabitants; E is tolerated error; p is the portion of the variable in the population" [39].

Questionnaire structure

The questionnaire was structured in order to evaluate four main dimensions: the main users of the area; the historical relevance of the place (Items 1–3), the building size and harmonious insertion in the urban context (Items 4–7), and the importance of developing green areas compared to other options (Items 8–9). In particular, considering Item 9, three alternatives in the use of the area were indicated, and respondents could express a preference value. This last aspect was developed after taking inspiration from Directive 2001/42/EC and national law (D.lgs 152/2006), which provide for the analysis of different planning hypotheses in the drafting of the Strategic Environmental Assessment (SEA) to identify the best possible solution [51,52].

The questions were formulated with a simple and clear structure to be filled out quickly by ticking the preference box. Specifically, respondents were asked to express a preference using a five-point for the second section and a ten-point for the third section. In this way, respondents indicate their level of agreement to a statement. In the present survey, the ten-point scale was used when respondents could express their preference for the different design solutions proposed (e.g., [53–56]).

The data analysis was performed with the descriptive statistics for the first seven items, while the last items were analyzed through tables of contingency. The contingency tables are used to represent and to analyze the relationships between two or more variables, through the study of their combined frequencies [57].

Before the survey, a pilot study was conducted, and five questionnaires were administered to people to verify whether the proposed items were adequate and easily understood.

2.5. Top-Down Activity: Sharing the Questionnaire Results with the Main Decision-Makers

As the final part of this work, we discussed the results of the questionnaires with the decision-makers, who, at that moment, had a different vision for the use of urban space. In this way, we try planning a new hypothesis of the shared urban space transformation.

This represents a simple exercise to try to design the new PUUC considering citizens' visions.

This was developed with informal appointments with decision-makers like the designers of the University and Superintendence.

In these appointments, starting from the original PUUC, we discussed potential new solutions of the PUUC and analyzed how the main results of the questionnaires could be incorporated in the new planning of the urban space. After that, we developed an illustrative conceptual graphic of one of the possible new urban space transformations. For practical reasons of this research, we have not considered the turnover between the various decision makers in time.

3. Results and Discussion

3.1. Bottom-Up Activities

3.1.1. Stakeholders Characterization

For the study area, the minimum sample size is 382 individuals. Therefore, our 624 questionnaires can be considered as representative of the population that characterizes the study area. Mainly, 42% of the questionnaires were compiled in the ex CRA area, 33% in the parking area, and 25% in the urban park "Belloluogo" (Figure 4).

Table 1 shows the main socio-demographic characteristics of the sample interviewed. It is possible to note that the main users of the area are young people and students. In particular, most of the interviewees were students under 25. This result could be strongly influenced by the presence of two university centers.

Table 1. Socio-demographic characteristics of the sample.

Characteristic.	Category	%
Gender	Male	35
	Female	62
No answer		3
Age	From 19 to 25	70
	Over 25	30
Occupation	University students	71
	Other activity	29
Residence	Municipality in the Province of Lecce	68
	Outside	32

3.1.2. Historical Relevance of the Place

Analyzing the historical relevance of the study area, it is clear how almost all of the respondents considered this area relevant to the historical and cultural point of view. Only 19% said that the area is "not very important" and only 3% that it is "not at all important". However, the surveyed people did not know the history of the ex CRA, or where precisely this building was located. Therefore, the surveyed respondents consider this building historically important for the urban context in which it is located, rather than for the specific history of the structure. The distribution of answers obtained is similar in the different sampled areas (Figure 5).

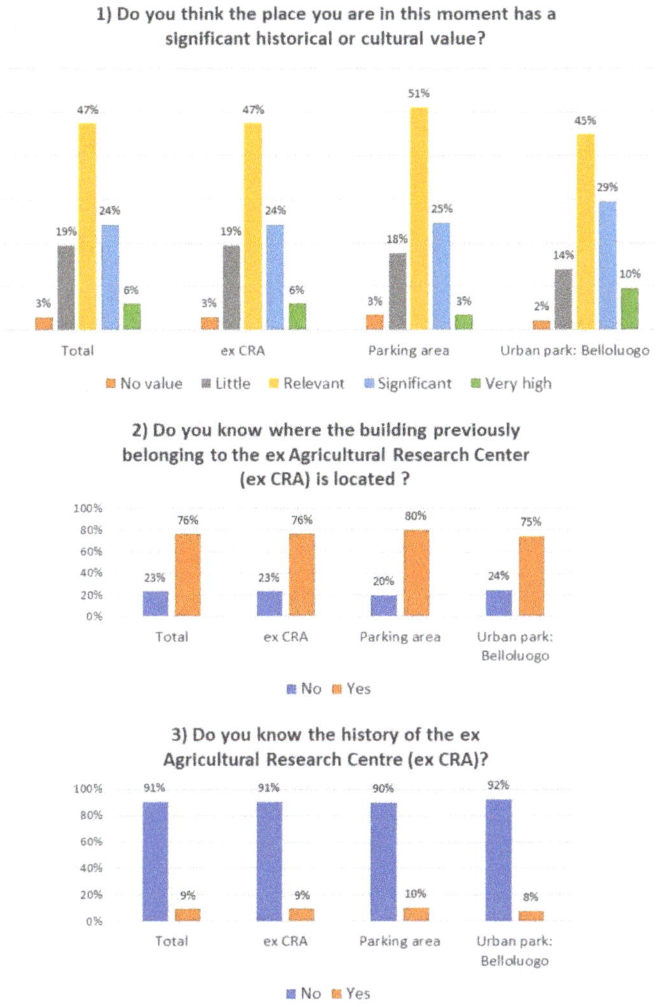

Figure 5. Answers to Questions 3, 4, and 5, concerning the historical relevance of the study area.

3.1.3. New construction in urban space

Subsequently, the opinion of the surveyed respondents was investigated with reference to the construction of a new building for teaching purposes within the former CRA. In particular, 43% of the respondents answered that the possibility of a new building altering the architectural harmony between existing buildings depends on the type of building that will be built. However, this building will not be a disturbance to the livability of the area (Figure 6).

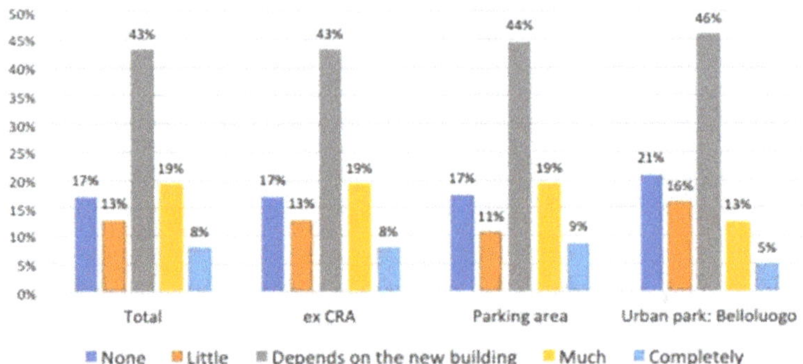

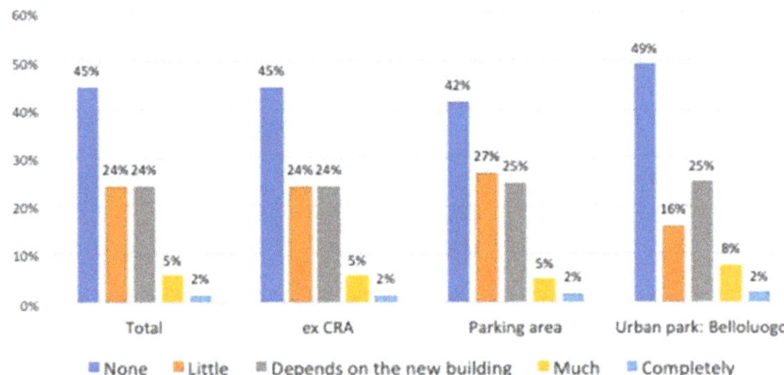

Figure 6. Answers to questions on the possibility of a new building inside the former CRA.

According to the surveyed respondents, the width and height of a new building could also be the same as the size of the central building existing within the former CRA, or even wider and higher. In fact, these two options were the second most selected choices in all the three sampling areas (Figure 7).

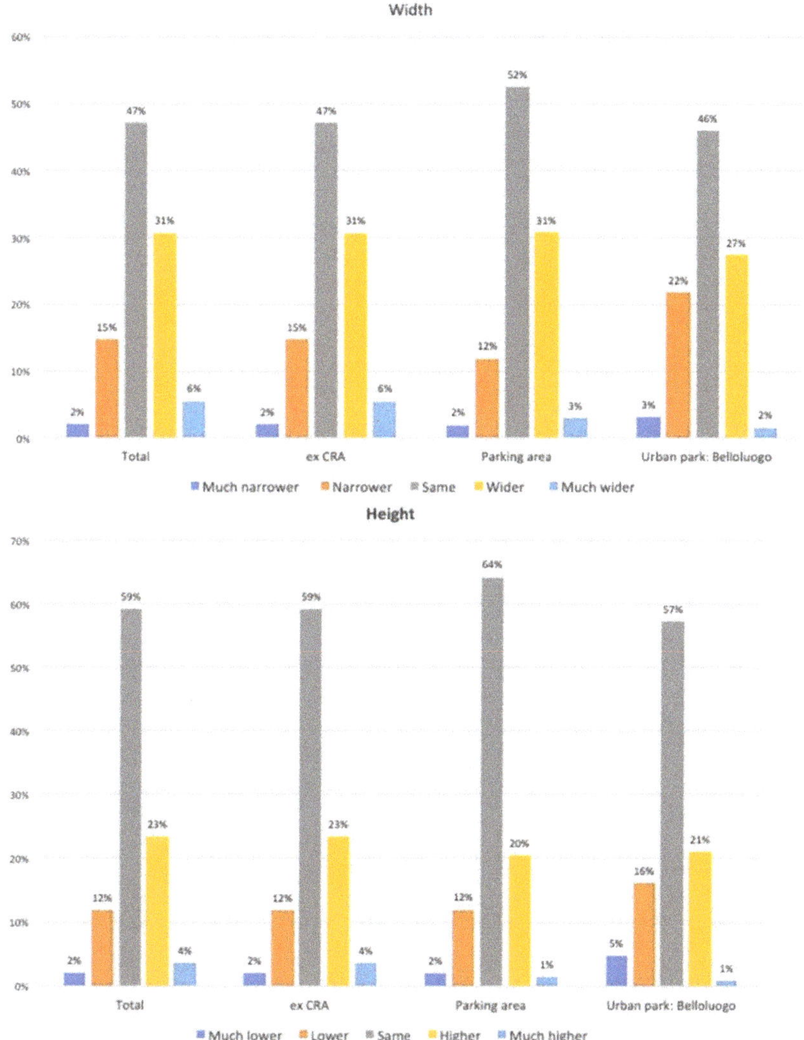

Figure 7. Answers to questions on the possible width and height of the new building that is to be realized within the former CRA.

However, the construction of the new building should not lead to the elimination of the currently unused greenhouse. According to the surveyed respondents, the greenhouse should be used for social activities (Figure 8).

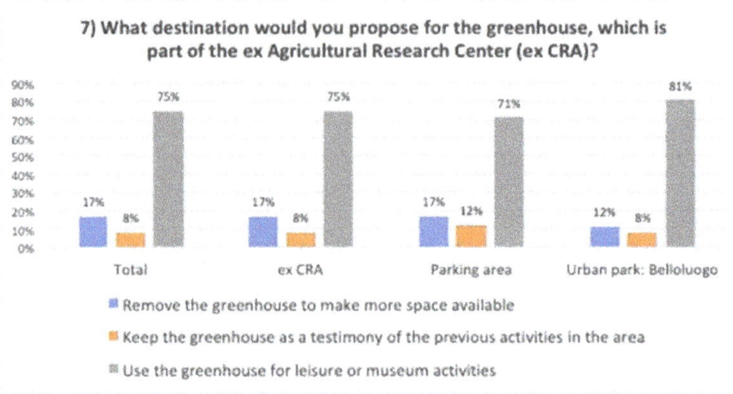

Figure 8. The question related to the greenhouse.

3.1.4. Potential actions that can be developed in the urban space

In the final questions of the questionnaire, we investigated the main actions of the plan to be launched within the former CRA (Figure 9). Surprisingly, it emerged that "improving the quality of urban green areas" is not among the main prerogatives indicated by the surveyed respondents. Overall, the main preferences were "finding a right harmony between use, green spaces, and economic activities" and "favoring the development of social activities (concerts, parties, and shows)". However, the surveyed respondents within the ex CRA expressed their preference "to create new classrooms and laboratories to enhance the university's educational offerings and accommodate a larger number of students" and "favoring the development of social activities" (Figure 9).

From the analysis of the contingency tables, statistically significant differences emerge. Mainly, there are differences between the alternatives "a", "b", "d", and "i". ($p < 0.01$). However, considering three sampling points (Figure 9), there are no statistically significant differences in the choice for the alternatives "a", "d", and "i", while there are statistically significant differences for the choice obtained for the alternative "b" in the different areas. Furthermore, the results show statistically significant differences in the choice of the solutions proposed considering the age of the surveyed respondents. The respondents aged 25 years and under expressed their main preferences as "i", "d", and "b", while those over 25 expressed as main preferences "a", "i", and "d" (Figure 9).

The numbers in the cells represent the frequencies relative to the appreciation value for each proposed alternative. The answers were analyzed, considering both the single survey areas and the whole (Figure 9a). In Figure 9b, the answers were analyzed considering two age groups: under 25 years and over 25 years.

8) What actions would you perform in a transformation project of the ex CRA?

a) Enhancing the city green areas

b) Realizing new classrooms and laboratories to enhance the educational activities and serve a higher number of students

c) Eliminating the perimetral fencing to allow a higher connection with other elements of the area: bus parking, car parking, park Belloluogo, the Olivetani monastery, shops in the surroundings, cemetery, Carlo Pranzo area and the fence walls

d) Finding a harmony among the fruition, the realization of green areas and the creation of economic activities

e) Realization of footpath and bus platforms

f) Realization of an area always accessible to the residents

g) Restoration of the greenhouse

h) Increasing the number of parking lots

i) Promotion of social activities

Table a

Total

Frequency of values / Appreciation values	a	b	c	d	e	f	g	h	i
0	2	5	25	3	22	5	32	41	20
1	3	3	6	2	8	3	10	17	2
2	8	5	19	4	20	5	14	16	7
3	13	9	29	8	23	12	23	20	6
4	10	16	23	10	18	9	16	30	8
5	39	25	34	21	60	40	44	56	34
6	29	36	42	47	54	43	53	48	34
7	67	69	58	65	58	78	81	71	61
8	123	113	86	101	87	134	86	73	95
9	61	80	70	79	78	56	60	53	62
10	[illegible]	207	176	[illegible]	140	183	149	143	[illegible]

Ex Cra

Frequency of values / Appreciation values	a	b	c	d	e	f	g	h	i
0	1	1	9	2	9	2	10	8	15
1	1	1	3	2	2	2	6	5	1
2	4	1	8	2	6	4	6	8	4
3	3	2	11	2	7	2	11	5	1
4	4	3	10	3	6	6	7	9	5
5	13	7	17	14	27	23	18	21	17
6	9	10	14	20	25	17	19	25	17
7	31	21	24	24	27	33	32	28	25
8	53	51	30	42	32	51	33	27	35
9	28	34	35	36	32	17	28	27	23
10	87	[illegible]	73	87	61	77	64	71	[illegible]

Parking area

Frequency of values / Appreciation values	a	b	c	d	e	f	g	h	i
0	1	0	9	1	6	0	13	10	4
1	2	0	1	0	2	1	3	3	1
2	3	3	7	2	8	1	5	4	3
3	7	4	12	5	10	10	8	8	3
4	5	6	10	3	5	1	7	12	3
5	16	5	11	5	19	12	20	25	9
6	14	12	14	13	14	19	16	11	13
7	23	29	23	24	18	22	33	26	22
8	47	40	32	37	42	54	30	35	38
9	18	35	22	26	29	21	20	15	24
10	65	67	60	[illegible]	48	60	46	52	[illegible]

Urban park: Belloluogo

Frequency of values / Appreciation values	a	b	c	d	e	f	g	h	i
0	0	4	7	0	7	3	9	23	1
1	0	2	2	0	4	0	1	9	0
2	1	1	4	0	6	0	3	4	0
3	3	3	6	1	6	0	4	7	2
4	1	7	3	4	7	2	2	9	0
5	10	13	6	2	14	5	6	10	8
6	6	14	14	14	15	7	18	12	4
7	13	19	11	17	13	23	16	17	14
8	23	22	24	22	13	29	23	11	22
9	15	11	13	17	17	18	12	11	15
10	[illegible]	37	43	[illegible]	31	46	39	20	[illegible]

Table b

Age≤25 years

Frequency of values / Appreciation values	a	b	c	d	e	f	g	h	i
0	2	1	16	2	11	[illegible]	23	17	16
1	3	2	5	2	6	1	9	12	1
2	8	4	13	4	14	4	11	11	7
3	11	7	21	4	13	8	19	14	3
4	9	11	18	7	13	8	14	23	8
5	25	14	24	17	48	31	36	42	27
6	27	24	32	36	45	36	42	43	22
7	52	48	47	47	44	64	68	51	47
8	98	89	60	76	71	105	61	64	72
9	47	68	58	61	60	40	43	42	53
10	[illegible]	[illegible]	128	[illegible]	97	120	96	103	[illegible]

Age≥26 years

Frequency of values / Appreciation values	a	b	c	d	e	f	g	h	i
0	0	4	9	0	10	2	8	23	4
1	0	1	1	0	2	0	1	5	1
2	0	1	5	0	6	1	2	5	0
3	1	1	7	3	9	3	3	5	2
4	1	5	5	3	5	1	2	6	0
5	12	10	9	4	9	7	8	13	7
6	2	11	8	11	8	6	11	5	12
7	12	18	9	17	12	12	12	15	12
8	21	23	21	21	12	24	20	8	20
9	14	9	12	16	16	15	14	8	8
10	[illegible]	[illegible]	47	44	55	41	[illegible]	49	37

Figure 9. Heat map for the analysis of possible planning actions considering the different survey areas. The colors from red to green represent a gradient of preference from low preference to high preference in relation to the number reported in the cells of the table. (**a**) The answers were analyzed, considering both the single survey areas and the whole. (**b**) The answers were analyzed considering two age groups: under 25 years and over 25 years.

The last question analyzed the preference of the surveyed respondents with respect to three project alternatives. Option 2, which involves a multifunctional use of the area and the realization of a new building, prevailed over the other two alternative projects. This preference did not show differences, considering the different sampling points or the age of the surveyed (Figure 10).

Land **2020**, _9_, 98

9) <u>Evaluate from 0 to 10</u> the following three project ideas regarding the ex CRA:

➤ Option 0: leave the area as it is

➤ Option 1: Enhance the historical and cultural aspects of the area: valorization of the greenhouse for recreational activities and improvement of the green areas inside the ex CRA garden

➤ Option 2: Enhance the fruition of the area: in addition to the activities foreseen in "option 1", this option includes the realization of a new building to host classrooms and laboratories, the removal of the perimetral wall to make the area always accessible to all the residents, realization of footpaths and bus platforms as well as other useful structures.

Table a

Appreciation values	Total			ex CRA			Parking area			Urban park: Belloluogo		
	option 0	option 1	option 2	option 0	option 1	option 2	option 0	option 1	option 2	option 0	option 1	option 2
0	[illegible]	13	9	[illegible]	6	4	[illegible]	4	3	[illegible]	3	2
1	38	5	2	18	2	0	14	1	0	6	2	2
2	49	10	5	23	7	2	17	2	0	9	1	3
3	46	14	11	15	3	4	20	7	3	11	4	4
4	37	23	15	10	6	5	14	11	6	13	6	4
5	58	62	28	30	23	8	15	24	12	13	15	8
6	16	59	25	8	25	10	4	24	7	4	10	8
7	18	78	53	4	34	18	10	25	21	4	19	14
8	5	118	146	1	51	53	3	37	61	1	30	32
9	1	56	83	0	26	33	1	17	31	0	13	19
10	5	145	206	1	57	103	3	51	59	1	37	44

(Row label column: Frequency of values)

Table b

Appreciation values	Age ≤25 years			Age ≥26 years		
	option 0	option 1	option 2	option 0	option 1	option 2
0	[illegible]	11	8	[illegible]	2	2
1	33	3	0	8	2	2
2	41	10	5	8	2	0
3	33	10	5	14	3	4
4	27	18	13	12	5	2
5	44	43	19	14	19	9
6	15	49	19	3	14	11
7	12	58	42	6	20	16
8	5	86	114	2	38	34
9	1	43	61	0	17	25
10	3	102	147	2	48	65

(Row label column: Frequency of values)

Figure 10. Heat map for the analysis of the results for the three project alternatives, considering the different survey areas and the age of respondents. The colors from red to green represent a gradient of preference from low preference to high preference in relation to the number reported in the cells of the table. (**a**) The answers were analyzed, considering both the single survey areas and the whole. (**b**) The answers were analyzed considering two age groups: under 25 years and over 25 years.

The numbers in the cells represent the frequencies relative to the satisfaction value for each option. In the first table, the answers were analyzed, considering both the single survey areas and the whole. In the second table, the answers were analyzed considering two age groups: under 25 years and over 25 years.

3.1.5. Relevant elements for the development of the new PUUC

The results showed that the urban space has to be planned as a social element of the context connected to ecological and economic aspects. Indeed, according to the opinions of the interviewed, in the study area, the green urban spaces were not the principal development vision of urban areas highlighted by citizens. The results showed how the main users emphasized the need to develop an integrated plan between new construction and green area that should favor economic and social development. The urban green space has to be planned as an element integrated within the development of new structures that can increase the social life of the area, including the creation of potential new buildings for economic or educative activities, creating a multifunctional center able to give vitality to the urban context of reference in different moments of the day.

From the analysis of the questionnaire, it emerged that the preservation of cultural-historical aspects should not be interpreted as prejudicial to urban transformation. What is important is to create harmony between the new urban elements that will be developed and the cultural–historical value of the area. In this perspective, according to the surveyed respondents, a new building within the former CRA should be designed as a multi-purpose center, without compromising the urban identity in which university activities, but also other social activities, can be developed, e.g., libraries, playrooms, and leisure centers.

Through participation, it was also possible to collect suggestions and specific proposals to plan the use and design of the urban space. In the initially proposed PUUC, the elimination of the greenhouse that is currently in a state of neglect was planned so as to make room for the new building. The surveyed respondents expressed their willingness to preserve the greenhouse and use it for recreational activities (Figure 11). An interesting aspect is that many surveyed respondents were unaware of the presence of the greenhouse. Even many students who attend the area have not noticed its presence or thought that the structure was indeed a greenhouse. This probably shows how limiting the use of this area can also compromise the awareness regarding the place and its identity. Therefore, creating a green area without increasing the usability of the place or the ability of people to move within it can limit the cultural value of the area.

Figure 11. Photos of the greenhouse.

To make this area more active from a social point of view, it is fundamental to ensure greater use of the spaces and movement of individuals within the area. The possibility of an individual to perceive an area of social value and obtain benefits depends on the experience that an individual has in moving within the space, creating occasions for social interactions and enjoyable time [16,18]. The imposing enclosure wall (h = 2.50 m) that surrounds the entire area of the CRA is an obstacle, isolating this urban space from the surrounding area (Figure 12).

Figure 12. Photos of the enclosure wall. (**a**,**b**) are to the south–west side; (**c**,**d**) are to the north–west side; (**e**) is to the east side.

3.2. Top-Down Activities: Reflective Analysis Using Bottom-Up Information

This work represents an experiment of a combination of the top-down and bottom-up planning processes. In particular, in this study, questionnaire activities allowed for the identification of the interests of the main citizens who frequent the area and their social needs that will stimulate the development of the new PUUC of the ex CRA ("bottom-up"). The results of the questionnaire were shared informally with decision-makers. The results were food for thought to hypothesize the development of a plan, which, starting from different visions, could produce a shared urban transformation process. Figure 13 shows an example of possible urban visions that can emerge from the comparison between decision-makers by analyzing the results of the questionnaires (top-down). The representation in Figure 13 is a simple example of potential PUUC visions that can be developed by this process. Other design solutions can also be developed because the focus is the social functionality that has to develop in the urban space independently of the urban elements such as buildings or other elements.

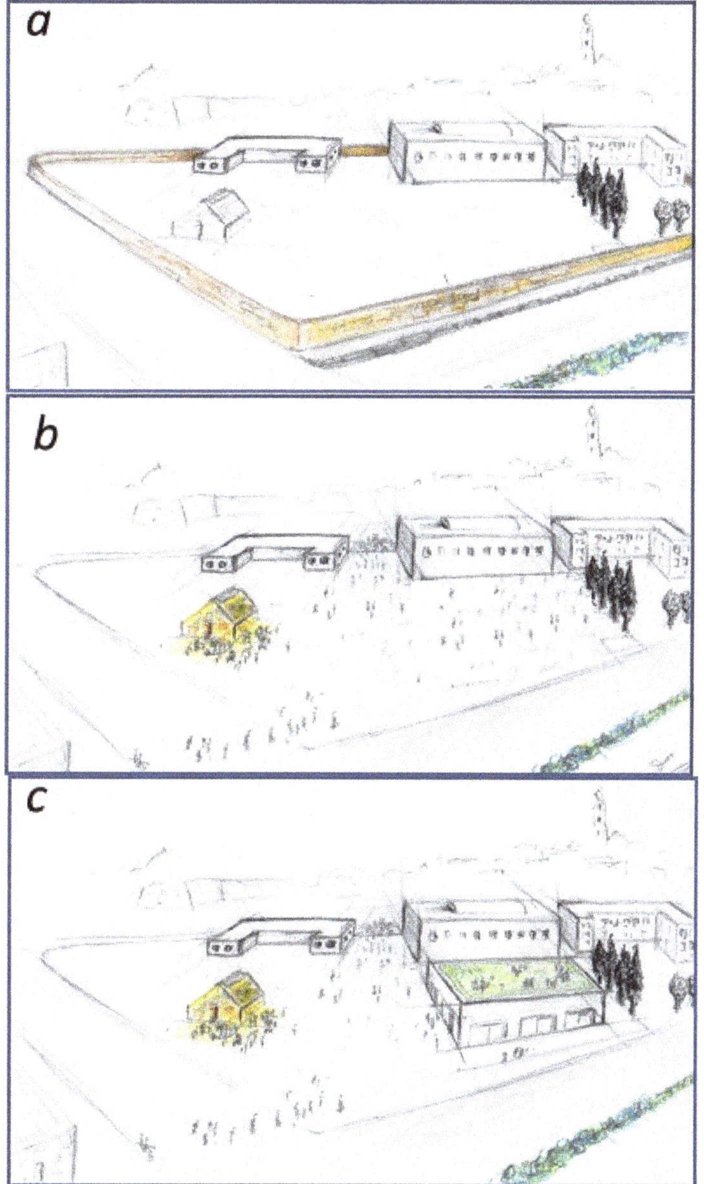

Figure 13. (**a**) schematic representation of the actual state of the area; (**b and c**) draft of the potential evolution of the area following and interpreting the indications of the respondents and discussion with the citizens, designers, and decision-makers. This figure represents a simple scientific exercise and does not have legal or institutional value or what this area will be in the future.

Considering these results, it emerged in these appointments that, in the new PUUC, the demolition of the wall should be further emphasized in order to encourage the development of social activities that may not necessarily be directly related to the university, but also related to leisure of a temporary nature, and, therefore, not pre-planned. Moreover, a greater permeability of this space would facilitate the circulation of people within the urban context of reference, to better harmonize the existing valuable elements and make the buildings of the ex CRA more visible. In addition, the redefinition of the ex CRA perimeter would allow the creation of pedestrian and cycle paths—currently somewhat absent—and new bus stops supporting public transportation The urban space quality is strongly connected with the quality of public transport and the promotion of healthy patterns of walking and cycling as daily activities [58,59]. This solution would increase the type of users of the site, which is currently limited to students, to enhance the cultural and historical value of the area, now restricted within four walls. In this case, the new use of the urban space can increase the contact between citizens and urban elements that influence the cultural identity of urban space, opening the opportunity to new knowledge and new relationships between stakeholder and urban space that was limited from the no-use of urban space. The new PUUC should include the recovery of the greenhouse. The greenhouse could host indoor cultural and social activities, student associations, or a place to set up co-working. Therefore, in the new PUUC, the greenhouse will play a fundamental role in implementing the social activity of the area, increasing the fruition and experiences that the inhabitants can have within the area (Figure 13b).

Another point of discussion is in regard to the possible use of the roof of a potential new building as a new social element of the PUUC. In this case, the roof can be designed as a green roof that can be used as an outdoor study area for students or small public events. Green roofs are an important tool used in residential, commercial, government, and public buildings to increase sustainability and biodiversity and decrease energy consumption, urban heat island impacts, and greenhouse gas generation in the city [60]. In this case, it helps to plan the multifunctional use of space. Therefore, green roofs can increase the quality of the buildings by reducing their impact on the urban landscape while also replacing some functions of natural areas and, therefore, assuming an important role in mental health [61] (Figure 13c). However, the realization of a new building would require more discussion among decision-makers.

The solutions highlighted can make the area more dynamic, which is an important element of the human experience in urban space that incorporates both the "relationship between the person and the person-to-place relationship", improving the perception of the identity of the urban space [62].

4. Discussion

The main inspiration in the urban regeneration of the urban spaces or degraded areas is the realization of green areas, as they are now widely recognized and documented in guaranteeing ecosystem services useful for the wellbeing of the population [62–65]. Indeed, the type of urban space use was the main focus of the conflict between some decision-makers: implementation of the university urban center vs. green areas.

As argued by such scholars as Bourdieu, Lefebvre, and Gans [66–68], the result of the bottom-up participation showed that the urban space must be thought of as a "social space", considering the main users and producing transitions able to support good quality of life. Therefore, the challenge going forward will be to apply an increasingly advanced and nuanced understanding of urban ecology in the practice of planning and designing urban ecosystems [69,70]. Urban spaces have to be planned as dynamic areas [71], giving the opportunity to develop new structures and functions to adapt them to new social needs and economic opportunities without upsetting urban identity and ecology quality. The implementation of a university urban center vs. green area is not the main new question of the PUUC, but how this aspect can be combined to implement the social use of the area going beyond the different vision or position of the decision-makers.

In socio-ecological systems like an urban ecosystem, the bottom-up and top-down participation approaches can give both a contribution to encourage the evolution in systems and increase the resilience of the area, understood as the ability to adapt their functions and structures to social changes. In particular, the bottom-up approach allows for the identification of the main stakeholders of the area and their social needs, which, in turn, will stimulate the development of the new urban plan of the ex CRA. This is important because it allows us to have a vision of the development of the area that is not conditioned only by the cultural background of the decision-maker, but of those who use the territory to meet the needs of everyday life. The important aspect of this approach is to actively connect the knowledge and information of bottom-up participation to decision-makers that manage the urban space at higher institutional levels.

Top-down participation, using the bottom-up information, in this case, can drive the choice and help decision-makers overcome an excessively deep-rooted view of conservation of the urban space that administratively slows down the urban regeneration process.

This would arguably help speed up the decision-making process by helping decision-makers become more aware of the transformations that are introduced in the urban context: "doing the right thing in the right place". This can be useful to produce a better acceptance of urban plans reducing the likelihood of conflicts between different experts or people that participate in the processes of planning development [31,72–74].

In this case study, participation activities started after the drafting of a first project by the university in relation to the use of the area that led the decision-makers to express a negative opinion on the project or highlighted specific critical points. From the meetings held with the decision-makers, it emerged that this approach could be useful in the initial phase of the project. However, this approach is also conditioned by the timing of urban planning. Long urban planning times can make this approach difficult to use because it can be conditioned by the turnover in management related to the decision-makers (for instance, the University, the municipality) that may require the need to restart the top-down phase. The vision of the use of urban space by current and future decision-makers can differ from previous decision-makers present when this work started.

Of course, the techniques used in this work can be improved. An important limitation of this work lies in the number of stakeholders involved in the bottom-up participation. This can affect the outcome of participation. To be representative, this method requires a great deal of effort in administering the questionnaires. Questionnaires are not the only tools to carry out the participation activity and cannot be defined as the best tool with respect to other methodologies because this depends on the scale of the investigation and type of activities. On a local level, however, questionnaires can represent valuable tools because they allow for the creation of a face-to-face relationship between planners and stakeholders. Therefore, questionnaires were used as a way to start the dialogue with the population and also to raise the interest of the interlocutors. Regarding the analysis of small transformations, it can be useful to involve citizens and their specific needs and vision in the reference context and try to put them in the final urban space planning.

Many participation approaches use the creation of thematic meetings or focus groups and tools such as online questionnaires that can open participation to a larger audience [22–24,37–40]. These actions may attract different citizens who are not necessarily users of the area, and, therefore, they may express a judgment based only on their preference or training and not on needs. This can also open a debate on the weight of the judgments expressed by those who are not familiar with or frequent the urban context of reference. We reckon that the chosen three main areas in which the questionnaires were submitted were the most functional to represent and characterize the typology of users of the urban reference ecosystem and, therefore, to analyze the social, economic, and ecological needs to be developed in the urban plan of the university (Figure 4). However, the characterization of social needs and stakeholders is the main issue of this study that needs improvement. As an alternative, using mixed methods would be an ideal solution, but this is not always feasible because of economic and time issues. For instance, in this work, the top-down activities were conducted with single appointments with the decision-makers. Therefore, in this work, we harmonized the response of decision-makers to social issues. Figure 13 is an example of the result of these activities. In the future, it can be interesting to plan the top-down participation activities by organizing focus groups between all decision-makers that participate in the authorization processes using the social issues derived from bottom-up participation as the starting point of discussion. In a similar focus group, the decision-makers should not limit themselves to expressing an opinion based on their skills and background but should try to produce a draft project, similar to the example in Figure 13, interacting directly with each other.

Even if the questionnaire did not provide open answers where citizens could freely express their thoughts, mainly in order not to weigh down the interview, an important aspect during the compilation of the questionnaire was the dialogue established with the interviewee. Often the surveyed was not limited to the simple answers to the questions, but to an open dialogue that went beyond the structure of the questionnaire. In this way, unforeseen or planned information, consideration, and opinions were obtained.

The majority of the questionnaires were filled out by young people aged between 18 and 25 years old. The results can be influenced by the greater presence of users under 25 years and students. Probably, this expresses the main current vocation of this area. However, we want to express some "reflections" that did not emerge from the analysis of data but were derived from the dialogue with all respondents during the interviews. People from 18 to 25 years old were more interested in addressing issues concerning the development of the territory in which they live. We can state that no young people refused to complete the questionnaire, a problem that has been found in other age groups, especially men. The students were interested and encouraged to make a contribution, providing their opinion, often critical, on the issues addressed in the questionnaire to try to improve the territory in which they live and project it in the future. In addition, during the interviews, a different attitude among the interviewees emerged. Specifically, the respondents between 18 and 25 years old seemed more cooperative and willing to make a contribution to improve urban quality. Older people sounded more pessimistic about the possibility of the political class favoring actions that could produce a change and were, therefore, less constructive in providing suggestions. The questionnaires showed a different view between young people and adults (Figure 10). In the context of urban planning, those who made the final decisions are mainly adults who occupy managerial positions. Such experts would make final decisions in terms of territorial sustainability, therefore, they are thinking about future generations. However, these managers sometimes have a different vision compared to the generations that they should be protecting. For this reason, participation is an effective tool for bridging this gap.

5. Conclusions and Recommendation

The combination of bottom-up and top-down participation methods can be a tool through which urban planning can drive the transformation or evolution of urban spaces at different institutional levels. It can increase the interactions between citizens in a vision that "unites and inspires" to develop urban quality space helping the decision-makers to identify hypotheses of territorial development that is more suitable on the basis of present and future scenarios of economic, environmental, and social evolution.

The study shows that in the socio-ecological system, the bottom institutional levels can introduce innovation or new vision in the use of free urban space and, therefore, bottom-up participation can push or trigger the evolution of the urban ecosystem, while the top institutional level drives the change from the top-down using the bottom-up participation information in planning actions between decision-makers. Therefore, considering the adaptive cycle in social–ecological systems, the bottom-up activity can be considered as a "revolt" process that affects the "urban space", changing it from a "conservation (K)" phase, where the urban space can be kept in a state of no-social use contributing in the degradation of identities of urban area, into a "collapse or release" (Ω) phase that stimulates new social uses of urban space. Therefore, the bottom-up can support the "reorganization" (α) phase to create situations able to drive innovation, considering economic and social processes. Top-down, in this case, plays a crucial role in determining and designing the new pattern of the urban space. The system will jump into a new adaptive cycle and therefore in new urban use with new environmental, social, and economic characteristics without losing the urban identity (Figure 14).

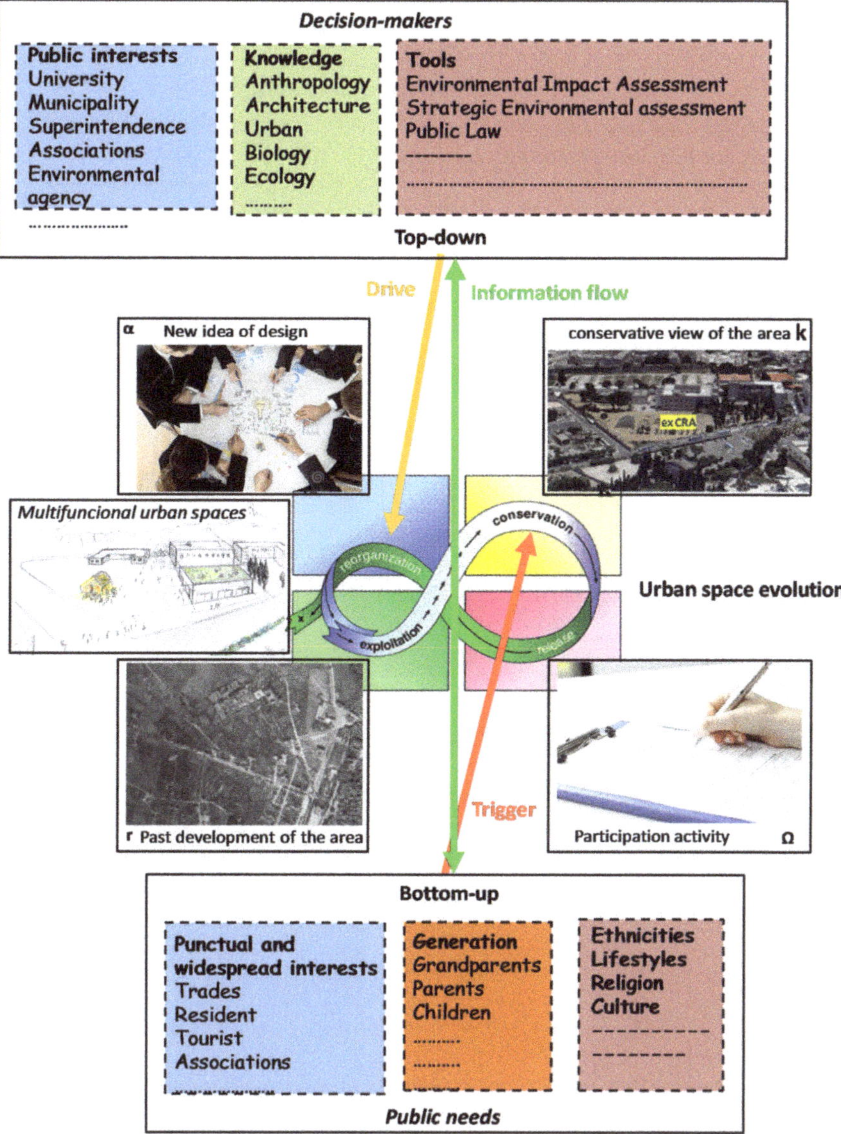

Figure 14. Contribution of the combination of bottom-up and top-down participation approaches in transdisciplinary planning and design of the urban space evolution considered as the socio-ecological system [10–13].

Therefore, the main aspect for the success of the bottom-up and top-down approaches is the creation of feedback between scientific knowledge derived from experiences and studies not directly connected to the characteristics of the study area and non-scientific visions deriving from those who live in the area, who express their opinions and advice based on their own life experiences. The bottom-up and top-down participation approaches can represent the base for transdisciplinary planning and design as they are useful to identify and correlate the ecological urban level and institutional levels integrating different cultural, knowledge, and generational needs, allowing the development of a holistic vision of the evolution of the urban space (Figure 14). This is important because it allows for a vision of the development of the area that is not conditioned only by the cultural background of the decision-makers, but of those who use the territory, in an effort to meet the needs of everyday life. (Figure 14). Therefore, this approach can be useful to harmonize the differences that can emerge at different institutional levels of the urban space, going from the single individual, the community that uses the area, and the different administrative levels that make the decisions [33,34].

In this way, the participation activities were not seen as an instrument for obtaining maximum consensus, but primarily as an opportunity to take into consideration the different stakeholders' interests and to better deal with urban issues that are not yet well defined. In this paper, the bottom-up and top-down participation approaches are important to combine the need of many stakeholders (single individuals) with the vision of urban development of fewer stakeholders that take the decision (decision-makers). The effectiveness of this approach lies precisely in the ability of the decision-makers to review their own position according to the different visions without remaining in a pre-decided position. Without the flexibility of the decision-makers, this approach can fail. Therefore, the main aspect of this approach is not primarily in the techniques used, but in the ability to acquire information and knowledge and make it turn transversely, creating synergy between the various stakeholders who often act and make decisions in isolation.

This approach allows for the creation of an urban plan with more "accountability", capable of reinforcing the responsibility of choices, guaranteeing greater insurance to the citizens about the proposed transformations, and giving an account of the choices made to combat the prejudices that accompany urban transformations and making the transformation process more reliable.

An aspect of strength of this process is the possibility to analyze conflicts to start an institutional dialogue between decision-makers and final users. In this case, the Superintendence had a central role in starting a productive dialogue, changing critical issues into strong points of the urban plan.

Author Contributions: Conceptualization, T.S.; methodology, T.S., R.A., and N.Z., A.L.; software, T.S.; validation, N.Z., A.L., and R.A.; formal analysis, T.S. and N.Z.; investigation, T.S., A.L., and R.A.; resources, T.S. and R.A.; data curation, T.S. and N.Z.; writing—original draft preparation, T.S., R.A., and A.L.; writing—review and editing, T.S., R.A., A.L., and F.S.C.; visualization, T.S., R.A., and A.L.; supervision, F.S.C. and N.Z.; project administration, T.S. All authors have read and agreed to the published version of the manuscript.

Funding: This research received no external funding.

Acknowledgments: I thank all the guys who have given their time to this project. I thank the researcher Sara Invitto for the corrections and suggestions that she made to the paper. I also thank Francesco Dieni for the graphic representation of Figure 13. I thank the Antonie De Vitis and Niko Antonelli for the professionalism and passion with which they worked on the PUUC project.

Conflicts of Interest: The authors declare no conflict of interest.

References

1. European Commission 2013. Green Infrastructure (GI) Enhancing Europe's Natural Capital. COM249. Available online: http://ec.europa.eu/environment/nature/ecosystems/docs/green_infrastructures/1_EN_ACT_part1_v5 (accessed on 25 November 2019).
2. Rebele, F. Urban Ecology and Special Features of Urban Ecosystems. *Glob. Ecol. Biogeogr. Lett.* **1994**, *4*, 173–187. [CrossRef]
3. Kourdounouli, C.; Jönsson, A.M. Urban ecosystem conditions and eco system services—A comparison between large urban zones and city cores in the EU. *J. Environ. Plan. Manag.* **2019**, *63*, 798–817. [CrossRef]
4. Lambin, E.F.; Meyfroidt, P. Land use transitions: Socio-ecological feedback versus socio-economic change. *Land Use Policy* **2010**, *27*, 108–118. [CrossRef]
5. Maggiore, G.; Semeraro, T.; Aretano, R.; De Bellis, L.; Luvisi, A. GIS Analysis of Land-Use Change in Threatened Landscapes by Xylella fastidiosa. *Sustainability* **2019**, *11*, 253. [CrossRef]

6. Leemans, R. *Modelling of Global Land Use: Connections, Causal Chains and Integration Inaugural Lecture*; Department of Plant Production Systems, Wageningen University: Wageningen, The Netherlands, 2000; p. 85.

7. Terry, H.Y.; Li, S.; Thomas, N.; Skitmore, M. Conflict or consensus: An investigation of stakeholder concerns during the participation process of major infrastructure and construction projects in Hong Kong. *Habitat Int.* **2012**, *36*, 333e342.

8. Peter, W.; de Langen, D. *Port Governance and Port Performance Research in Transportation Economics*; Elsevier Ltd.: Amsterdam, The Netherlands, 2007; Volume 17, pp. 457–477.

9. Piracha, A.L.; Marcotullio, P.J. *Urban Ecosystem Analysis Identifying Tools and Methods*; United Nations University Institute of Advanced Studies: Tokyo, Japan, 2003.

10. Gunderson, L.; Holding, C.S. *Panarchy: Understanding Transformations in Human and Natural Systems*; Island Press: Washington, DC, USA, 2002.

11. Cartier, C. Scale relations and China's spatial administrative hierarchy. In *Restructuring the Chinese City: Changing Society, Economy and Space*; Ma, L.J.C., Wu, F., Eds.; Routledge: London, UK, 2005; pp. 21–38.

12. Ferguson, J.; Gupta, A. Spatializing states: Toward an ethnography of neoliberal governmentality. *Am. Ethnol.* **2002**, *29*, 981–1002. [CrossRef]

13. Smith, N.R. Beyond top-down/bottom-up: Village transformation on China's urban edge. *Cities* **2014**, *41*, 209–220. [CrossRef]

14. Rienicts, T. Shrinking cities: Causes and effects of urban population losses in the twentieth century. *Nat. Cult.* **2009**, *3*, 231–254. [CrossRef]

15. Hollander, J.; Németh, J. The bounds of smart decline: A foundational theory for planning shrinking cities. *Hous. Policy Debate* **2011**, *3*, 349–367. [CrossRef]

16. Jiirgen, F. A Theory of Urban Decline: Economy, Demography and Political Elites. *Urban Stud.* **1993**, *30*, 907–917. [CrossRef]

17. Haase, D. Urban ecology of shrinking cities: An unrecognised opportunity? *Nat. Cult.* **2008**, *3*, 1–8. [CrossRef]

18. Masys, A. *Disaster Management: Enabling Resilience*; Springer: Berlin, Germany, 2008; Volume 3, pp. 1–8. [CrossRef]

19. Walker, B.; Anderies, J.; Kinzig, A.; Ryan, P. Exploring resilience in socialecological systems through comparative studies and theory development: Introduction to the special issue. *Ecol. Soc.* **2006**, *11*, 12–16. [CrossRef]

20. U.N.-HABITAT. *Objetivos de Desarrollo Sostenible*; ONU-HABITAT: Geneva, Switzerland, 2015.

21. Giddings, R.; Hopwood, W. *A Critique of Masterplanning as a Technique for Introducing Urban Quality into British Cities*; Available online: https://stuffit.org/bnb/research/regeneration/pdf/masterplan.pdf (accessed on 23 November 2019).

22. Asmara, J.-P.E.; Ebohonb, J.O.; Taki, A. Bottom-up approach to sustainable urban development in Lebanon: The case of Zouk Mosbeh. *Sustain. Cities Soc.* **2012**, *2*, 37–44. [CrossRef]

23. Judith, I.; Booher, D. *Planning with Complexity: An Introduction to Collaborative Rationality for Public Policy*; Routledge: New York, NY, USA, 2010.

24. Sabatier, P.A.; Focht, W.; Lubell, M.; Trachtenberg, Z.; Vedlitz, A.; Matlock, M. *Swimming Upstream: Collaborative Approaches to Watershed Management*; Sabatier, P.A., Focht, W., Lubell, M., Trachtenberg, Z., Vedlitz, A., Matlock, M., Eds.; MIT Press: Cambridge, MA, USA, 2005.

25. Wolcha, J.R.; Byrneb, J.; Newellc, J.P. Urban green space, public health, and environmental justice: The challenge of making cities 'just green enough'. *Landsc. Urban Plan.* **2013**, *125*, 234–244. [CrossRef]

26. Elbakidze, M.; Dawson, L.; Andersson, K.; Axelsson, R.; Angelstam, P.; Stjernquist, I.; Thellbro, C.; Schlyter, P.; Thellbro, C. Is spatial planning a collaborative learning process? A case study from a rural-urban gradient in Sweden. *Land Use Policy* **2015**, *48*, 270–285. [CrossRef]

27. Qu, Z.; Lu, Y.; Jiang, Z.; Basset, E.; Tan, T. A Psychological Approach to Public Perception of Land-Use Planning: A Case Study of Jiangsu Province, China. *Sustainability* **2018**, *10*, 3056. [CrossRef]

28. Fan, Y. A Research about Public Participation in the Process of Master-Urban-Planning in China—An Introduction of Practice of Gathering Public Opinion via Questionnaire Survey. *Adv. Appl. Sociol.* **2015**, *5*, 13–22. [CrossRef]

29. Hanan, A.; Harry, S. Users' perceptions about planning and design of public open spaces: A case study of Muscat. In *EAEA-11 Conference*; Lodz University of Technology: Lodz, Poland, 2014.

30. Berke, P.R.; Godschalk, D.R.; Kaiser, E.J.; Rodriguez, D.A. *Urban Land Use Planning*, 5th ed.; University of Illinois Press: Urbana, IL, USA, 2006.

31. Gauthier, M.; Simard, L.; Waaub, J.P. Public participation in strategic environmental assessment (SEA): Critical review and the Quebec (Canada) approach. *Environ. Impact Assess. Rev.* **2011**, *31*, 48–60. [CrossRef]

32. Liepa-Zemeša, M.; Hess, D.B. Effects of public perception on urban planning: Evolution of an inclusive planning system during crises in Latvia. *Town Plan. Rev.* **2016**, *87*, 71–92. [CrossRef]

33. Walker, H.; Sinclair, J.; Spaling, H. Public participation in and learning through SEA in Kenya. *Environ. Impact Assess. Rev.* **2014**, *45*, 1–9. [CrossRef]

34. Soma, K.; Dijkshoorn-Dekker, M.W.C.; Polman, P.N.B. Stakeholder contributions through transitions towards urban sustainability. *Sustain. Cities Soc.* **2018**, *37*, 438–450. [CrossRef]

35. Gavrilidis, A.A.; Ciocanea, C.M.; Nita, M.R.; Onose, D.A.; Nastase, I.I. Urban Landscape Quality Index—Planning Tool for Evaluating Urban Landscapes and Improving the Quality of Life. *Procedia Environ. Sci.* **2016**, *32*, 155–167. [CrossRef]

36. Derak, M.; Cortina, J.; Taiqui, L. Integration of stakeholder choices and multi-criteria analysis to support land use planning in semiarid areas. *Land Use Policy* **2015**, *64*, 414–428. [CrossRef]

37. Timms, P. Urban transport policy transfer: Bottom-up and top-down perspectives. *Transp. Policy* **2011**, *18*, 513–521. [CrossRef]

38. Tomas, M.K.; Newig, J. From Planning to Implementation: Top-Down and Bottom-Up Approaches for Collaborative Watershed Management. *Policy Stud. J.* **2014**, *42*, 3.

39. Suadnya, W.; Yanuartati, Y.; Handayani, T.; Habibi, P.; Puspadi, K.; Bou, N.; Vaghelo, D.; Rochester, W. Integrating Top-Down and Bottom-Up Adaptation Planning to Build Adaptive Capacity: A Structured Learning Approach. *Coast. Manag.* **2015**, *43*, 346–364. [CrossRef]

40. Sherman, M.; Ford, J. Stakeholder engagement in adaptation interventions: An evaluation of projects in developing nations. *Clim. Policy* **2013**, *14*, 417–441. [CrossRef]

41. UNDP. *Designing Climate Change Adaptation Initiatives: A UNDP Toolkit for Practitioners*; United Nations Development Programme: New York, NY, USA, 2010.

42. Massey, D. For Space. London: SAGE. Available online: https://selforganizedseminar.files.wordpress.com/2011/07/massey-for_space.pdf (accessed on 20 November 2019).

43. Fan, L.; Lei, C. Da chengshi bianyuanqu yanhua fazhan zhong de maodun ji duice—Jiyu guangzhou shi anli de tantao Contradiction and governance research in the developing urban fringe area of large cities—A discussion based on the case of Guangzhou Municipality. Chengshi Fazhan Yanjiu. *Urban Stud.* **2009**, *12*.

44. Wei, L.; Yan, X. Da chengshi jiaoquhua zhong shehui kongjian de fei junheng pocuihua—Yi guangzhou shi wei lie The unbalanced fragmentation of social space in the context of the suburbanization of large cities—Taking Guangzhou Municipality as an example. *Chengshi Guihua City Plan. Rev.* **2006**, *5*, 55–60.

45. Martín-Lópeza, B.; Felipe-Lucia, M.R.; Bennette, E.M.; Norström, A.; Peterson, G.; Plieningerg, T.; Hicksh, C.C.; Turkelboom, F.; García-Llorente, M.; Jacobs, S.; et al. A novel telecoupling framework to assess social relations across spatial scales for ecosystem services research. *J. Environ. Manag.* **2019**, *241*, 251–263. [CrossRef]

46. Holling, C.S.; Gunderson, L.; Peterson, G. Sustainability and Panarchies. In *Panarchy: Understanding Transformations in Human and Natural Systems*; Gunderson, L.H., Holling, C.S., Eds.; Island Press: Washington, DC, USA, 2002; pp. 63–102.

47. Hasan, M.A.; Nahiduzzaman, K.M.; Aldosary, A.S. Public participation in EIA: A comparative study of the projects run by government and non-governmental organizations. *Environ. Impact Assess. Rev.* **2018**, *72*, 12–24. [CrossRef]

48. Sharifi, A.; Yamagata, Y. Resilience-Oriented Urban Planning. In *Resilience-Oriented Urban Planning*; Sharifi, A., Yagamata, Y., Eds.; Springer: Berlin, Germany, 2018; pp. 3–28.

49. Lara, A.; Saurı', D.; Ribas, A.; Pavo'n, D. Social perceptions of floods and flood management in a Mediterranean area (Costa Brava, Spain). *Nat. Hazards Earth Syst. Sci.* **2010**, *10*, 2081–2091. [CrossRef]

50. Sierra, R. *Técnicas de Investigación Social: Teoría y Ejercicios*, 14th ed.; Paraninfo Thomson Learning: Madrid, Spain, 2001.

51. Fischer, T.B. Theory and practice of Strategic Environmental Assessment. In *Towards a More Systematic Approach*; Earthscan: London, UK, 2007.

52.	European Parliament and Council of the European Union. Directive 2001/42/EC of the European Parliament and of the Council of 27 June 2001 on the assessment of the effects of certain plans and programmes on the environment. *Off. J.* **2001**, *197*, 30–37.

53.	Aretano, R.; Parlagreco, L.; Semeraro, T.; Zurlini, G.; Petrosillo, I. Coastal dynamics vs beach users attitudes and perceptions to enhance environmental conservation and management effectiveness. *Mar. Pollut. Bull.* **2017**, *123*, 142–155. [CrossRef] [PubMed]

54.	Aretano, R.; Petrosillo, I.; Zaccarelli, N.; Semeraro, T.; Zurlini, G. People perception of landscape change effects on ecosystem services in small Mediterranean islands: A combination of subjective and objective assessments. *Landsc. Urban Plan.* **2013**, *112*, 63–73. [CrossRef]

55.	Nagarale, V.; Harpale, D. An assessment of environmental impact on Bhimashankar and Lonavala with the help of Likert scale. *Trans. Inst. Indian Geogr.* **2012**, *34*, 127–136.

56.	Rega, C.; Baldizzone, G. Public participation in Strategic Environmental Assessment: A practioners' perspective. *Environ. Impact Assess. Rev.* **2015**, *50*, 105–115. [CrossRef]

57.	Tsumoto, S. Contingency matrix theory: Statistical dependence in a contingency table. *Inf. Sci.* **2009**, *179*, 1615–1627. [CrossRef]

58.	Gehl, J. Open Space: People Space. In *Public Spaces for a Changing Public Life*; Thompson, C.W., Travlou, P., Eds.; Taylor & Francis: Oxon, UK, 2007.

59.	Gehl, J. *Cities for People*; Island Press: London, UK, 2010.

60.	Korola, E.; Shushunova, N. Benefits of A Modular Green Roof Technology. *Procedia Eng.* **2016**, *161*, 1820–1826. [CrossRef]

61.	Semeraro, T.; Aretano, R.; Pomes, A. Green Roof Technology as a Sustainable Strategy to Improve Water Urban Availability. *IOP Conf. Ser. Mater. Sci. Eng.* **2019**, *471*, 092065. [CrossRef]

62.	Clementsen, A. Experiencing and Reacting Upon Social Diversity in Urban Spaces. International Conference on The Ideal City: Between myth and reality. In Proceedings of the Representations, Policies, Contradictions and Challenges for Tomorrow's Urban Life, Urbino, Italy, 27–29 August 2015.

63.	Dinnie, E.; Brown, M.B.; Morris, B. Community, cooperation and conflict: Negotiating the social well-being benefits of urban greenspace experiences. *Landsc. Urban Plan.* **2013**, *112*, 1–9. [CrossRef]

64.	Kothencz, G.; Kolcsar, R.; Cabrera-Barona, P.; Szilassi, P. Urban Green Space Perception and Its Contribution to Well-Being. *Int. J. Environ. Res. Public Health* **2017**, *14*, 766. [CrossRef] [PubMed]

65.	Farahani, L.M.; Maller, C. Perceptions and Preferences of Urban Greenspaces: A Literature Review and Framework for Policy and Practice. *Landsc. Online* **2018**, *6*, 1–22. [CrossRef]

66.	Bourdieu, P. The Social Space and the Genesis of Groups. *Theory Soc.* **1985**, *14*, 723–744. [CrossRef]

67.	Lefebvre, H. *The Production of Space*; Blackwell Publishing: Hoboken, NJ, USA, 1991.

68.	Gans, H. The Sociology of Space—A Use-centered view. *City Community* **2002**, *1*, 329–339. [CrossRef]

69.	Pataki, D.E. Grand challenges in urban ecology. *Front. Ecol. Evol.* **2016**, *3*, 5–7. [CrossRef]

70.	Semeraro, T.; Gatto, E.; Buccolieri, R.; Vergine, M.; Gao, Z.; De Bellis, L.; Luvisi, A. Changes in Olive Urban Forests Infected by Xylella fastidiosa: Impact on Microclimate and Social Health in urban areas. *Int. J. Environ. Res. Public Health* **2019**, *16*, 2642. [CrossRef]

71.	Seamon, D. Place Attachment and Phenomenology. In *Place Attachment*; Manzo, L.C., Devine-Wright, P., Eds.; Routledge: London, UK, 2014.

72.	Piga, B.; Morello, E. Environmental design studies on perception and simulation: An urban design approach. *Ambiances Nternational J. Sens. Environ. Archit. Urban Space* **2015**, 2–20. [CrossRef]

73.	Seeliger, L.; Turok, I. Towards Sustainable Cities: Extending Resilience with Insights from Vulnerability and Transition Theory. *Sustainability* **2013**, *5*, 2108–2128. [CrossRef]

74.	COWI. *Study Concerning the Report on the Application and Effectiveness of the SEA Directive (2001/42/EC)-Final report to the European Commission*; COWI: Kongens Lyngby, Denmark, 2009.

 *land*

Article

Monitoring of Urban Landscape Ecology Dynamics of Islamabad Capital Territory (ICT), Pakistan, Over Four Decades (1976–2016)

Hammad Gilani [1],*, Sohail Ahmad [1], Waqas Ahmed Qazi [1], Syed Muhammad Abubakar [2] and Murtaza Khalid [1]

[1] Geospatial Research & Education Lab (GREL), Department of Space Science, Institute of Space Technology, Islamabad 44000, Pakistan; sohailahmedsings@gmail.com (S.A.); waqas.qazi@grel.ist.edu.pk (W.A.Q.); murtazakhalid96@gmail.com (M.K.)

[2] Freelance Journalist, Lahore 54000, Pakistan; s.m.abubakar@hotmail.com

* Correspondence: hammad.gilani@grel.ist.edu.pk; Tel.: +92-51-907-5767

Received: 25 March 2020; Accepted: 16 April 2020; Published: 20 April 2020

Abstract: In the late 1960s, the Islamic Republic of Pakistan's capital shifted from Karachi to Islamabad, officially named Islamabad Capital Territory (ICT). In this aspect, the ICT is a young city, but undergoing rapid expansion and urbanization, especially in the last two decades. This study reports the measurement and characterization of ICT land cover change dynamics using Landsat satellite imagery for the years 1976, 1990, 2000, 2010, and 2016. Annual rate of change, landscape metrics, and urban forest fragmentation spatiotemporal analyses have been carried out, along with the calculation of the United Nations Sustainable Development Goal (SDG) indicator 11.3.1 Land Consumption Rate to the Population Growth Rate (LCRPGR). The results show consistent increase in the settlement class, with highest annual rate of 8.79% during 2000–2010. Tree cover >40% and <40% canopy decreased at an annual rate of 0.81% and 0.77% between 1976 to 2016, respectively. Forest fragmentation analysis reveals that 'core forests of >500 acres' class decreased from 392 km^2 (65.41%) to 241 km^2 (55%), and 'patch forest' class increased from 15 km^2 (2.46%) to 20 km^2 (4.54%), from 1976 to 2016. The LCRPGR ratio was 0.62 from 1976 to 2000, increasing to 1.36 from 2000 to 2016.

Keywords: forest fragmentation; sustainable development goal (SDG); land consumption rate to the population growth rate (LCRPGR)

---

1. Introduction

The conversion of one land cover type to another is one of the most visible and rapid changes that the earth is experiencing and these conversions have profound social and environmental impacts at multiple scales. In the past, the drivers of land cover changes have been classified into two categories: proximate and distant, or indirect [1]. Recently, the phenomenon of tele-coupling between places has been recognized as key to understanding how distant drivers relate to proximate drivers and influence local landscape changes [2–4]. Some of the key drivers of landscape changes include economic, technological, institutional and policy, cultural, and demographic factors [5].

The growth of urban areas leads to land cover change in many parts of the world, especially in developing countries [3,6]. Intense urbanization and increase in anthropogenic activities reflect the scope, intensity, and frequency of human interference, and the changes they cause in ecological processes and systems in the urbanized areas [7]. The urban areas consist of only 1–6% of the earth's land surface, yet they have enormous impacts on the functioning and service of local and global ecosystems, by modifying local climate conditions, eliminating and fragmenting native habitats, generating anthropogenic pollutants, etc. [8]. The spatial pattern of an urban landscape is a result of the interaction between various driving forces including natural and socioeconomic factors [9]. Increasing trends in industrialization and urbanization, along with the migration from rural to urban areas, are the most dominant factors influencing the land cover transformation. In the rural areas, employment opportunities and income are insufficient, which contributes to large differences in income and facility levels between urban and rural areas [10]. In developing countries, new cities are being developed due to human migration, infrastructure development, and growing job opportunities [10,11].

The United Nations (UN) Sustainable Development Goals (SDGs) are a collection of 17 global goals, which include 232 indicators set by the UN General Assembly for the year 2030. These goals are an urgent call for action by all countries—developed and developing—in a global partnership. Pakistan is a signatory to the UN SDGs, and this study analyzes the land use change dynamics of Islamabad Capital Territory (ICT) in the context of the relevant SDGs and indicators. The Goal 11 of the SDGs, "Make cities and human settlements inclusive, safe, resilient and sustainable", with indicator number 11.3.1 "Ratio of Land Consumption Rate to Population Growth Rate (LCRPGR)" [12] is an important parameter for analyzing sustainability of land use and land change with population growth. The LCRPGR parameter is actually based on the previously defined parameters of LCR [13,14] and PGR [15], and several studies of urban areas have been carried out before on the basis of these parameters. The LCRPGR parameter and terminology has been given prime importance now after being linked with the UN defined SDGs. LCRPGR is vital to understand the rate of land change as compared to the population boom, to understand historical land consumption traditions, and to guide decision and policy makers on the planned expansion of the city along with the protection of environmental, social, and economic assets. Another SDG, Goal 15, "Protect, restore and promote sustainable use of terrestrial ecosystems, sustainably manage forests, combat desertification, and halt and reverse land degradation and halt biodiversity loss" emphasizes the protection of tree cover and sustainable management of forests. The SDG indicator number 15.2.1, "Progress towards sustainable forest management" [16] calls for sustainable management strategies for conserving the forest cover, enhancing environmental education, and engaging a wide range of stakeholder institutions, policies, regulations, and considerations that promote sustainability and utilization of natural resources at multiple spatial scales [17].

Assessment and monitoring of land cover dynamics is essential for the sustainable management of natural resources, environmental protection, biodiversity conservation, and developing sustainable livelihoods. Therefore, the development of applicable and systematic methods for producing and updating land cover databases are considered an urgent need [18]. Around the globe, urban land expansion rates are higher than or equal to urban population growth rates [19–21]. Many research studies focus on big cities and metropolises, where increase in population through analysis of census statistics is directly linked with the urban land expansion; however, these statistics do not provide information regarding spatial distribution, pattern, and scale of urban land use change. Multi-temporal land cover change analysis and simulation based on coarse to very high resolution satellite remote sensing images is becoming a well established technique for quantifying changes occurring on the earth's surface, and multi-temporal aerial and satellite datasets are now widely and continuously being used for urban growth mapping, monitoring, and modeling with a focus on the spatial dimension and structures [22–25]. In urban expansion studies, spatiotemporal analysis of land cover and land use changes has helped towards understanding the underlying natural and socio-economic factors and drivers. For instance, Seto et al. [26] have presented a meta-analysis of 326 studies which used temporal satellite images to map urban land conversion. A total of 58,000 km^2 increase in urban land area was reported in thirty years (1970 to 2000) and by 2030, global urban land cover is expected to increase between 430,000 km^2 and 12,568,000 km^2, with an estimate of 1,527,000 km^2 more likely. According to Seto et al. [26], across all regions and for all three decades, urban land expansion rates are higher than or equal to urban population growth rates. Yang et al. [27] studied and reported the evidence of urban agglomerations through satellite images in four major bay areas of US (San Francisco and New York), China (Hong Kong-Macau), and Japan (Tokyo), from 1987 to 2017.

Clarke et al. [28] proposed a framework to combine remote sensing and spatial metrics for improved understanding and representation of urban dynamics to come up with alternative conceptions of urban spatial structure and change. Particularly with regards to urban forestry, a few studies have focused towards the assessment, mapping, and monitoring of urban forest parameters in fast-growing cities of developing countries. For example, Gong et al. [29] carried out a 30-year forest fragmentation study over Shenzhen Special Economic Zone (SEZ), a city which was established in 1979 in Southern China. Huang et al. [30] utilized satellite images of 77 metropolitan areas in Asia, US, Europe, Latin America, and Australia to calculate and analyze seven spatial metrics (area weighted mean shape, area weighted mean patch fractal dimension index, centrality, compactness index, compactness index of the largest patch, ratio of open space, and density). According to their analysis, the compactness, density, and regularity of urban areas in developing regions generally exceeded the levels reported throughout developed countries [30]. Dewan et al. [31] studied the dynamics of land use/cover changes though landscape fragmentation analysis in Dhaka Metropolitan, Bangladesh, computing and analyzing the following metrics: Number of patches, Patch density, Landscape shape index, Largest patch index, Mean patch size, Area-weighted mean fractal dimension, Interspersion and juxtaposition, Contagion, and Shannon's diversity index.

In Pakistan, like other developing countries, most urban development is haphazard, typically lacking appropriate planning strategies [10]. Pakistan's urbanization rate is the highest in South Asia, and by 2030, Pakistan will have more people in cities than in rural areas. Growing population and rapid development is causing prime agricultural land to be encroached and also causing loss of tree cover [32–36]. In the late 1960s, the capital of the Islamic Republic of Pakistan was shifted from Karachi to Islamabad (officially named Islamabad Capital Territory (ICT)). The masterplan of ICT was developed by the famous Greek architect and town planner C. A. Doxiadis [37]. In terms of a planned new capital, ICT is similar to planned new post-colonial capitals/relocations as in the cases of Brasilia (Brazil), Nur-Sultan (named as Astana from 1998 to 2019, and Akmola previously) in Kazakhstan, and Canberra (Australia) [38–40]. In the recent decades, with various ongoing development activities, ICT has been struggling with rapid urbanization and gigantic levels of pollution from industrial, residential, and transportation sources. In terms of population, ICT is considered as the most diverse city of Pakistan with a large percentage of immigrants and foreigner population [41]. Unprecedented influx of migrants and population increase has resulted in urban sprawl and conversion of fertile agricultural land and green cover into concrete—a clear deviation from the original ICT master plan [39,40]. Uncontrolled population growth in ICT due to rapid urbanization has deteriorated the living environment, and increased the adverse ecological impacts on human health, flora, and fauna [42].

1.1. Literature Review—ICT Mapping and Monitoring

In the last 20 years, several studies have been conducted on the ICT, regarding land cover change, biomass estimation, water quality monitoring, and temperature increase using satellite datasets. In this section, we present a synthesis of published work regarding land cover change dynamics in the ICT.

Adeel [43] identified urban growth potential through land use for ICT zone IV (Figure 1), based on SPOT-5 2.5 m panchromatic dataset and population census data, and found that nearly 63% of zone IV carries a 'High' to 'Very High' future growth potential, which is mainly located close to Islamabad Expressway. This work uses satellite imagery and field data from one year (2007) and does not report spatiotemporal change dynamics. Butt et al. [35] studied the metropolitan development in ICT, based on growth direction and expansion trends from the city center, for the period 1972–2009 using Landsat satellite images. Using Principal Components Analysis (PCA), band ratios, and supervised classification methods, they found that the urban development had expanded by 87.31 km^2 in 38 years. Butt et al. [44] conducted a study on land cover change analysis over Simly dam watershed, ICT; the results derived from maximum likelihood supervised classification showed tree cover loss of up to 26% and 6% increase in settlements from 1992–2012, based on Landsat 5 TM and SPOT-5 imagery, respectively. Similarly, the watershed analysis of Rawal dam, ICT using Landsat 5 TM imagery, showed

3% degradation of tree cover and 2% gain of settlement from 1992–2012 [45]. Another ICT land cover change dynamics study conducted by Hassan et al. [46] utilized 30 m Landsat 5 TM data for 1992 and 2.5 m SPOT-5 data for 2012, using the maximum likelihood algorithm for image classification. The study revealed a decrease in forest cover of approximately 49% and over 213% gain of settlement area from 1992–2012. Sohail et al. [47] conducted a study to assess the water quality index and analyze the major change in land cover types, vegetation cover, rate of urbanization and its possible impact on groundwater resources, vegetation, and barren land. They used Landsat images for the years 1993, 1997, 2002, 2007, 2013, and 2017 for the assessment and mapping of land cover dynamics; according to their findings, from 1993 to 2017, vegetation areas decreased by 101.77 km^2, surface water was reduced by 1.10 km^2, barren land was reduced by 2.90 km^2, while built-up lands expanded by 105.77 km^2.

A comparison of Beijing, China and ICT for the role of vegetation in "controlling the eco environmental conditions for sustainable urban environment" was performed by Naeem et al. [42], where they used Gaofen-1 (GF-1) and Landsat-8 Operational Land Imager (OLI) satellite imagery with 8 m and 30 m spatial resolution, respectively. They evaluated various scenarios and models for future development to predict future spatial patterns in both cities. Another study was conducted by Naeem et al. [48] to study the association between green space characteristics, analyzed through landscape metrics, and land surface temperature for sustainable urban environments comparing Beijing, China and ICT.

Khalid et al. [49] conducted a study to quantify the decline of forest reserves and associated temperature variations in a relatively unexplored biodiversity hotspot of ICT, the Margalla Hills National Park (MHNP). In this work, Landsat satellite imagery from 1992, 2000, and 2011 was used to monitor the changes in forest cover and statistical significance tests were used to determine the significance of temperature variation associated with a shift in land cover classes. The study finds that deforestation and forest degradation by local communities is an ongoing practice in MHNP; this necessitates the promotion of conservation practices to minimize ecological disturbances here [49]. Batool and Javaid [50] carried out a study on the assessment of Margalla Hills forest by using Landsat imagery for years 2000 and 2018, and report that the forest cover has decreased from 87% in 2000 to 74% in 2018, whereas built-up area has increased from 5% in 2000 to 7% in 2018, and open land in the study area increased from 2% in 2000 to 7% in 2018. Mannan et al. [51] conducted a study using Landsat imagery, Markov Chain, and Cellular Automata on Margalla Hills, focusing on the quantitative assessment of spatiotemporal land use and land cover changes during 1998, 2008, 2018, and a simulation of 2028. In addition, a forest inventory survey was conducted for biomass and carbon sink estimations. This work shows that the forest area has reduced from 409.36 km^2 to 392.31 km^2 and settlement area has increased from 14.97 km^2 to 39.66 km^2 from 1998 to 2018. The average yearly biomass and carbon losses were 50.34 Gg/ha/yr and 31.33 Gg C/ha/yr, respectively.

The ICT is a relatively new and spatially heterogeneous city surrounded by the Himalaya mountainous dense forest as compared to other fast growing and expanding cities like Dhaka, Bangladesh [31,52–55], New Delhi, India [56], Beijing, China [42,48,57,58], Shanghai, China [59–61], Tokyo, Japan [27,62], etc. Based on the literature review, we observed that most studies of the ICT land cover dynamics have used different remotely sensed data, methods, definitions, and classification schemes, and have provided diverse results. Most studies which have analyzed the land change dynamics in the ICT focus on the overall analysis of land-cover and land-use change, and a detailed analysis of the landscape ecology and urban forestry characteristics is missing. There has further been very little focus in these studies towards urban landscape metrics and indicators of sustainable urban growth such as LCRPGR.

1.2. Study Objectives

In this paper, well established, proven, and articulated research methodology, satellite datasets, and definitions of features were adopted with the goal to systematically achieve the following defined objectives:

- Detection, measurement, and characterization of land cover features using Landsat medium resolution freely available satellite data (1976, 1990, 2000, 2010, and 2016) and determine the annual rate of change in land cover classes at 10 years interval.
- Landscape metrics and forest fragmentation spatiotemporal analysis, to estimate and report changes in the ICT urban ecosystem over forty years.
- Calculation of SDG indicator number 11.3.1 "Land Consumption Rate to the Population Growth Rate (LCRPGR)."

2. Study Area

The ICT is the capital city of Islamic Republic of Pakistan (Figure 1) located in the Potohar plateau. It comprises an area of 906 km^2 including mountains and uneven plains exceeding 1,175 m in height above the mean sea level [63]. According to the 2017 national population census, the total population of ICT is approximately two million, which makes it the ninth largest city of Pakistan. It has a humid and sub-tropical climate with four distinct seasons: autumn, spring, summer, and winter. The temperatures vary from 13 °C in January to 38 °C in June. ICT consists of five planning zones: Zone I, II, and V are reserved for planned urban development, while the remaining two zones, III and IV, are managed as National Park and the rural fringes. Marble and chemical factories, steel mills, flour mills, oil units, pigments, paints, and pharmaceutical manufacturing plants are some of the main industries in the city [34].

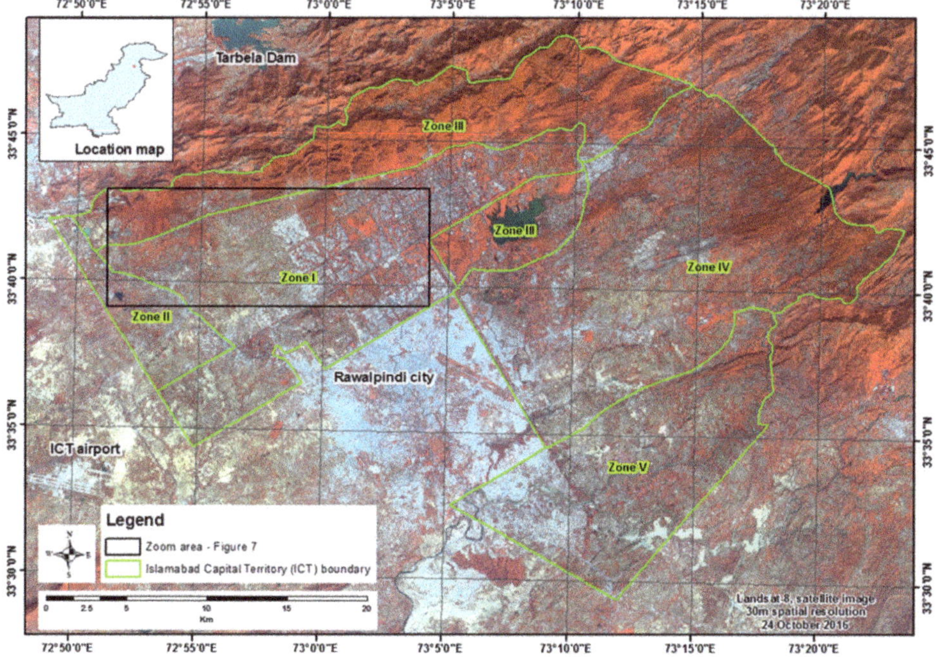

Figure 1. Study area map—Islamabad Capital Territory (ICT), Pakistan. The bounded box inset covering Zone I and Zone II partially is representing the area shown in the Figure in Section 5.

3. Materials and Methods

3.1. Datasets

To study the land cover dynamics and spatial analysis of ICT, various datasets have been collected from primary and secondary sources. The data collected from the primary sources include Landsat medium resolution satellite imagery and field observations using Geographical Positioning System (GPS) receiver, while the secondary or ancillary data consist of population census and ground truth data from Very High Resolution Satellite (VHRS) imagery.

3.1.1. Landsat Satellite and Digital Elevation Model (DEM) Data

For land cover mapping, we used 30 m spatial resolution orthorectified and cloud free Landsat Multispectral Scanner System (MSS), Thematic Mapper (TM), Enhanced Thematic Mapper Plus (ETM+), and Operational Land Image (OLI) sensor images. The TM, ETM+, and OLI sensors have the same 30 m spatial resolution, while the MSS has a spatial resolution of 57 m. The overall ICT region is covered within one Landsat scene (185 × 185 km). A total of five images for years 1976, 1990, 2000, 2010, and 2016 were downloaded from the United States Geological Survey (USGS) Earth Resources Observation and Science (EROS) archive. All satellite data was collected for the months of October and November, to avoid cloud cover (Table 1).

Table 1. Landsat images used for land cover mapping.

Acquisition Date	Satellite	Sensor	Path/Row	Spatial Resolution
19 October 1976	Landsat 3	MSS	161/37	57 m
01 October 1990	Landsat 5	TM	150/37	30 m
05 November 2000	Landsat 7	ETM+	150/37	30 m
09 November 2010	Landsat 5	TM	150/37	30 m
24 October 2016	Landsat 8	OLI	150/37	30 m

A 30 m spatial resolution Shuttle Radar Topography Mission (SRTM) Digital Elevation Model (DEM) from USGS was used to understand the topography of the area and to prepare the study area maps.

3.1.2. Population Census Data

For the calculation of SDG Goal 11 indicator 11.3.1 LCRPGR, we used the data collected from secondary sources including demographic data (primary census abstracts for the year 1972, 1998, and 2017) from the Pakistan Bureau of Statistics. We used actual census data of 1972 and 1998 to determine the shape of the population curve, which was exponential, as expected. Using the fitted exponential regression function, we derived the growth rate through the exponential function derivative—Equation (1), and then Equation (2) was utilized to determine the approximate population for the year 1976. The population data used for the calculation of LCRPGR is given in Table 2.

$$\text{Growth rate} = (ab)e^{ac} \tag{1}$$

$$\text{Pop}_{1976} = \text{Pop}_{1972} + 4 \times \text{Growth rate} \tag{2}$$

where a and b are regression coefficients, c represents the population variable, POP_{1976} = Estimated Population in 1976, and POP_{1972} = Past Population in 1972.

Table 2. Population data of ICT for 1972, 1976, 1998/2000, and 2016/2017.

Year	1972	1976	1998/2000	2016/2017
Population	238,000	291,888	805,235	2,006,572

3.1.3. Ground Truth Data

In this study, we collected ground truth data from two sources: the primary ground data was collected using a Global Positioning System (GPS) receiver in October 2016 during field survey at various locations of ICT, while for the secondary ground data we relied on sub-meter VHRS imagery from Google Earth. A total of 205 samples were gathered, which were utilized as reference data for the satellite image classification and accuracy assessment of results. Using the Google Earth temporal VHRS data, we visually identified and geo-tagged the sites witnessing land degradation, urban development, and tree loss. The geo-tagged dataset helped to validate Landsat-based land cover changes.

3.2. Methodology

Our methodology (Figure 2) on land cover assessment and spatial analysis of ICT consists of (i) pre-processing of Landsat images, (ii) classification of Landsat images, (iii) accuracy assessment of land cover products, (iv) temporal land cover assessment, (v) urban landscape metrics analysis, (vi) urban forest fragmentation analysis, and (vii) LCRPGR calculation.

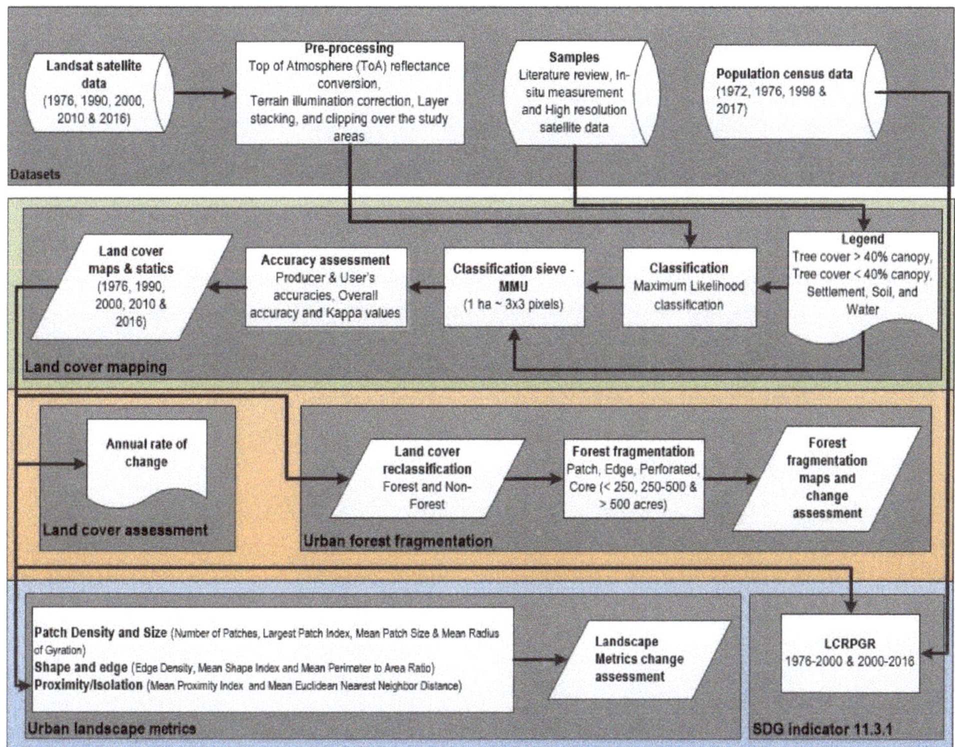

Figure 2. Methodological flow chart.

3.2.1. Pre-Preprocessing Landsat Images

Pre-processing steps for Landsat images included Top of Atmosphere (ToA) reflectance conversion, terrain illumination correction, and layer stacking of spectral bands. A reduction in between-scenes variability was accomplished through normalization for solar irradiance with a two-step process. First, we converted all pixels' Digital Numbers (DNs) to radiance values using the bias and gain values, which were scene-specific and given in the metadata file of the respective scene. Second, we converted radiance data to ToA reflectance [64]. Terrain illumination correction was conducted using an empirical rotation model proposed by Tan et al. [65]. Layer stacking was performed to construct multi-spectral images at 30 m spatial resolution. The false color composite (FCC) of bands 542 from the multi-spectral images was used to visually interpret land cover features.

3.2.2. Classification of Landsat Satellite Images

The image classification procedure is to instinctively categorize all pixels in an image into land cover classes [66]. In this study, the maximum likelihood classification method was used for land cover-mapping from Landsat images. The maximum likelihood classification method is a supervised classification technique, which works on the basis of multivariate normal probability density function of categories, and utilizes both the variance and co-variance of the spectral response of each unknown pixel to assign it to a particular category [67,68]. In this study, five land cover classes were specified: Tree cover >40% canopy, Tree cover <40% canopy, Settlement, Soil, and Water. Almost 35 training samples were collected for each land cover class to classify the various Landsat images. The classified objects with an area smaller than the Minimum Mapping Unit (MMU) (i.e., 1 ha ~ 3 × 3 pixels) were fused with the neighboring land cover classes [69]. We adopted on-scene digitization technique for land cover change detection. First, we overlaid 2016 base land cover map on the multi-spectral Landsat image of 2010. We traced patches where land changes had occurred, leaving unchanged patches unmodified for consistency [70]. We followed the similar approach to detect land cover change between 2000 and 2010, using 2010 land cover map as a base layer. This process was repeated to generate land cover change maps between 1976–1990, 1990–2000, 2000–2010, and 2010–2016.

3.2.3. Accuracy Assessment of Land Cover Products

Assessment of classification accuracy of 1976, 1990, 2000, 2010, and 2016 land cover maps was carried out to determine the quality of information derived from the data. Random sampling method was adopted to assess the accuracy of satellite derived land cover products. Accuracy was assessed using 125 reference points, based on ground truth data and satellite visual interpretation. The comparison of classification results and reference data was carried out statistically using error matrices. In addition, the non-parametric kappa statistic was computed for each classified map to measure the accuracy of results, as it not only accounts for diagonal elements but also for all elements in the confusion matrix [71].

3.2.4. Temporal Land Cover Assessment

In this study, within the 906 km^2 total area of ICT, the land cover maps (1976, 1990, 2000, 2010, and 2016) were compared in terms of the area. According to Puyravaud [72], the annual rate of change is based on the compound interest law, considering non-linear change across the timeline to estimate the percentage change per year. In this study, the annual rate of change was calculated using Equation (3) proposed by Puyravaud [72].

$$r = \left(\frac{1}{t_2 - t_1}\right) \times \left(\ln \frac{A_2}{A_1}\right) \times 100 \tag{3}$$

where r is the annual rate of change in percentage, and A_1 and A_2 is the area at earlier time t_1 and later time t_2, respectively, and ln denotes the natural logarithm function.

3.2.5. Urban Landscape Metrics Analysis

Landscape metrics are used in this study to quantify the spatial patterns of land cover categories. The landscape metrics can be defined as quantitative and comprehensive measurements showing spatial diversity at a specific scale and resolution [73,74]. In this study, landscape metrics analysis was performed at land cover class level to quantitatively analyze spatial structures and patterns of the topographically and biophysically heterogeneous and diverse landscape of the ICT [74].

Under the three habitat categories Patch Density and Size, Shape and Edge, and Proximity/Isolation, a total of nine landscape metrics were used to quantify structural changes: Number of patches (NP), Mean Patch Size (MPS), Largest Patch Index (LPI), Mean Radius of Gyration (MRG), Mean Shape Index (MSI), Edge Density (ED), Mean Perimeter to Area Ratio (MPAR), Mean Euclidean Nearest Neighbor Distance (MED), and Mean Proximity Index (MPI) (Table 3). The landscape metrics were calculated using the FRAGSTATS v4 software tool. The output of the landscape metric analysis depends upon the spatial resolution of the data [75]; in this particular study, 30 m spatial resolution data has been chosen from 1990 to 2015 and 60 m spatial resolution for 1976.

Table 3. Land cover class level description of landscape metrics used in this study.

Category	Metrics	Unit	Description
Patch size and density	Number of Patches (NP)		Gives the total number of patches in the landscape of a particular category.
	Largest Patch Index (LPI)	Percentage (%)	Ratio of area of largest patch to total landscape area.
	Mean Patch Size (MPS)	Square Kilometer (km^2)	Sum of area of all the patches divided by the number of patches of the class.
	Mean Radius of Gyration (MRG)	Meter (m)	Mean distance for each cell in one patch to the patch centroid. It measures connectivity inside habitat patches.
Shape and edge	Edge Density (ED)	Meter per Square Kilometer (m/km^2)	Measures total length of edge per unit area. It explains complexity of patch shape.
	Mean Shape Index (MSI)		Measures complexity of patch shape compared to a standard shape of the same size. Its value increases with complexity of shape.
	Mean Perimeter to Area Ratio (MPAR)		Patch shape complexity, based on perimeter length to patch area. It explains shape complexity without standardization to a standard Euclidean shape (square).
Proximity/Isolation	Mean Proximity Index (MPI)		Measure of connectedness of a habitat class. Considers size and proximity of all patches with the same habitat type inside a specified search radius.
	Mean Euclidean Nearest Neighbor Distance (MED)	Meter (m)	Measures minimum edge-to-edge distance to the nearest neighboring patch of the same type. It explains connectedness or isolation in landscape or habitat class.

3.2.6. Urban Forest Fragmentation Analysis

Forest fragmentation is the splitting up of large contiguous forest fields into smaller or less contiguous areas. A number of events or activities can lead to forest fragmentation including road formations, woodcutting, forest conversion to agriculture, forest fires, and human conflict over forest patches [25]. To assess forest fragmentation in ICT, the land cover map is divided into two major categories: forest and non-forest. Forest class consists of tree cover greater and less than 40% canopy cover while non-forest class comprises of settlement, soil, and water land cover classes. The outcome of forest fragmentation analysis was represented into six categories: Patch, Edge, Perforated, Core (<250 acres), Core (250–500 acres), and Core (>500 acres). These categories are signs of forest ecosystem quality and can be used to estimate the amount of fragmentation present in a landscape and the potential habitat impacts [25,76].

3.2.7. Ratio of Land Consumption Rate to the Population Growth Rate (LCRPGR) Calculation

One of the goals of this study is the calculation of the SDGs indicator 11.3.1 Land Consumption Rate to the Population Growth Rate (LCRPGR), which aims at monitoring and measuring urban development by comparing the urban expansion rate with the population growth rate on similar temporal and spatial scales [12,77]. If the LCRPGR ratio value lies between $0 \leq LCRPGR \leq 1$, it shows the simultaneous increase of population growth rate (PGR) and land consumption rate (LCR), but the land consumption rate is much slower than the population growth rate. On the other hand, if LCRPGR > 1, it reflects the simultaneous increase of PGR and LCR, with a faster LCR than the PGR. To estimate the LCRPGR, satellite data derived land cover maps were used for the years 1976, 1998/2000 and 2016/2017, and census data was used for the specific years (Table 2). The indicator 11.3.1 was assessed at the local level for ICT using the mathematical expressions currently proposed by UN-Habitat, given below in Equations (4)–(6) [77]:

$$LCRPGR = \frac{Land\ Consumption\ Rate}{Population\ Growth\ Rate} \tag{4}$$

$$Land\ Consumption\ Rate = \frac{\ln\left(\frac{Urb_{t+n}}{Urb_t}\right)}{n} \tag{5}$$

$$Population\ Growth\ Rate = \frac{\ln\left(\frac{Pop_{t+n}}{Pop_t}\right)}{n} \tag{6}$$

where, ln = Natural logarithm, Urb_{t+n} = Surface occupied by urban areas at the final year (t + n). Urb_t = Surface occupied by urban area at the initial year (t), Pop_{t+n} = Population living in urban areas at the final year (t + n), Pop_t = Population living in urban areas at the initial year (t), and n = Number of years between the two time intervals.

4. Results

4.1. Accuracy Assessment of Land Cover Products

According to the confusion matrix analysis, 90% overall accuracy and kappa coefficient value of 0.85 was attained for the 2016 classified map. Similarly, overall classification accuracy levels of 88% with kappa coefficient of 0.84 for 2010, 86% with kappa coefficient of 0.82 for 2000, 85% with kappa coefficient of 0.81 for 1990, and 83% with kappa coefficient of 0.79 for 1976 image classifications were achieved (Table A1 in Appendix A).

4.2. Temporal Land Cover Assessment

The overall settlement in the ICT increased while tree cover classes (greater and less than 40% tree canopies) decreased in forty years (1976 to 2016). The estimated land cover area for each image is summarized in Table 4 and spatial distributions are presented in Figure 3.

Table 4. Land cover 1976, 1990, 2000, 2010, and 2016 assessment based on satellite images.

Land Cover Classes	1976	1990	2000	2010	2016
	km² (%)				
Tree cover >40% canopy	182.87 (20.18)	192.19 (21.21)	178.56 (19.70)	140.76 (15.53)	132.27 (10.86)
Tree cover <40% canopy	417.03 (46.02)	342.94 (37.85)	399.31 (44.07)	360.79 (39.82)	306.53 (35.72)
Settlement	29.99 (3.31)	44.51 (4.91)	59.73 (6.60)	143.82 (15.87)	170.40 (18.80)
Soil	270.27 (29.83)	317.91 (35.08)	257.00 (28.37)	254.34 (28.02)	289.45 (33.82)
Water	5.86 (0.65)	8.44 (0.94)	10.51 (1.16)	6.29 (0.69)	6.51 (0.76)
Total			906 (100)		

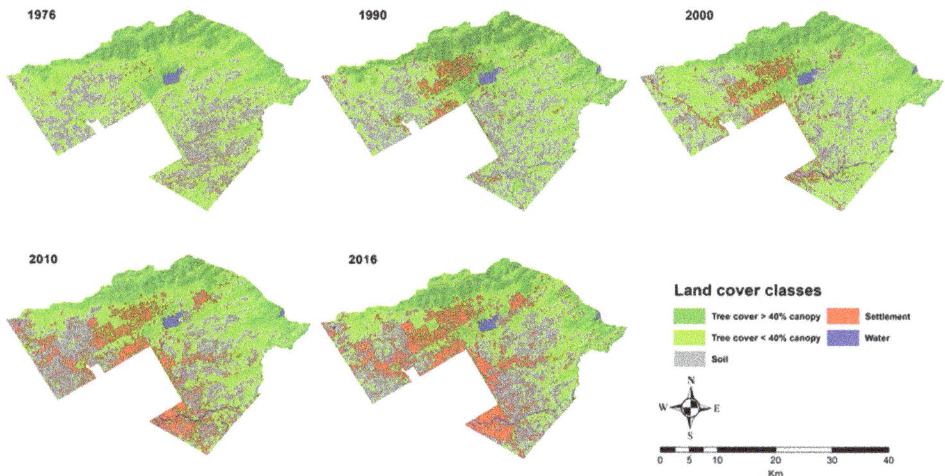

Figure 3. Land cover maps for 1976, 1990, 2000, 2010, and 2016.

The settlement class in ICT increased from 29.99 km^2 (3.31%) in 1976 to 170.40 km^2 (18.80%) in 2016. The tree cover >40% canopy class showed a slight increase, 182.87 km^2 (20.18%) in 1976 to 192.19 km^2 (21.21%) in 1990 and then declined to 132.27 km^2 (14.59%) in 2016. Tree cover <40% canopy land cover class faced overall decline from 1976 to 2016: 417.03 km^2 (46.02%) in 1976 to 342.94 km^2 (37.85%) in 1990, increase from 1990 to 2000 to 399.31 km^2 (44.07%), while again decreased from 2000 to 2016 to 306.53 km^2 (35.72%). The other two land cover classes, soil and water, observed fluctuating trends in the forty year period. The soil area of 270.27 km^2 (29.83%) in 1976 increased to 317.91 km^2 (35.08%) in 1990 and then reduced to 257 km^2 (28.37%) in 2000, then further decreased to 254.34 (28.02%) in 2010, and again increased to 289.45 km^2 (33.82%) in 2016. The change in water area was not significant though it has shown fluctuations during the study period; it increased from 5.86 km^2 (0.64%) in 1976 to 8.44 km^2 (0.94%) in 1990, and to 10.51km^2 (1.16%) in 2000, then decreased to 6.29 km^2 (0.69%) in 2010, and then increased slightly to 6.51 km^2 (0.76%) in 2016 (Table 4).

Based on the temporal land cover data in the ICT, urban landscape settlement increased at an annual rate of 4.34% since 1976, with the highest annual rate of 8.79% during 2000–2010. The tree cover >40% canopy decreased at an annual rate of 0.81% between 1976 to 2016 and tree cover >40% canopy declined at an annual rate of 0.77% in forty years (1976–2016). The tree cover >40% canopy witnessed the highest loss rate at 2.38% per annum between 2000–2010, and tree cover <40% canopy experienced tree cover loss at 2.72% during 2010–2016 (Table 5).

Table 5. Land cover annual rate of change (% change per year) from 1976–1990, 1990–2000, 2000–2010, 2010–2016, and 1976–2016.

Land Cover Classes	1976–1990	1990–2000	2000–2010	2010–2016	1976–2016
Tree cover >40% canopy	0.36	−0.74	−2.38	−1.04	−0.81
Tree cover <40% canopy	−1.40	1.52	−1.01	−2.72	−0.77
Settlements	2.82	2.94	8.79	2.83	4.34
Soil	1.16	−2.09	−0.14	2.16	0.17
Water	2.61	2.19	−5.13	1.59	0.42

4.3. Urban Landscape Matrix Analysis

Landscape matrix analysis was performed at class level, and the results for each class are reported in Figure 4 and Table A2 in Appendix A. The parameters Number of patches (NP), Largest Patch Index

(LPI), Edge Density (ED), and Mean Proximity Index (MPI) for the settlement class overall increased from 1976 to 2016. Largest Patch Index (LPI) increased to 16% in 1990 from 13% in 1976, and then reduced to less than 10% in 2000, 2010, and 2016. For tree cover greater than and less than 40% canopy, Number of patches (NP) and Edge Density (ED) initially increased from 1976 to 2000, and then declined from 2000–2016. For tree cover greater than and less than 40% canopy, soil, and settlement classes, the Mean Patch Size (MPS) and Mean Euclidean Nearest Neighbor Distance (MED) sharply declined from 1976 to 1990 and remained lower. In the Mean Euclidean Nearest Neighbor Distance (MED), tree cover greater than and less than 40% canopy, soil and settlement classes remained less than 250 m throughout from 1976 to 2016. From 1976 to 2010, the Mean Radius of Gyration (MRG) decreased for tree cover greater than and less than 40% canopy, and then slightly increased during the years 2010 to 2016. For tree cover >40%, the canopy Mean Proximity Index (MPI) slightly increased between 1976 to 2016 while for tree cover >40%, the canopy MPI decreased from 1976 to 1990, grew between 1990 to 2010, and then again sharply reduced from 2010 to 2016. In all the land cover classes, the Mean Shape Index (MSI) value remained greater than 1 with slight fluctuations. In the Mean Perimeter to Area Ratio (MPAR), the settlement class increased from 640 in 1976 to 1000 in 1990 with slight increase till 2016.

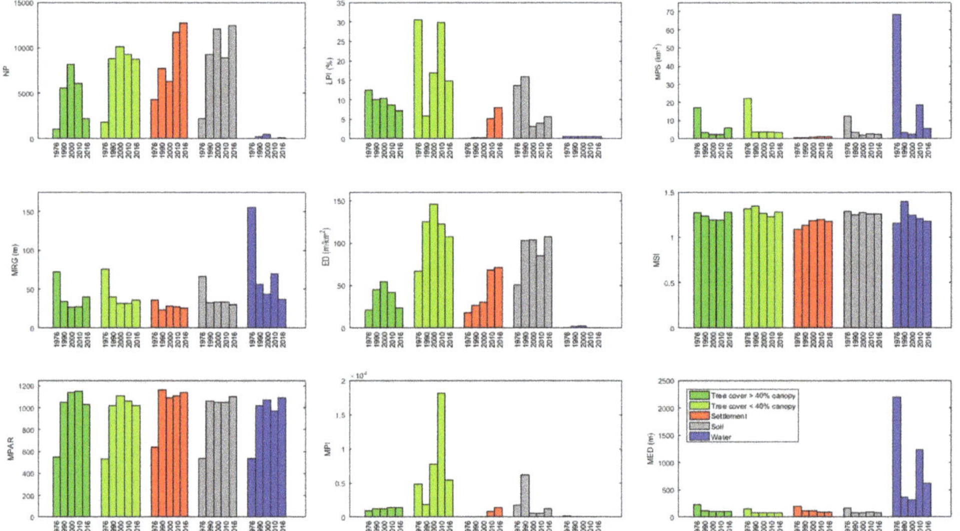

Figure 4. Land cover class level landscape matrix parameters graphs for 1976, 1990, 2000, 2010, and 2016.

4.4. Urban Forest Fragmentation Analysis

The forest fragmentation analysis results are presented in Figures 5 and 6, and Table A3 in Appendix A. The analysis shows that the overall core forests of >500 acres decreased from 391.98 km^2 (65.41%) in 1976 to 241.44 km^2 (40.29%) in 2016. In the forty years' time span, the patch class increased to 20 km^2 (4.54%) in 2016 from 14.74 km^2 (2.46%) in 1976. The perforated forest fragmentation class was 35.21 km^2 (5.88%) in 1976, which increased 41.79 km^2 (7.23%) in 2000, and then again declined to 22.74 km^2 (5.18%) in 2016. For the edge class, a minor increase was observed from 1976 to 2000 and then a drop in 2016.

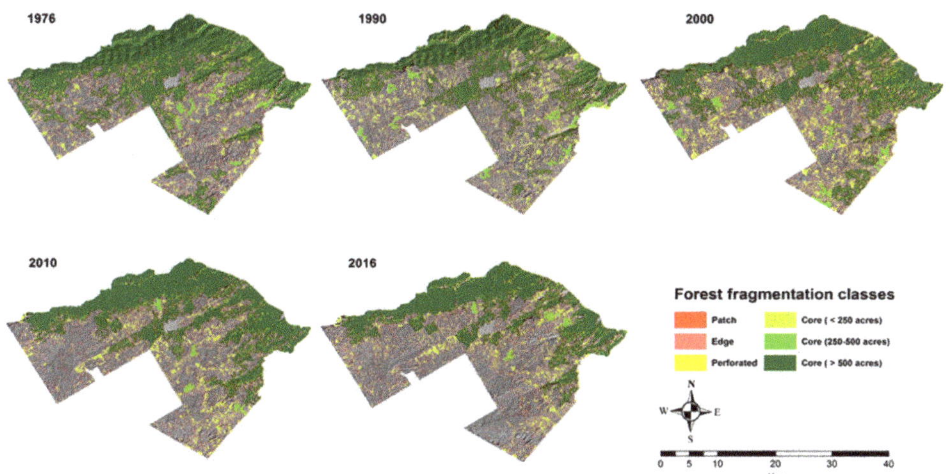

Figure 5. Forest fragmentation maps for 1976, 1990, 2000, 2010, and 2016.

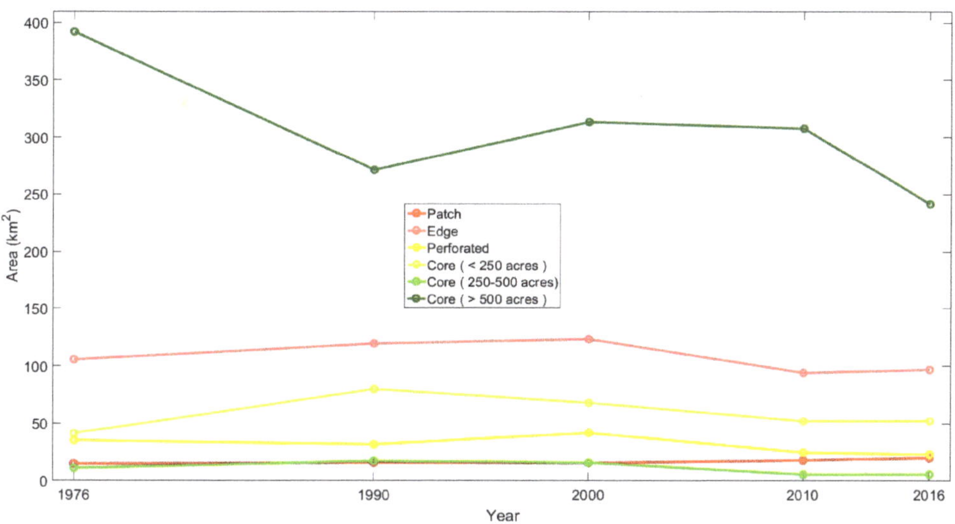

Figure 6. Forest fragmentation graphs for 1976, 1990, 2000, 2010, and 2016.

4.5. Ratio of Land Consumption Rate to the Population Growth Rate (LCRPGR)

For the years 1976–2000, the LCRPGR ratio (SDG indicator 11.3.1) was obtained to be 0.62, which highlights the simultaneous increase of Population Growth Rate (PGR) and Land Consumption Rate (LCR), but the LCR is much slower than the PGR. For the years 2000–2016, the LCRPGR ratio was obtained to be 1.36, which reflects the simultaneous increase of PGR and LCR, with a faster LCR than PGR.

5. Discussion

The present study was conducted in order to quantitatively analyze the landscape change dynamics in the ICT from 1976 to 2016 using medium spatial resolution Landsat images. In this study, we observed an increase in settlements over the last forty years. The dramatic land cover change in settlements is exerting severe pressure on other land cover classes, particularly tree cover and soil. Already existing urban areas can be seen expanding through rapid construction of residential blocks in the form of housing societies, industrial blocks and road expansions, leading to horizontal and vertical developments in the city and development of lavish farm houses in the vicinity areas (Figure 7). This rapid increase in urbanization is linked also with migrations. Urban growth may have positive or negative impacts on the environment but unplanned growth of urban areas has negative effects. For the economic development of the country, necessary planning is required to make urbanization helpful, as social, health, and environmental issues often accompany the process of urbanization.

In the time span of forty years, an overall decrease has been observed in the area of tree cover classes (i.e., greater than and less than 40% tree canopy) in the ICT, and most of the tree loss has occurred after the year 2000 with a corresponding increase in built-up areas. Apart from tree cutting, another important reason behind rapid tree loss or degradation is forest fires. In the Margalla Hills, forest fires usually take place during dry hot climate conditions when there is no rain for months and temperature goes up to 45 °C. According to Khalid and Ahmad [78], a total of 320 forest fires were recorded from 2002 to 2012 and approximately 8 km^2 area got burnt as a result. In the ICT, due to uncontrolled urbanization and lack of awareness, huge tree loss has been observed in the last sixteen years, i.e., 2000–2016. Another factor for forest degradation is uncontrolled grazing of livestock [78]. As such, there are no adequate plans or manageable methods to stop grazing activities. For ecotourism and public awareness drives, a number of jogging and hiking trails have been formed in Margalla Hills National Park, and a large number of visitors has severely affected the Margalla Hills National Park by dumping waste. These illegal and unmonitored activities in the forested area cause threats to the forest ecosystem [50,79].

Massive migrations have occurred over the past few years from rural to urban areas, mostly due to low cultivated land output, landlessness, sub-division of land, poor economy, and better educational and health opportunities in urban areas. The rapid increase in population has contributed towards natural resource depletion and rapid deforestation close to settlements [80].

The LCRPGR parameter is an indicator of urban sustainable development, whether urban expansion is in balance with population growth or not. According to literature review, limited scientific peer reviewed studies have been reported on the monitoring and mapping of LCRPGR. Under each SDG, a number of targets and indicators have been defined, which countries have to quantify, but most of the developing countries do not have comprehensive databases through which they can compute, quantify, and report the SDGs indicators. To the best of the authors' knowledge, only Nicolau et al. [77] and Wang et al. [81] have computed and reported scientific results of SDG proposed LCRPGR over urban areas. Nicolau et al. [77] based their study over the mainland of Portugal, while Wang et al. [81] carried out their study over mainland China, using earth observation and population census data, and reported the increase of LCRPGR value from 1.69 in 1990–2000 to 1.78 in 2000–2010. The LCRPGR related research findings from these studies show that in most cities, both horizontal and vertical urban expansions are carried out in an unplanned manner which has already effected the equilibrium of land consumption versus population increase to attain effective development goals by 2030 [77,81]. In this study over the ICT, the LCRPGR ratio was 0.62 from 1976 to 2000, which increased to 1.36 from 2000 to 2016. Based on studies conducted on the global scale, in the most of the cases LCR is higher than or equal to PGR due to high demand of luxurious occupancies in the urban areas [26,30].

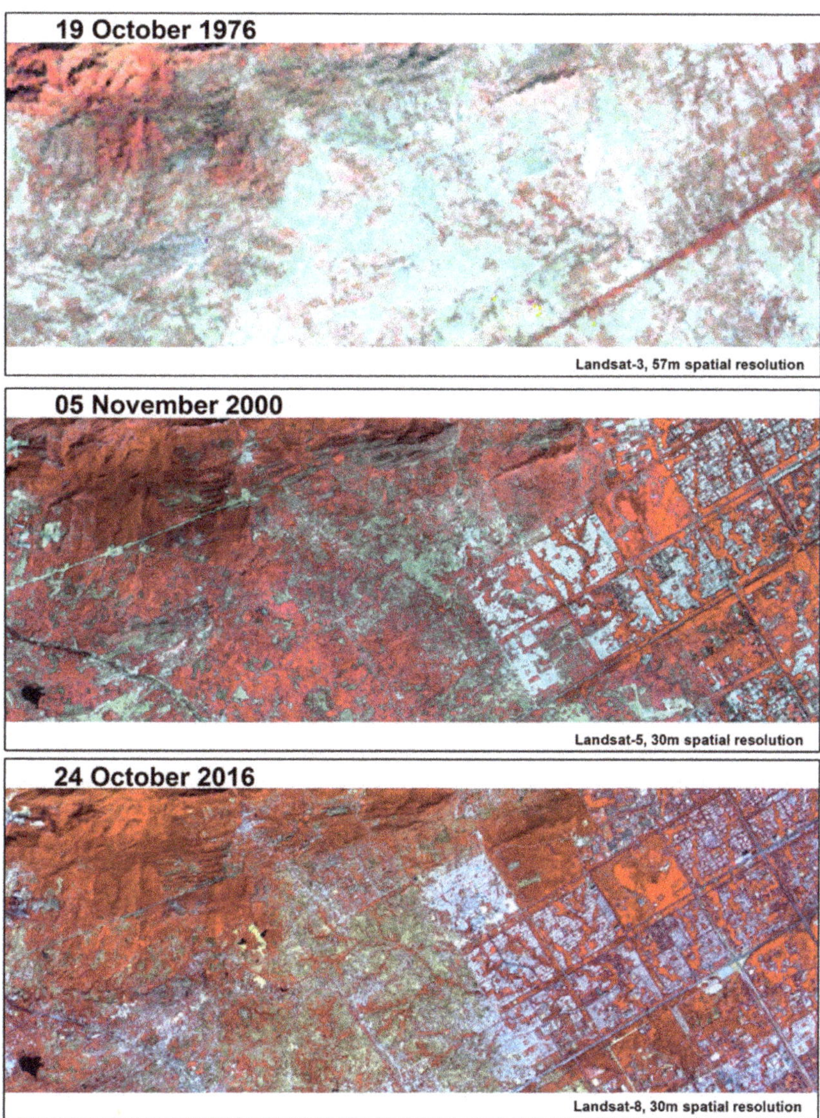

Figure 7. Temporal Landsat satellite images of 1976, 2000, and 2016 in standard False Color Composite (FCC), illustrating an overall increase in built-up areas over the years in the ICT. The area of the ICT shown here represents the bounded box inset covering Zone I and Zone II partially in Figure 1.

The Shenzhen SEZ and ICT cities were both developed in late 1970s. In terms of urban forest cover change, Shenzhen SEZ had been restored to ~85% (1973–2005) [29] while in this study, we detected urban forest loss of ~27% (1976–2016) in the ICT. In both cities (SEZ and ICT), urban forest fragmentation results revealed losses in forest patches. In South Asia, the landscape of the ICT resembles strongly with the cities of Kathmandu (the capital of Nepal) and Thimphu (the capital of Bhutan), as they are all cities topographically surrounded by the pine trees. As compared to ICT, Kathmandu and Thimphu cities are older but highly migrant resident populated, vastly encroached, and over grazed. Although many studies have investigated the land cover dynamics of Kathmandu and Thimphu cities using temporal satellite data [82–85], detailed analysis using parameters such as landscape metrics, forest fragmentation, and LCRPGR has not been performed. In this aspect, this study serves as a methodological framework for application and analysis over other similar cities in developing countries.

Due to rapid tree loss and urbanization, the ICT has observed rapid spells of dust storms and soil erosion [86]. Soil erosion, dry temperatures, and consequent dust storms have a negative impact in the form of air pollution [87] and land degradation. The soil in ICT and the surrounding areas is shallow and has a clay composition [63]. The alluvial lands and terraces in the area tend to have low agricultural productivity and in the southern and western parts of the Potohar plateau, the soil is thin and infertile [88]. Streams and ravines cut the loose plain and cause erosion and steep slopes. This land is generally unsuitable for cultivation. However, large patches of deep, fertile soil are found in the depressions and sheltered parts of the plateau and these support small forests and agriculture [89]. Butt et al. [45] described the most significant land degradation issues in the Rawal watershed as soil erosion and loss of soil nutrients. They further showed that, between 1992 to 2012, the majority of land that was previously vegetation, bare soil, or water bodies was converted to agriculture and settlements, suggesting increased pressure on natural resources in the Rawal watershed [45].

The results of landscape analysis in this study reveal that the ICT urban landscape has become more heterogeneous, disproportional and diverse, and tree patches have declined. Alarmingly, core forests of >500 acres have declined almost 15% in forty years. Although at the individual level, the residents of ICT, civil society, and the local government are trying to recover tree cover loss by planting trees, but these initiatives should be continuous and ongoing on a regular basis to monitor the growth of trees without damaging the existing mature standing trees. The temporal forest fragmentation analysis shows that due to tree cover loss, the three categories of core forest fragmentations (i.e., <250 acres, 250–500 acres, and >500 acres) have decreased in forty years (1976–2016). The loss in forest fragmentation negatively influences the habitat and terrestrial biodiversity of ICT [79,90,91]. Based on landscape metrics analysis over the Dhaka metropolitan, a similar Asian developing city, Dewan et al. [31] revealed that cultivated areas and vegetated lands became highly fragmented with increasing anthropogenic disturbances and urban built up category became aggregated and convoluted.

While our analysis of ICT land cover change dynamics in this study is important and unique, there are a few caveats which are worth mentioning. First of all, for the 1976 land cover map, we have relied on approximately 57 m spatial resolution Landsat 3 Multispectral Scanner System (MSS) sensor data which is relatively coarser spatially as compared to Landsat 5 and 8 (i.e., 30 m), which may affect the spatial heterogeneity and accuracy of developed land cover maps. Second, as Landsat is a pioneer Earth Observation (EO) satellite program initiated in 1972, so acquiring remote sensing imagery of the years before that is not possible, and thus we cannot derive land cover maps before the 1960s, when the Islamic Republic of Pakistan's capital shifted from Karachi to Islamabad. Third, from Landsat 30 m medium spatial resolution satellite data, we cannot detect and delineate the boundaries of built-up areas as well as can be done from sub-meter VHRS images available from the 2000s onward. However, the VHRS datasets come with a high cost, especially when acquisitions at multiples times have to be acquired, and may not be therefore feasible for study sites in developing countries. Of course, research in this domain is ongoing with regards to utilization of VHRS data for study of urban areas, using methods like object-based image analysis, machine learning, etc. Fourth, there may be some level of uncertainty for the field measurement data, as there is always a potential for human error especially when ground truth is collected over larger areas. Fifth, in this study, we utilized only temporal optical satellite data, which may cause optical signal saturation in closed canopy forests, and atmospheric effects (i.e., cloud coverage, haze, and smog).

6. Conclusions

This study sheds new light on systematic land cover dynamics of ICT over the period of four decades, underlining a major decrease in tree cover along with significant increase in settlements and soil. The findings from the analysis of land cover changes, landscape metrics, forest fragmentation, and LCRPGR, are aligned well and agree overall with previous published work on the regional and global scales. The computation of LCRPGR is vital to understand the rate of land change as compared to the population boom, to understand historical land consumption traditions, and to guide decision and policy makers on planned urban expansion while ensuring the protection of environmental, social, and economic assets. It is indeed a scientifically accepted phenomenon that in most of the cities, the land consumption rate overall is much faster than the population growth rate, due to high demand of luxurious occupancies in the urban areas.

Both natural and anthropogenic activities are responsible for land cover changes in ICT. In terms of landscape and fragmentations, the ICT urban landscape ecology is continuously under threat due to the encroachment of housing societies, and influential business community. The findings from urban landscape matrix and forest fragmentation analysis in this study could help ICT Capital Development Authority (CDA) and Islamabad Wildlife Management Board (IWMB) in making strategic decisions to prevent tree loss, forest degradation, and encroachment in the urban landscape of ICT, and also plan future urban growth keeping with the use of remote sensing imagery and geospatial analysis. The methodology outlined in this study is cost-effective and could be easily replicable to other countries/regions, especially in developing countries, through integrating freely available satellite images with a 5–10 years interval.

Author Contributions: Conceptualization, H.G., W.A.Q.; data analysis, S.A., M.K.; supervision, H.G., visualization, H.G., W.A.Q., S.A., M.K.; writing—original draft preparation, H.G., S.A., W.A.Q., S.M.A.; writing—review and editing, H.G., W.A.Q., S.M.A. All authors have read and approved the final version of this paper. All authors have read and agreed to the published version of the manuscript.

Funding: This research received no external funding.

Acknowledgments: This work is an extension of an Undergraduate Final Year Project, which was completed in 2018, and we would like to express our thanks to Institute of Space Technology (IST) management and academic staff for their support. The authors also would like to acknowledge the editors and three anonymous reviewers for their extensive and critical comments on the paper, which helped to improve the paper immensely.

Conflicts of Interest: The authors declare no conflict of interest. The views and interpretations in this publication are those of the authors, and they are not necessarily attributable to their organizations.

Appendix A

Table A1. Accuracy assessment of land covers classification for 1976, 1990, 2000, 2010, and 2016.

	Tree cover > 40% canopy	Tree cover < 40% canopy	Settlement	Soil	Water	Total	User's accuracy (%)
2016's Land cover accuracy assessment report							
Tree cover >40% canopy	22	3	0	0	0	25	88
Tree cover <40% canopy	3	21	0	1	0	25	84
Settlement	0	0	23	3	0	26	88
Soil	0	1	2	21	0	24	88
Water	0	0	0	0	25	25	100
Total	25	25	25	25	25	125	
Producer's accuracy (%)	88	84	92	84	100		
Overall accuracy: 0.90, Kappa: 0.85							
2010's Land cover accuracy assessment report							
Tree cover >40% canopy	22	5	0	0	1	28	79
Tree cover <40% canopy	3	20	0	0	0	23	87
Settlement	0	0	23	4	0	27	85
Soil	0	0	2	21	0	23	91
Water	0	0	0	0	24	24	100
Total	25	25	25	25	25		
Producer's accuracy (%)	88	80	92	84	96		
Overall accuracy: 0.88, Kappa: 0.84							
2000's Land cover accuracy assessment report							
Tree cover >40% canopy	22	3	0	0	2	27	81
Tree cover <40% canopy	3	20	1	1	0	25	80
Settlement	0	0	22	3	0	25	88
Soil	0	2	2	21	0	25	84
Water	0	0	0	0	23	23	100
Total	25	25	25	25	25	125	
Producer's accuracy (%)	88	80	88	84	92		
Overall accuracy: 0.86, Kappa: 0.82							

Table A1. *Cont.*

		1990's Land cover accuracy assessment report					
Tree cover >40% canopy	21	2	0	0	1	24	88
Tree cover <40% canopy	4	20	1	1	0	26	77
Settlement	0	0	21	4	0	25	84
Soil	0	3	3	20	0	26	77
Water	0	0	0	0	24	24	100
Total	25	25	25	25	25	125	
Producer's accuracy (%)	84	80	84	80	96		
		Overall accuracy: 0.85, Kappa: 0.81					
		1976's Land cover accuracy assessment report					
Tree cover >40% canopy	21	3	0	0	1	25	84
Tree cover <40% canopy	4	21	1	2	0	28	75
Settlement	0	0	20	5	0	25	80
Soil	0	1	4	18	0	23	78
Water	0	0	0	0	24	24	100
Total	25	25	25	25	25	125	
Producer's accuracy (%)	84	84	80	72	96		
		Overall accuracy: 0.83, Kappa: 0.79					

Table A2. Land cover class level landscape matrix parameters results for 1976, 1990, 2000, 2010, and 2016.

Land Cover Class	Year	Patch Density and Size				Shape and Edge			Proximity/Isolation	
		NP	LPI (%)	MPS (km^2)	MRG (m)	ED (m/km^2)	MSI	MPAR ×10^3	MPI	MED (m)
Tree cover >40% canopy	1976	1072	12.50	16.98	71.58	21.08	1.27	0.55	869.65	228.47
	1990	5590	9.95	3.44	34.13	44.78	1.24	1.05	1233.12	109.63
	2000	8195	10.31	2.18	26.89	54.25	1.19	1.14	1247.77	98.59
	2010	6121	8.59	2.30	27.34	41.94	1.20	1.15	1401.38	104.15
	2016	2223	7.10	5.95	40.30	23.54	1.28	1.03	1407.03	99.17
Tree cover <40% canopy	1976	1872	30.62	22.29	75.52	66.35	1.32	0.53	4809.76	145.76
	1990	8856	5.89	3.87	40.00	125.30	1.35	1.02	1837.19	77.20
	2000	10145	16.92	3.94	31.18	146.14	1.26	1.11	7773.81	73.00
	2010	9253	29.83	3.89	31.39	122.42	1.23	1.06	18114.32	79.52
	2016	8728	14.97	3.51	36.00	107.08	1.28	1.02	5386.70	83.08
Settlement	1976	4296	0.02	0.70	35.83	17.73	1.09	0.64	1.06	193.66
	1990	7752	0.22	0.57	22.92	26.69	1.13	1.16	12.70	116.02
	2000	6310	0.30	0.95	28.28	30.58	1.19	1.09	33.80	115.14
	2010	11694	5.16	1.23	27.39	68.28	1.20	1.11	784.86	86.70
	2016	12754	8.07	1.34	25.56	71.35	1.18	1.14	1382.81	87.29
Soil	1976	2180	13.76	12.40	65.86	50.43	1.29	0.54	1763.86	159.37
	1990	9271	15.94	3.42	32.34	102.90	1.25	1.06	6172.97	81.38
	2000	12060	3.15	2.13	33.12	103.57	1.27	1.05	564.47	80.48
	2010	8885	4.02	2.86	33.49	84.81	1.26	1.05	530.05	85.99
	2016	12436	5.72	2.33	30.23	107.68	1.26	1.1	1197.42	76.53
Water	1976	8	0.60	68.27	154.90	0.25	1.15	0.54	162.66	2192.65
	1990	242	0.62	3.49	55.79	1.90	1.39	1.02	29.98	362.05
	2000	449	0.61	2.35	43.54	2.93	1.25	1.07	26.36	322.78
	2010	34	0.59	18.51	69.64	0.47	1.21	0.97	65.20	1230.49
	2016	120	0.58	5.43	36.36	0.71	1.18	1.09	46.78	620.00

Table A3. Forest fragmentation results for 1976, 1990, 2000, 2010, and 2016 and percentage assessment based on satellite images).

Forest Fragmentation Classes	1976	1990	2000	2010	2016
	km^2				
Patch	14.74 (2.46)	15.67 (2.93)	15.46 (2.68)	17.84 (3.56)	19.92 (4.54)
Edge	105.58 (17.62)	119.48 (22.33)	123.42 (21.36)	94.06 (18.75)	96.83 (22.07)
Perforated	35.21 (5.88)	31.52 (5.89)	41.79 (7.23)	24.39 (4.86)	22.74 (5.18)
Core (<250 acres)	41.4 (6.91)	79.94 (14.94)	68.04 (11.77)	52.05 (10.38)	52.24 (11.91)
Core (250–500 acres)	10.99 (1.83)	17.18 (3.21)	15.74 (2.72)	5.54 (1.10)	5.63 (1.28)
Core (>500 acres)	391.98 (65.41)	271.34 (50.71)	313.42 (54.24)	307.67 (61.34)	241.44 (55.02)
Total	585.16 (100)	535.13 (100)	577.87 (100)	501.55 (100)	438.80 (100)

References

1. Lambin, E.F.; Meyfroidt, P. Land use transitions: Socio-ecological feedback versus socio-economic change. *Land Use Policy* **2010**, *27*, 108–118. [CrossRef]
2. Balooni, K.; Lund, J.F. Forest rights: The hard currency of REDD+. *Conserv. Lett.* **2014**, *7*, 278–284. [CrossRef]
3. Belal, A.A.; Moghanm, F.S. Detecting urban growth using remote sensing and GIS techniques in Al Gharbiya governorate, Egypt. *Egypt. J. Remote Sens. Sp. Sci.* **2011**, *14*, 73–79. [CrossRef]
4. Congalton, R.; Gu, J.; Yadav, K.; Thenkabail, P.; Ozdogan, M. Global Land Cover Mapping: A Review and Uncertainty Analysis. *Remote Sens.* **2014**, *6*, 12070–12093. [CrossRef]
5. Rasul, G. Ecosystem services and agricultural land-use practices: A case study of the Chittagong Hill Tracts of Bangladesh. *Sustain. Sci. Pract. Policy* **2009**, *5*, 15–27. [CrossRef]
6. Addae, B.; Oppelt, N. Land-Use/Land-Cover Change Analysis and Urban Growth Modelling in the Greater Accra Metropolitan Area (GAMA), Ghana. *Urban Sci.* **2019**, *3*, 26. [CrossRef]
7. Redman, C.L. Human Dimensions of Ecosystem Studies. *Ecosystems* **1999**, *2*, 296–298. [CrossRef]
8. Xu, C.; Liu, M.; Zhang, C.; An, S.; Yu, W.; Chen, J.M. The spatiotemporal dynamics of rapid urban growth in the Nanjing metropolitan region of China. *Landsc. Ecol.* **2007**, *22*, 925–937. [CrossRef]
9. Burgi, M.; Hersperger, A.M.; Schneeberger, N. Driving forces of landscape change - current and new directions. *Landsc. Ecol.* **2005**, *19*, 857–868. [CrossRef]
10. Bilsborrow, R.E.; McDevitt, T.M.; Kossoudji, S.; Fuller, R. The Impact of Origin Community Characteristics on Rural-Urban Out-Migration in a Developing Country. *Demography* **1987**, *24*, 191–210. [CrossRef]
11. Magidi, J.; Ahmed, F. Assessing urban sprawl using remote sensing and landscape metrics: A case study of City of Tshwane, South Africa (1984–2015). *Egypt. J. Remote Sens. Sp. Sci.* **2019**, *22*, 335–346. [CrossRef]
12. Melchiorri, M.; Pesaresi, M.; Florczyk, A.J.; Corbane, C.; Kemper, T. Principles and applications of the global human settlement layer as baseline for the land use efficiency indicator—SDG 11.3.1. *ISPRS Int. J. Geo-Inf.* **2019**, *8*, 96. [CrossRef]
13. Sharma, L.; Pandey, P.C.; Nathawat, M.S. Assessment of land consumption rate with urban dynamics change using geospatial techniques. *J. Land Use Sci.* **2012**, *7*, 135–148. [CrossRef]
14. Salvati, L.; Carlucci, M. Distance matters: Land consumption and the mono-centric model in two southern European cities. *Landsc. Urban Plan.* **2014**, *127*, 41–51. [CrossRef]
15. Sibly, R.M.; Hone, J. Population growth rate and its determinants: An overview. *Philos. Trans. R. Soc. B Biol. Sci.* **2002**, *357*, 1153–1170. [CrossRef]
16. Philippe, D.; Jean-Marie, O.-B.; Marta, G.-D.; Thomas, C. Global immunization: Status, progress, challenges and future. *BMC Int. Health Hum. Rights* **2009**, *9*, S2. [CrossRef]
17. Ghazanfari, H.; Namiranian, M.; Sobhani, H.; Mohajer, R.M. Traditional forest management and its application to encourage public participation for sustainable forest management in the northern Zagros mountains of Kurdistan Province, Iran. *Scand. J. For. Res. Suppl.* **2004**, *19*, 65–71. [CrossRef]
18. Turner, A.B.L.; Meyer, W.B.; Skole, D.L. Global land-use/land-cover change: Towards an integrated study. *Ambio* **1994**, *23*, 91–95.
19. Seto, K.C.; Guneralp, B.; Hutyra, L.R. Global forecasts of urban expansion to 2030 and direct impacts on biodiversity and carbon pools. *Proc. Natl. Acad. Sci.* **2012**, *109*, 16083–16088. [CrossRef] [PubMed]
20. Grimm, N.B.; Faeth, S.H.; Golubiewski, N.E.; Redman, C.L.; Wu, J.; Bai, X.; Briggs, J.M. Global Change and the Ecology of Cities. *Science.* **2008**, *319*, 756–760. [CrossRef] [PubMed]
21. Angel, S.; Parent, J.; Civco, D.L.; Blei, A.; Potere, D. The dimensions of global urban expansion: Estimates and projections for all countries, 2000–2050. *Prog. Plann.* **2011**, *75*, 53–107. [CrossRef]
22. Boggs, G.S. Assessment of SPOT 5 and QuickBird remotely sensed imagery for mapping tree cover in savannas. *Int. J. Appl. Earth Obs. Geoinf.* **2010**, *12*, 217–224. [CrossRef]
23. Wulder, M.A.; Coops, N.C.; Roy, D.P.; White, J.C.; Hermosilla, T. Land cover 2.0. *Int. J. Remote Sens.* **2018**, *39*, 4254–4284. [CrossRef]
24. Dong, J.; Xiao, X.; Menarguez, M.A.; Zhang, G.; Qin, Y.; Thau, D.; Biradar, C.; Moore, B. Mapping paddy rice planting area in northeastern Asia with Landsat 8 images, phenology-based algorithm and Google Earth Engine. *Remote Sens. Environ.* **2016**, *185*, 142–154. [CrossRef]

25. Uddin, K.; Shrestha, H.L.H.L.; Murthy, M.S.R.S.R.; Bajracharya, B.; Shrestha, B.; Gilani, H.; Pradhan, S.; Dangol, B. Development of 2010 national land cover database for the Nepal. *J. Environ. Manage.* **2015**, *148*, 82–90. [CrossRef]

26. Seto, K.C.; Fragkias, M.; Güneralp, B.; Reilly, M.K. A Meta-Analysis of Global Urban Land Expansion. *PLoS ONE* **2011**, *6*, e23777. [CrossRef]

27. Yang, C.; Li, Q.; Hu, Z.; Chen, J.; Shi, T.; Ding, K.; Wu, G. Spatiotemporal evolution of urban agglomerations in four major bay areas of US, China and Japan from 1987 to 2017: Evidence from remote sensing images. *Sci. Total Environ.* **2019**, *671*, 232–247. [CrossRef]

28. Herold, M.; Couclelis, H.; Clarke, K.C. The role of spatial metrics in the analysis and modeling of urban land use change. *Comput. Environ. Urban Syst.* **2005**, *29*, 369–399. [CrossRef]

29. Gong, C.; Yu, S.; Joesting, H.; Chen, J. Determining socioeconomic drivers of urban forest fragmentation with historical remote sensing images. *Landsc. Urban Plan.* **2013**, *117*, 57–65. [CrossRef]

30. Huang, J.; Lu, X.X.; Sellers, J.M. A global comparative analysis of urban form: Applying spatial metrics and remote sensing. *Landsc. Urban Plan.* **2007**, *82*, 184–197. [CrossRef]

31. Dewan, A.M.; Yamaguchi, Y.; Ziaur Rahman, M. Dynamics of land use/cover changes and the analysis of landscape fragmentation in Dhaka Metropolitan, Bangladesh. *GeoJournal* **2012**, *77*, 315–330. [CrossRef]

32. Mandelas, E. A fuzzy cellular automata based shell for modeling urban growth–a pilot application in Mesogia area. In Proceedings of the 10th AGILE International Conference on Geographic Information Science, Allborg, Denmark, 8–11 May 2007; pp. 1–9.

33. Farhat, K.; Waseem, L.A.; Khan, A.A.; Baig, S. Spatiotemporal Demographic Trends and Land Use Dynamics of Metropolitan Lahore. *J. Hist. Cult. Art Res.* **2018**, *7*, 92. [CrossRef]

34. Shah, M.H.; Shaheen, N. Seasonal behaviours in elemental composition of atmospheric aerosols collected in Islamabad, Pakistan. *Atmos. Res.* **2010**, *95*, 210–223. [CrossRef]

35. Butt, M.J.; Waqas, A.; Iqbal, M.F.; Muhammad, G.; Lodhi, M.A.K. Assessment of Urban Sprawl of Islamabad Metropolitan Area Using Multi-Sensor and Multi-Temporal Satellite Data. *Arab. J. Sci. Eng.* **2012**, *37*, 101–114. [CrossRef]

36. Shahzad, N.; Saeed, U.; Gilani, H.; Ahmad, S.R.; Ashraf, I.; Irteza, S.M. Evaluation of state and community/private forests in Punjab, Pakistan using geospatial data and related techniques. *For. Ecosyst.* **2015**, *2*, 7. [CrossRef]

37. Doxiadis, C.A. Islamabad, the creation of a new capital. *Town Plan. Rev.* **1965**, *36*, 1. [CrossRef]

38. Stephenson, G.V. Two Newly Created Capitals: Islamabad and Brasilia. *Town Plan. Rev.* **1970**, *41*, 317. [CrossRef]

39. Kreutzmann, H. Islamabad, living the plan. *Südasien-Chronik-South Asia Chron.* **2013**, *3*, 2013.

40. Beacco, D. Urban Planning in Islamabad: From the Modern Movement to the Contemporary Urban Development Between Formal and Informal Settlements. In *Sustainable Urban Development and Globalization*; Springer: Cham, Switzerland, 2018; pp. 65–74.

41. Bokhari, S.A.; Saqib, Z.; Ali, A.; Haq, M.Z. Perception of Residents about Urban Vegetation: A Comparative Study of Planned Versus Semi-Planned Cities of Islamabad and Rawalpindi, Pakistan. *J. Ecosyst. Ecography* **2018**, *8*, 1–8. [CrossRef]

42. Naeem, S.; Cao, C.; Waqar, M.M.; Wei, C.; Acharya, B.K. Vegetation role in controlling the ecoenvironmental conditions for sustainable urban environments: A comparison of Beijing and Islamabad. *J. Appl. Remote Sens.* **2018**, *12*, 016013. [CrossRef]

43. Adeel, M. Methodology for identifying urban growth potential using land use and population data: A case study of Islamabad Zone IV. *Procedia Environ. Sci.* **2010**, *2*, 32–41. [CrossRef]

44. Butt, A.; Shabbir, R.; Ahmad, S.S.; Aziz, N. Land use change mapping and analysis using Remote Sensing and GIS: A case study of Simly watershed, Islamabad, Pakistan. *Egypt. J. Remote Sens. Sp. Sci.* **2015**, *18*, 251–259. [CrossRef]

45. Butt, A.; Shabbir, R.; Ahmad, S.S.; Aziz, N.; Nawaz, M.; Shah, M.T.A. Land cover classification and change detection analysis of Rawal watershed using remote sensing data. *J Biodivers Env. Sci* **2015**, *6*, 236–248.

46. Hassan, Z.; Shabbir, R.; Ahmad, S.S.; Malik, A.H.; Aziz, N.; Butt, A.; Erum, S. Dynamics of land use and land cover change (LULCC) using geospatial techniques: A case study of Islamabad Pakistan. *Springerplus* **2016**, *5*, 812. [CrossRef] [PubMed]

47. Sohail, M.T.; Mahfooz, Y.; Azam, K.; Yat, Y.; Genfu, L.; Fahad, S. Impacts of urbanization and land cover dynamics on underground water in Islamabad, Pakistan. *Desalin. WATER Treat.* **2019**, *159*, 402–411. [CrossRef]

48. Naeem, S.; Cao, C.; Qazi, W.; Zamani, M.; Wei, C.; Acharya, B.; Rehman, A. Studying the association between green space characteristics and land surface temperature for sustainable urban environments: An analysis of Beijing and Islamabad. *ISPRS Int. J. Geo-Inf.* **2018**, *7*, 38. [CrossRef]

49. Khalid, N.; Ullah, S.; Ahmad, S.S.; Ali, A.; Chishtie, F. A remotely sensed tracking of forest cover and associated temperature change in Margalla hills. *Int. J. Digit. Earth* **2019**, *12*, 1133–1150. [CrossRef]

50. Batool, R.; Javaid, K. Spatio-temporal assessment of Margalla hills forest by using Landsat imagery for year 2000 and 2018. In Proceedings of the International Archives of the Photogrammetry, Remote Sensing and Spatial Information Sciences, ISPRS TC III Mid-term Symposium "Developments, Technologies and Applications in Remote Sensing", Beijing, China, 7–10 May 2018; Volume XLII–3, pp. 69–72.

51. Mannan, A.; Liu, J.; Zhongke, F.; Khan, T.U.; Mukete, B.; Chaoyong, S.; Saeed, S.; Yongxiang, F.; Ahmad, A.; Amir, M.; et al. Application of land-use/land cover changes in monitoring and projecting forest biomass carbon loss in Pakistan. *Glob. Ecol. Conserv.* **2019**, *17*, e00535. [CrossRef]

52. Dewan, A.M.; Yamaguchi, Y. Land use and land cover change in Greater Dhaka, Bangladesh: Using remote sensing to promote sustainable urbanization. *Appl. Geogr.* **2009**, *29*, 390–401. [CrossRef]

53. Dewan, A.M.; Yamaguchi, Y. Using remote sensing and GIS to detect and monitor land use and land cover change in Dhaka Metropolitan of Bangladesh during 1960–2005. *Environ. Monit. Assess.* **2009**, *150*, 237–249. [CrossRef]

54. Pramanik, M.M.A.; Stathakis, D. Forecasting urban sprawl in Dhaka city of Bangladesh. *Environ. Plan. B Plan. Des.* **2016**, *43*, 756–771. [CrossRef]

55. Gruebner, O.; Sachs, J.; Nockert, A.; Frings, M.; Khan, M.M.H.; Lakes, T.; Hostert, P. Mapping the Slums of Dhaka from 2006 to 2010. *Dataset Pap. Sci.* **2014**, *2014*, 1–7. [CrossRef]

56. Sharma, R.; Joshi, P.K. Monitoring Urban Landscape Dynamics Over Delhi (India) Using Remote Sensing (1998–2011) Inputs. *J. Indian Soc. Remote Sens.* **2013**, *41*, 641–650. [CrossRef]

57. Naeem, S.; Cao, C.; Fatima, K.; Najmuddin, O.; Acharya, B.K. Landscape greening policies-based land use/land cover simulation for Beijing and Islamabad—An implication of sustainable urban ecosystems. *Sustainability* **2018**, *10*, 1049. [CrossRef]

58. Huang, H.; Chen, Y.; Clinton, N.; Wang, J.; Wang, X.; Liu, C.; Gong, P.; Yang, J.; Bai, Y.; Zheng, Y.; et al. Mapping major land cover dynamics in Beijing using all Landsat images in Google Earth Engine. *Remote Sens. Environ.* **2017**, *202*, 166–176. [CrossRef]

59. Cui, L.; Shi, J. Urbanization and its environmental effects in Shanghai, China. *Urban Clim.* **2012**, *2*, 1–15. [CrossRef]

60. Zhao, S.; Da, L.; Tang, Z.; Fang, H.; Song, K.; Fang, J. Ecological consequences of rapid urban expansion: Shanghai, China. *Front. Ecol. Environ.* **2006**, *4*, 341–346.

61. Yue, W.; Fan, P.; Wei, Y.D.; Qi, J. Economic development, urban expansion, and sustainable development in Shanghai. *Stoch. Environ. Res. Risk Assess.* **2014**, *28*, 783–799. [CrossRef]

62. Bagan, H.; Yamagata, Y. Landsat analysis of urban growth: How Tokyo became the world's largest megacity during the last 40 years. *Remote Sens. Environ.* **2012**, *127*, 210–222. [CrossRef]

63. Sheikh, I.M.; Pasha, M.K.; Williams, V.S.; Raza, S.Q.; Khan, K.S.A.A. Environmental Geology of the Islamabad-Rawalpindi Area. In *Regional Studies of the Potwar Plateau Area, Northern Pakistan*; 2007; pp. 1–32. Available online: https://pubs.usgs.gov/bul/2078/ (accessed on 5 February 2020).

64. Teillet, P.; Barker, J.; Markham, B.; Irish, R.; Fedosejevs, G.; Storey, J. Radiometric cross-calibration of the Landsat-7 ETM+ and Landsat-5 TM sensors based on tandem data sets. *Remote Sens. Environ.* **2001**, *78*, 39–54. [CrossRef]

65. Tan, B.; Masek, J.G.; Wolfe, R.; Gao, F.; Huang, C.; Vermote, E.F.; Sexton, J.O.; Ederer, G. Improved forest change detection with terrain illumination corrected Landsat images. *Remote Sens. Environ.* **2013**, *136*, 469–483. [CrossRef]

66. Congalton, R.G. Remote Sensing and Image Interpretation. 7th Edition. *Photogramm. Eng. Remote Sens.* **2015**, *81*, 615–616. [CrossRef]

67. Kasereka, K. Remote Sensing and Geographic Information System for Inferring Land Cover and Land Use Change in Wuhan (China), 1987-2006. *J. Sustain. Dev.* **2010**, *3*, 221–229. [CrossRef]

68. Khatami, R.; Mountrakis, G.; Stehman, S.V. A meta-analysis of remote sensing research on supervised pixel-based land-cover image classification processes: General guidelines for practitioners and future research. *Remote Sens. Environ.* **2016**, *177*, 89–100. [CrossRef]

69. Saura, S. Effects of minimum mapping unit on land cover data spatial configuration and composition. *Int. J. Remote Sens.* **2002**, *23*, 4853–4880. [CrossRef]

70. Roy, P.S.; Roy, A.; Joshi, P.K.; Kale, M.P.; Srivastava, V.K.; Srivastava, S.K.; Dwevidi, R.S.; Joshi, C.; Behera, M.D.; Meiyappan, P.; et al. Development of decadal (1985-1995-2005) land use and land cover database for India. *Remote Sens.* **2015**, *7*, 2401–2430. [CrossRef]

71. Rosenfield, G.H.; Fitzpatrick-Lins, K. A coefficient of agreement as a measure of thematic classification accuracy. *Photogramm. Eng. Remote Sens.* **1986**, *52*, 223–227.

72. Puyravaud, J.P. Standardizing the calculation of the annual rate of deforestation. *For. Ecol. Manage.* **2003**, *177*, 593–596. [CrossRef]

73. McGarigal, K. Landscape Pattern Metrics. In *Wiley StatsRef: Statistics Reference Online*; John Wiley & Sons, Ltd.: Chichester, UK, 2014; Volume 45, pp. 1–18.

74. Nichol, J.E.; Abbas, S.; Fischer, G.A. Spatial patterns of degraded tropical forest and biodiversity restoration over 70-years of succession. *Glob. Ecol. Conserv.* **2017**, *11*, 134–145. [CrossRef]

75. McGarigal, K.; Marks, B.J. FRAGSTATS: Spatial pattern analysis program for quantifying landscapes structure. *Gen. Tech. Rep. PNW-GTR-351. U.S. Dep. Agric. For. Serv. Pacific Northwest Res. Station. Portland, OR* **1994**, *97331*, 134.

76. Vogt, P.; Riitters, K.; Estreguil, C.; Kozak, J.; Wade, T.; Wickham, J. Mapping spatial patterns with morphological image processing. *Landsc. Ecol.* **2007**, *22*, 171–177. [CrossRef]

77. Nicolau, R.; Caetano, M.; David, J.; Caetano, M.; Pereira, J. Ratio of Land Consumption Rate to Population Growth Rate—Analysis of Different Formulations Applied to Mainland Portugal. *ISPRS Int. J. Geo-Inf.* **2018**, *8*, 10. [CrossRef]

78. Khalid, N.; Saeed Ahmad, S. Monitoring Forest cover change of Margalla Hills over a period of two decades (1992-2011): A spatiotemporal perspective. *J. Ecosyst. Ecography* **2016**, *06*, 1–8. [CrossRef]

79. Malik, I.U.; Faiz, A.H.; -i- Abbas, F. Biodiversity Assessment and its Effect on the Environment of Shakarparian Forest. *J. Bioresour. Manag.* **2014**, *1*, 4. [CrossRef]

80. Ahmed, K.; Jatoo, W.; Jing, F.C.; Saengkrod, W.; Mastoi, A.G. Urbanization in Pakistan: Challenges and Way Forward (Options) For Sustainable Urban Development. In Proceedings of the 4th International Conference on Energy, Environment and Sustainable Development, Jamshoro, Pakistan, 1–3 November 2016.

81. Wang, Y.; Huang, C.; Feng, Y.; Zhao, M.; Gu, J. Using Earth Observation for Monitoring SDG 11.3.1- Ratio of Land Consumption Rate to Population Growth Rate in Mainland China. *Remote Sens.* **2020**, *12*, 357. [CrossRef]

82. Rimal, B. Application of remote sensing and GIS, land use/land cover change in Kathmandu Metropolitan City, Nepal. *J. Theor. Appl. Inf. Technol.* **2011**, *23*, 80–86.

83. Thapa, B.R.; Ishidaira, H.; Pandey, V.P.; Shakya, N.M. A multi-model approach for analyzing water balance dynamics in Kathmandu Valley, Nepal. *J. Hydrol. Reg. Stud.* **2017**, *9*, 149–162. [CrossRef]

84. Giri, N.; Singh, O.P. Urban growth and water quality in Thimphu, Bhutan. *J. Urban Environ. Eng.* **2013**, *7*, 82–95. [CrossRef]

85. Khanal, N.; Uddin, K.; Matin, M.; Tenneson, K. Automatic Detection of Spatiotemporal Urban Expansion Patterns by Fusing OSM and Landsat Data in Kathmandu. *Remote Sens.* **2019**, *11*, 2296. [CrossRef]

86. Ellis, S.; Taylor, D.; Masood, K.R. Land degradation in northern Pakistan. *Geography* **1993**, *78*, 84–87.

87. Cavanaugh, R. Increased pollution and problems in respiratory health menace Pakistan's soaring population. *Lancet Respir. Med.* **2018**, *6*, 175–176. [CrossRef]

88. Rashid, K.; Rasul, G. Rainfall variability and maize production over the Potohar plateau of Pakistan. *Pakistan J. Meteorol.* **2011**, *8*, 63–74.

89. Pandey, V.P.; Manandhar, S.; Kazama, F. Climate change vulnerability assessment. In *Climate Change and Water Resources*; CRC Press: Boca Raton, FL, USA, 2014; pp. 183–208. ISBN 9781466594678.

90. Akram, A.; Rais, M.; Asadi, M.A.; Jilani, M.J.; Balouch, S.; Anwar, M.; Saleem, A. Do habitat variables correlate anuran abundance in arid terrain of Rawalpindi–Islamabad Areas, Pakistan? *J. King Saud Univ. Sci.* **2015**, *27*, 278–283. [CrossRef]
91. Shabbir, A.; Bajwa, R. Distribution of parthenium weed (Parthenium hysterophorus L.), an alien invasive weed species threatening the biodiversity of Islamabad. *Weed Biol. Manag.* **2006**, *6*, 89–95. [CrossRef]

 *land*

Article

Can Rock-Rubble Groynes Support Similar Intertidal Ecological Communities to Natural Rocky Shores?

Paul Holloway [1,2,*] and **Richard Field [3]**

[1] Department of Geography, University College Cork, T12K8AF Cork, Ireland
[2] Environmental Research Institute, University College Cork, T23XE10 Cork, Ireland
[3] School of Geography, University of Nottingham, University Park, Nottingham NG7 2RD, UK;
Richard.Field@nottingham.ac.uk
[*] Correspondence: paul.holloway@ucc.ie

Received: 30 March 2020; Accepted: 24 April 2020; Published: 28 April 2020

Abstract: Despite the global implementation of rock-rubble groyne structures, there is limited research investigating their ecology, much less than for other artificial coastal structures. Here we compare the intertidal ecology of urban (or semi-urban) rock-rubble groynes and more rural natural rocky shores for three areas of the UK coastline. We collected richness and abundance data for 771 quadrats across three counties, finding a total of 81 species, with 48 species on the groynes and 71 species on the natural rocky shores. We performed three-way analysis of variance (ANOVA) on both richness and abundance data, running parallel analysis for rock and rock-pool habitats. We also performed detrended correspondence analysis on all species to identify patterns in community structure. On rock surfaces, we found similar richness and abundance across structures for algae, higher diversity and abundance for lichen and mobile animals on natural shores, and higher numbers of sessile animals on groynes. Rock-pool habitats were depauperate on groynes for all species groups except for sessile animals, relative to natural shores. Only a slight differentiation between groyne and natural shore communities was observed, while groynes supported higher abundances of some 'at risk' species than natural shores. Furthermore, groynes did not differ substantially from natural shores in terms of their presence and abundance of species not native to the area. We conclude that groynes host similar ecological communities to those found on natural shores, but differences do exist, particularly with respect to rock-pool habitats.

Keywords: biodiversity; non-native species; protected species; range expansion; species distributions

1. Introduction

Climate change and anthropogenic pressures have fragmented and restricted the distribution of many species worldwide, with significant shifts documented in a vast array of ecological communities [1–4]. In coastal ecosystems, numerous studies have reported a decline in biodiversity [5–12]. For example, Sorte et al. [9] noted a 60% decline in the abundance of the blue mussel *Mytilus edulis* over the past 40 years along the coastline of Eastern USA and linked this decline with that of several other species within the intertidal community. With at least a billion people expected to live within the lower-elevation coastal zone by 2060 [13] and up to 12.5 million km^2 of natural habitat potentially replaced by 2030 [14,15], there persists a need to identify how novel artificial habitats impact coastal ecology.

Introducing hard-engineering structures can negate many of the perceived negative geomorphological and economic impacts of coastal erosion, particularly in urban environs; however, these structures can have significant implications for the configuration of intertidal habitats and biodiversity [16–18]. Studies have predominantly focused on determining whether the communities found across analogous natural shores (largely situated in rural areas) versus artificial structures (largely situated in urban or semi-urban areas) are comparable [18]. The consensus so far is that there is higher diversity on natural shores than analogous artificial rocky sea defences [19–26]. However, investigations have not been conclusive, identifying differences across tidal heights [27] and taxa [28,29], while others have found a strong similarity between structures [30,31].

Rock-rubble groynes are commonly implemented hard-engineering structures that run perpendicular to the shoreline, intercepting longshore transport of sediment. The intertidal habitat of these artificial structures offers a new rocky habitat that is typically located in predominantly sandy shorelines; however, there are very few studies investigating the ecology of rock-rubble groynes. Studies by Pinn and Rodgers [32] of a natural rocky shore and by Pinn et al. [21] of rock-rubble groynes in Dorset, UK, found higher species richness and abundances of most species on the natural rocky shore at Kimmeridge Bay than on the rock-rubble groynes at the nearby Sandbanks Peninsula. However, the ecological comparison between the structures was not the predominant focus of either study, and there was no statistical analysis beyond the simple comparisons of biodiversity. Similarly, in a study comparing the ecology of eight artificial structures (five of which were groynes) and eight natural rocky-shores in the UK, Firth et al. [23] found higher mean species richness on the natural rocky shores than the artificial structures, with no species unique to the artificial structures. Firth et al. [23] also compared habitats, finding that rock pools supported greater species richness than rock habitats, irrespective of structure. This contrasts with the results of Pinn et al. [21], who found more species on exposed rock than in the pools on the groynes at Sandbanks. The findings of these studies suggest that, while groynes support a lower level of biodiversity than their natural counterparts, they could provide a refuge for intertidal communities found on rocky shores that are under pressure from increasing urbanisation.

Conversely, a major criticism of locating hard-bottom artificial structures in soft-bottom areas of coastline is that they can contribute to the decline of barriers (impassable areas of soft-bottom coastline) which isolate distinct regions of rocky shores: removing barriers may enable the dispersal of larvae and propagules of invasive species beyond their natural limits [33–35]. For example, Airoldi et al. [35] found that non-indigenous species were 2–3 times more abundant on artificial structures in part of the North Adriatic Sea, and several other studies have also found artificial structures to support non-indigenous species [36–39]. These studies conclude that there are more non-indigenous species on artificial structures than on nearby rocky shores because they act as points of invasion for many of the non-indigenous taxa. Because of these points of invasion and the resulting creation of stepping-stone dispersal corridors, hard-bottomed artificial structures may pose a serious concern for biodiversity [40].

The UK coastline is one of the most highly human-impacted ecosystems in the world [41]. With the projected coastal urbanisation and climatic changes, there is a pressing need to identify how intertidal communities differ between rock-rubble groynes in urban environs and analogous natural rocky shores in rural areas. Theoretically, rock-rubble groynes provide a conservation dilemma. They offer an opportunity for the presence of novel rock habitat that could have beneficial implications for populations of under-pressure intertidal species predominantly in highly-impacted urban environments, yet they increase connectivity between isolated rocky shores, which may support populations of native species or increase the potential for non-native species invasion, or both. With only two studies [21,23] having compared the ecology of natural rocky shores with those of rock-rubble groynes (one of these only as an in-passing comment), and with contrasting results found in a single study area (Sandbanks peninsula, Dorset), the questions of whether rock-rubble groynes support ecological communities similar to natural rocky shores, and whether they represent a conservation opportunity or threat, remain open. Furthermore, we still lack basic knowledge of how diverse and abundant the rock-rubble groyne communities are.

Here we compare the ecological communities of both exposed rock and rock pools between rock-rubble groynes and nearby rocky shores, using a paired sampling design repeated in three locations around the coast of England. We focus on four main questions: do urban rock-rubble groynes and nearby rural rocky shores (hereafter we refer to the two collectively as rocky 'structure types') differ in terms of (1) their species richness and (2) their species' abundances? (3) Are there specific communities found on one of the structure types but not the other? (4) Do the rocky structure types differ in terms of presence and abundance of species not native to the area (considering native status at both a country and a within-country level)? In addressing these questions, we investigate the role of rock pools as well as rock surfaces and investigate all macro-organisms found, doing parallel analyses for algae, lichens, sessile animals and mobile animals.

2. Materials and Methods

2.1. Study Area and Data Collection

We selected three stretches of the English coastline that each contained a rural natural rocky shore and urban (or semi-urban) rock-rubble groynes in reasonably close proximity to each other (Figure 1, Table SI1). This gave us three pairs of study sites (Supplementary Information 1), which we sampled in summer 2008. In each site with artificial structures, we randomly selected three rock-rubble groynes of the same age to survey (Table SI1). To control for variations in tidal shore height, we separated each structure into low-, mid- and high-shore sections (we call this variable 'level') based on the mean low- and high-water spring tides. We did not identify tidal height boundaries by the limits of organisms, as often recommended [42] because both physical and biological components influence the zonation of organisms [43], meaning there is the possibility to introduce considerable cross-site error.

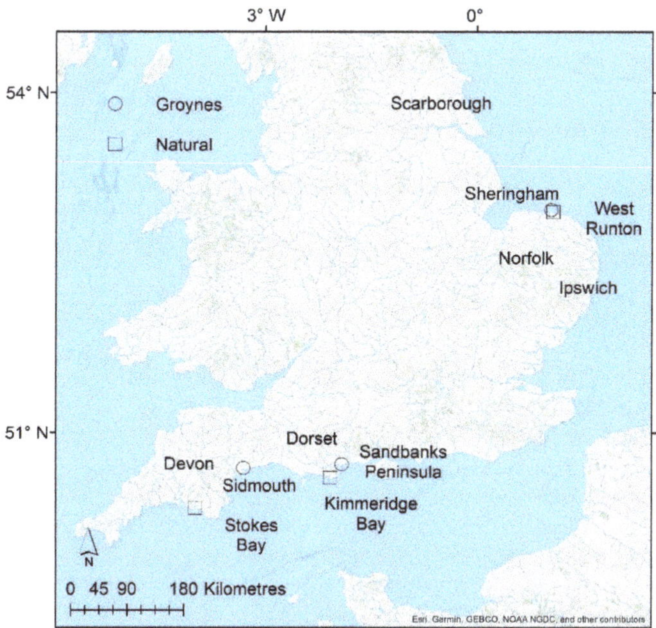

Figure 1. Location of study sites around the coast of England. Devon, Dorset, and Norfolk are counties. Ipswich and Scarborough are locations mentioned in the text.

We used a stratified random sampling technique. With 2 structure types (natural, groyne), 3 counties (Devon, Dorset, Norfolk), 3 replicates of each structure type in each county and 3 levels (high, mid, low) per structure, we had 54 sampling sections in total. Due to the uncertain influence of aspect and exposure on intertidal biological assemblages [21,44], we only surveyed the side of the groynes that did not face the dominant longshore current (therefore sheltered from wave action). Each section of the sampling area was divided up into 0.5 m^2 squares to reduce the chance of recording the same individual twice and selected using a random number generator [42]. On open rock surfaces, we randomly located 12 quadrats in each sampling section, each of 0.25 m × 0.25 m, making 648 rock-surface quadrats in total, and where necessary moved quadrats to avoid sampling very deep crevices or between boulders to allow for consistent surface areas [29]. In each sampling section, we also selected three rock pools (of similar size to the quadrats across structures) in the same way, or all rock pools in the section if fewer than four were present. Ten of the sampling sections (nine of them on groynes) contained no rock pools at all. Overall, we placed quadrats in 123 rock pools: 74 on natural rocky shores and 49 on groynes. In addition, we counted the total number of rock pools present on each entire groyne or equivalent area of natural shore. We did not measure the depth and perimeter of rock pools as is sometimes recommended [42] due to the time constraints associated with the survey, and we acknowledge the limitations of this approach on the analysis in the discussion.

We used a non-destructive sampling method, recording species as present if observed within the quadrat. We recorded species richness as the total number of species in each quadrat. For mobile organisms, we recorded abundance as the count of individuals in each quadrat or pool, and for lichens, algae and sessile animals, we recorded abundance as percentage cover using a grid. We grouped the twelve rock quadrats in each sampling section into four sets of three neighbouring quadrats and grouped the three rock-pool quadrats into one set of three rock-pool quadrats (where we had sufficient numbers), with each data value being the arithmetic mean of these quadrats (e.g. the mean number of animal species A across three quadrats). Thus, each unit in the analysis represented a small section of habitat sampled using three quadrats, rather than a single very small quadrat. This was to allow a better representation of the local community in each unit of analysis and to reduce any variation and uncertainty associated with slight differences in pool volume and surface area due to the uneven surface of the study areas. Overall, the data analysed included 216 rock surface samples and 46 rock pool samples.

2.2. Data Analysis

We used a three-way analysis of variance (ANOVA) to explore the differences between rock-rubble groynes and natural rocky shores (the 'structure type' explanatory variable). We also included the following factors to account for their expected influence: 'county' (Devon, Dorset, Norfolk—reflecting the pairing of the sites) and [tidal] 'level' (low, mid, high). We used Levene's test to examine the assumption of homogeneity of variance, and we visually examined the model residuals for patterning and tested them for normality using Kolmogorov–Smirnov tests. These diagnostics caused us to square-root transform the response variable in the analyses of species' abundances. We ran parallel analyses for rock and pool habitats. We then used detrended correspondence analysis (DCA) to identify ecological communities and significant environmental centroids within the full species dataset among sites which had species recorded within them. DCA reveals the dispersion of points in ordination space, which reflects species abundances within sampling sites [45]. Due to the high frequency of rare species in pools, analysis was undertaken for both rock and pool habitats together, but rare species were not down-weighted as sometimes suggested [46] due to minimal differences in the results when both habitats (rock and pool) were considered together and the importance of the rarer species to ecological communities on natural shores. We chose DCA over other ordination analyses due to the long gradient lengths [47] and the fact that DCA is based on the underlying unimodal model of species distributions [48]—a key foundation of our research questions. We explored the implementation of another method (nonmetric multidimensional scaling: nMDS), but found that the long computation time, coupled with the lack of model convergence and the impact of rare species, made manually exploring model options and subsequent interpretation of the results overly complex. Moreover, results from the nMDS were largely congruent with those of DCA, with the exception of the extreme impact of rare *Idotea* and *Gammarus* spp (results not shown). Environmental factors were passively projected on the ordination plot. All analysis was implemented in the open-source software R 3.6.2 [49], with the DCA implemented using the *vegan* package [46]; see Supplementary Information 2 for data and Supplementary Information 3 for R code.

We compared recorded presences in our study with species distribution maps from Gibson et al. [50] and the Marine Life Information Network (MarLIN; [51]) to determine whether we recorded species beyond their distributions, as judged by the two sources. Gibson et al. [50] provide a generalised range of each species, while MarLIN [51] maps are generated from published species records and verified sightings. We chose these sources to correspond with the period of data collection, rather than the most up-to-date distributional data.

3. Results

3.1. Species Richness and Abundance

We recorded 81 species in total: 27 algae, 6 lichens, 11 sessile animals and 37 mobile animals (Supplementary Information 4). We found 48 species on the groynes and 71 species on the natural rocky shores. Species richness was higher on the natural rocky shores than on the groynes; however, this varied by species group and habitat type (Figure 2a,b). On rock surfaces, species richness on natural shores was significantly higher than that for groynes (Figure 2a, Table 1); however, when species groups were disaggregated, the difference was not significant for algae and there were significantly more sessile animal species on groynes. In the rock pools, species richness was substantially higher on natural rocky shores than groynes for all species groups except sessile animals (no significant difference) and lichens (none recorded in pools) (Figure 2b, Table 1). Indeed, the rock pool habitat represents a key difference between the groynes and the natural rocky shores. Of the 81 species we recorded, 66 were present in rock pools, including 21 only recorded in rock pools. Of these 21, twelve were only in natural rock pools (5 algae, 7 mobile animals), three only in pools on groynes (all were mobile animals) and six in both (5 mobile animals, 1 sessile animal). Species richness varied significantly between counties and across tidal levels, and these factors interacted with structure type to varying degrees to affect the magnitude of species richness difference between groynes and natural rocky shores (Figure 3, Table 1).

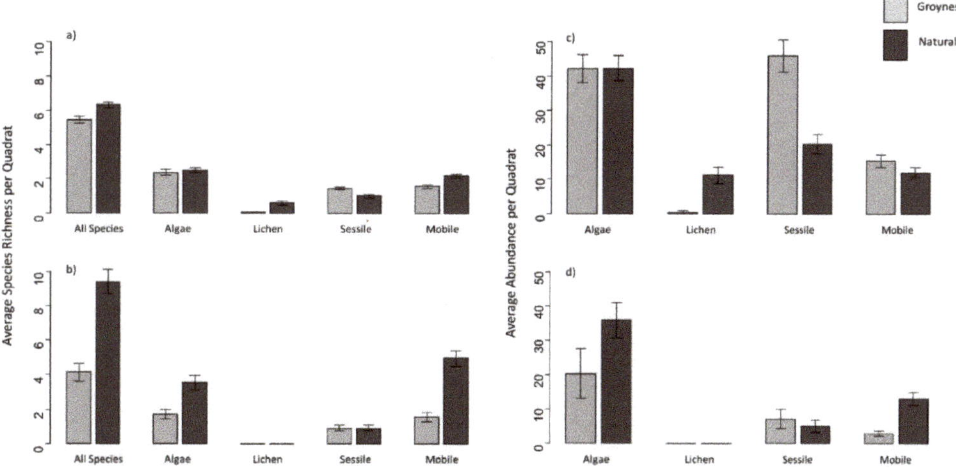

Figure 2. Comparison of species richness on (**a**) rock surfaces and (**b**) rock pools, and of species' abundance on (**c**) rock surface and (**d**) rock pools, between rock-rubble groynes and natural rocky shores. Error bars represent 1 standard error of the mean. Abundance for algae, lichen and sessile animals is shown as percentage cover, and for mobile animals is shown as the total count.

Table 1. Results of the analysis of variance (ANOVA) testing the effects of County, Type, and Level, on average species richness per quadrat, averaged over a set of up to three quadrats. 'Type' is the rocky structure type (natural vs. artificial). 'Level' is the tidal level (low, medium, high). 'County' refers to the three stretches of coastline (Devon, Dorset, Norfolk). ** $p < 0.01$, * $p < 0.05$.

Response Variable		Type	Level	County	Type * Level	Type * County	Level * County	Type * Level * County
	df:	1	2	2	2	2	4	4
total: Rock	F:	14.76 **	50.85 **	9.18 **	5.09 **	1.00	3.48 **	4.88 **
algae: Rock	F:	0.56	37.45 **	41.17 **	4.76 **	8.19 **	3.26 *	2.12
lichen: Rock	F:	58.89 **	50.06 **	13.79 **	33.92 **	9.32 **	8.49 **	5.07 **
sessile: Rock	F:	17.75 **	36.02 **	12.45 **	3.05 *	0.83	7.12 **	7.31 **
mobile: Rock	F:	24.27 **	14.71 **	22.98 **	7.18 **	8.24 **	6.11 **	9.15 **
total: Pool	F:	58.74 **	5.55 **	2.72	1.40	1.46	5.46 **	2.68
algae: Pool	F:	22.45 **	3.96 *	3.85 *	3.23	7.20 **	4.03 *	0.22
sessile: Pool	F:	0.007	8.90 **	3.15	1.10	5.48 *	0.45	1.77
mobile: Pool	F:	56.97 **	0.81	2.52	0.54	0.41	3.23 *	3.21 *

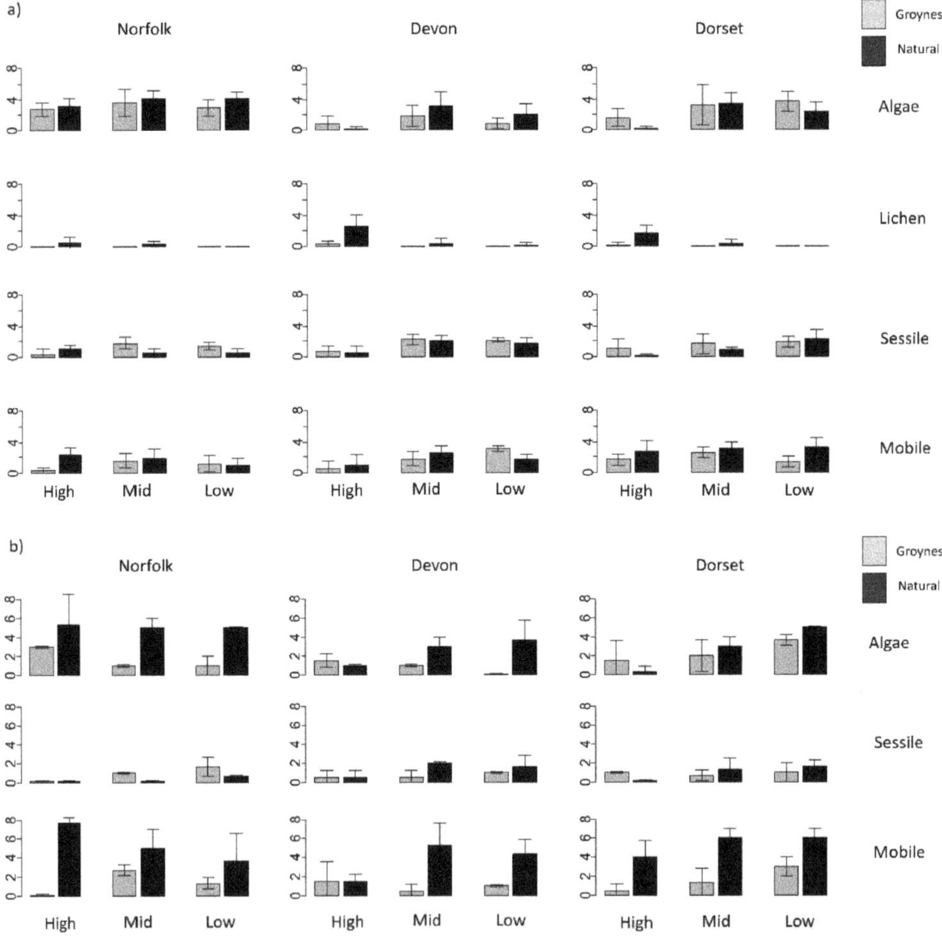

Figure 3. Comparison of species richness on (**a**) rock surfaces and (**b**) rock pools, among tidal levels and counties. Error bars represent 1 standard error of the mean.

Findings were different when we considered abundance (Figure 2c,d, Table 2). We found no significant differences in abundance between structure types for algae or mobile animals in the rock habitat, although in pools the abundances were significantly higher on natural shores. Again, we found a significantly higher abundance of sessile animals on the rock habitat of groynes than natural shores, although this pattern was not significant for pool habitats. Abundance also depended strongly on the tidal level, county, and habitat studied (Supplementary Information 5). For example, in Devon, algal abundances on natural shores increased towards low tide on the rock habitat while it decreased towards low tide in pool habitats. Differences also existed across structures, with algal abundances in Norfolk decreasing towards low tide on groynes in pools but increasing on natural structures in pools (Supplementary Information 5). Abundance differences between structures were notable for mobile species, particularly in rock pools (F = 56.97; Table 2). The abundance of mobile animals was generally greater on natural rocky shores, with this difference prominent in Dorset. However, we observed nearly the opposite pattern in Devon, with abundances of mobile animals not significantly different across structures in pools and on rocks at most tidal levels, and abundance higher at low tide for groynes than natural shores.

Table 2. Results of the analysis of variance (ANOVA) testing the effects of County, Type, and Level, on average species abundance per quadrat, averaged over a set of up to three quadrats. 'Type' is the rocky structure type (natural vs. artificial). 'Level' is the tidal level (low, medium, high). 'County' refers to the three stretches of coastline (Devon, Dorset, Norfolk). ** $p < 0.01$, * $p < 0.05$.

Response Variable		Type	Level	County	Type * Level	Type * County	Level * County	Type * Level * County
	df:	1	2	2	2	2	4	4
algae: Rock (sqrt)	F:	0.02	11.37 **	31.90 **	10.32 **	12.47 **	3.13 *	3.81 **
lichen: Rock (sqrt)	F:	57.52 **	47.30 **	12.74 **	31.16 **	8.53 **	8.90 **	5.53 **
sessile: Rock (sqrt)	F:	52.21 **	79.60 **	15.27 **	20.44 **	15.65 **	3.53 **	14.62 **
mobile: Rock (sqrt)	F:	1.05	17.13 **	53.42 **	16.90 **	8.60 **	8.59 **	2.16
algae: Pool (sqrt)	F:	11.25 **	1.88	0.01	8.03 **	5.56 **	3.16 *	1.60
sessile: Pool (sqrt)	F:	0.97	7.37 **	5.42 *	1.51	10.80 **	0.29	3.22 *
mobile: Pool (sqrt)	F:	54.02 **	2.49	3.80 *	3.98 *	9.13 **	2.67	1.66

3.2. Species- and Community-Level Analysis

Of the 21 species found only in rock pools, most were mobile organisms. Of these mobile species, *Idotea granulosa*, *Gibbula cineraria* and *Lacuna vincta* could potentially survive outside of pools; however, all are sensitive to rapid desiccation or feed on algal species that require water. The other mobile species were either shrimp or fish, which could not survive on open rock. The five species of algae that were only found in rock pools all require sublittoral habitats, except *Gelidium pusillum*, and were all found only on natural shores. The species found only in pools on groynes were generalist species, which were widely distributed and associated with sandy habitats, such as the brown shrimp *Crangon crangon*. Species only found on rock surfaces on groynes were the Greenleaf worm *Eulalia viridis* and the Dahlia anemone *Urticina felina*, with each only identified at one site: Sidmouth and Sandbanks, respectively. We identified only one Dahlia anemone at Sandbanks, while we found Greenleaf worms (in high abundances) in every low tide quadrat of one groyne at Sidmouth. Both species are habitat-specific, with the anemone requiring crevices and water, and the Greenleaf worm requiring mussel beds. Our dataset contains three species that are of conservation concern (Table 3), with the dog whelk *Nucella lapillus* occurring in higher abundances on the groynes and the other two at slightly higher abundances on the natural rocky shores.

DCA (Figure 4, Table SI4) revealed two main gradients in the differentiation of intertidal species composition. The first axis (DCA1) appears to be influenced by type and county and contains the most spread [eigenvalue 0.8114], while the second axis (DCA2) appears to be influenced by tidal level, with less spread [eigenvalue 0.5751]. Distinct communities were observable on the ordination plot. For example, communities found high on axis 1 are dominated by samples from natural shores at high tide and consist of predominantly lichens. Lichens favour areas of high stress and disturbance and are only found on bedrock or boulders [52], so were expected to be found on rocks in the high shore; however, their absence on groynes is notable and is perhaps related to the slow growth rates observed in marine lichens [53]. Species found low on axis 1 are those that occur on groynes in high abundance (e.g., Northern Acorn Barnacle *Semibalanus balanoides*, *Ulva intestinalis*) across all sites. Species found at either end of axis 2 include sponges and algae sensitive to desiccation at the lower end and fish and anemones at the higher end. These results suggest that distinct communities of species that are sensitive to desiccation exist at all tidal levels. Overall, this analysis suggested only a slight differentiation between communities on groynes versus natural rocky shores, part of which reflects the absence of lichens on the relatively new surfaces of the groynes.

Table 3. Total abundance of the three species of conservation concern. For *Padina pavonica*, the units of abundance are percentage cover averaged across a set of up to three quadrats; for *Nucella lapillus* and *Pomatoschistus minutus* they are the total number of individuals per quadrat averaged over a set of up to three quadrats (see Methods). Conservation conventions and legislation: UK Biodiversity Action Plan (UKBAP), the Convention for the Protection of the Marine Environment of the North-East Atlantic (OSPAR), and Bern Convention is the Bern Convention on the Conservation of European Wildlife and Natural Habitats.

		Species		
		Padina Pavonica	*Nucella Lapillus*	*Pomatoschistus Minutus*
Marine and Conservation Conventions and Legislation		**UK BAP**	**OSPAR**	**Bern Convention**
Structure Type	**Habitat**			
Groyne	Rock	X	102.3	X
Groyne	Pool	X	26.5	0.3
Natural	Rock	X	1.0	X
Natural	Pool	14.3	1.0	11.7

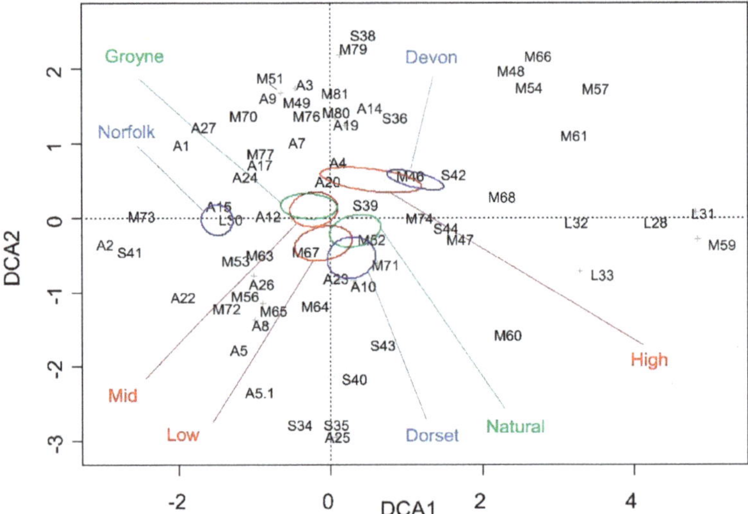

Figure 4. Ordination plot of detrended correspondence analysis (DCA) results on all species for rock and pool habitats combined. Labels represent species (see Supplementary Information 4 for a full list). Significant ($\alpha < 0.01$) environmental variables are represented as factor centroids with standard error ellipses.

3.3. Species Ranges

We observed only two species beyond the ranges recorded for them in the sources we used: the lined top shell *Phorcus lineatus* at Sandbanks (only just beyond its recorded range [54]) and the small periwinkle *Melarhaphe neritoides* at Sheringham (closest recorded presence near Ipswich, nearly 140 km away [55]). Our dataset contained only one species not native to Great Britain: *Austrominius modestus*, identified at Stokes, Sidmouth and West Runton (Table 4). Gibson et al. [50] identified this species as present at all three sites, while Avant [56] recorded this species as only present at Sidmouth and Stokes. When present on groynes, *A. modestus* was at higher abundances than on the paired natural rocky shores (Table 4).

Table 4. Mean abundance of *Austrominius modestus* on open rock at different shore levels. The unit is the percentage of cover.

Site	County	Structure Type	Shore Level		
			High	Mid	Low
Sidmouth	Devon	Groyne	5.9	49.8	60.8
Stokes	Devon	Natural	13.9	31.8	29.9
West Runton	Norfolk	Natural	0	2.3	0

4. Discussion

The main aim of this investigation was to ascertain the extent to which rock-rubble groynes can support similar ecological communities to natural rocky shores, which several studies have identified as being under pressure [32]. Our findings suggest that although rock-rubble groynes support fewer species of intertidal macro-organisms than natural rocky shores (when measured using small quadrats across groyne-sized extents), these differences are not consistent across habitat (i.e., rock or pool) and tidal level (i.e., high, mid, low). On the exposed rock surfaces, the relatively small difference in species richness between the groynes and the natural rocky shores may reflect no more than a species–area effect over the entire ecosystem, i.e., the total area of rock-surface habitat is relatively small on the groynes compared with the natural rocky shores when viewed at a beach level. On the rock surfaces, the abundance of macro-organisms is, if anything, higher for the groynes. In the pools, it is much lower. Species that only occur on the groynes are associated with the higher abundance of mussels and barnacles (e.g., the Greenleaf worm) or they utilise the sandy habitat that is not present on the rocky shores (e.g., brown shrimp). Alternatively, the species found only on or predominantly on the natural shores appear to be those most sensitive to desiccation (e.g. *Desmarestia aculeata*). Much of the difference between structures appears to be due to the paucity of rock pools on the groynes, especially complex pools (Table 5).

Differences in the complexity of the intertidal zone, particularly in relation to water-retention capacity, may therefore explain some of the results. Natural rocky shores typically have higher complexity, with more microhabitats providing shelter from desiccation and other stresses [43]—for example groves, gullies and cracks, all of which can maintain moisture at low tide. In algae, higher rates of water loss speed up the decline of photosynthesis and respiration [57], and when desiccation past a critical water content occurs, irreversible damage results [58]. Retention of water is also vital for mollusc survival and maintenance [59]. Incorporating rock pool features on artificial structures has been shown to increase the diversity of intertidal taxa, sometimes by four- to nine-fold [24,60–62]. The structure of rock pools on groynes was relatively simple. The predominant formation of pools on groynes was where substrate had become saturated and water had built up against the rocks, with a few pools in the higher tidal level having been formed by chemical weathering. The simplistic structure of these pools meant only certain species (e.g., *Palaemon serratus* with anti-predatory mechanisms of immobility and cryptic appearance [63]) could survive there.

Table 5. Mean numbers of rock pools in the sampling sections (±1 standard error).

	Groynes	Natural
High shore		
All counties	**0.9**	**9.7**
Devon	1.0 ±1.0	0.7 ±0.6
Dorset	2.0 ±1.7	7.3 ±1.2
Norfolk	0.3 ±0.6	18.7 ±7.6
Mid shore		
All counties	**2.3**	**14.8**
Devon	1.0 ±1.0	6.7 ±1.5
Dorset	3.3 ±0.6	13.7 ±5.1
Norfolk	3.0 ±1.0	26.3 ±3.2
Low shore		
All counties	**3.8**	**13.7**
Devon	0.3 ±0.6	8.7 ±2.5
Dorset	5.0 ±1.0	20.0 ±1.7
Norfolk	4.7 ±0.6	12.3 ±2.5

While clear delineations of communities exist across the tidal levels for rock and pool habitats combined (Figure 4), there was a difference between generalist species (e.g., *P. serratus*, *Anemonia viridis*) at the higher end of axis 2 and specialist species (e.g., *Halichondria panicea*, *Adocia cinerea)* at the lower end of axis 2 in the DCA. These results suggest that distinct communities of species that are sensitive to desiccation exist at all tidal levels, but currently the structure of rock pools on groynes only support generalist species. It should be noted that pool depth and perimeter were not measured within this study, following similar approaches used on groynes [21,23], and while precautions were made to minimise variations among pools (e.g., pools of similar size; grouping of quadrats), there is the possibility that differences between pools may be due to pool depth and complexity [64,65]. Thus, while our results identified lower biodiversity in rock pool habitats on groynes, there may be potential for artificial rock pool habitats to replicate those found on natural shores, and further research into which species benefit from rock pool presence, as well as which type of rock pools harbour the most species on groynes needs to be undertaken.

Furthermore, in times of low amplitude, a characteristic of the diurnal currents in the English Channel (in Dorset) causes the effects of the varying topography to become pronounced, causing shallow water harmonics which are responsible for a double-tide [66]. This means that there is a relatively long stand of high water, and for approximately 16 hours of the 24-h cycle, the water level is above the mid-water spring tide level [67]. Consequently, while Sandbanks may not be as complex as a natural shore, it retains water in other ways through this double-tide and the intertidal species are subject to reduced levels of desiccation, which could explain the higher biodiversity observed there (Figure 3, Supplementary Information 5). We felt it was important to include Sandbanks in our study due to the contrasting results found in previous research [21,23], but it should be noted that tidal regimes have a large impact on the ecological communities, and while pairing sites by geographic proximity removed some of this uncertainty, the tidal regime may be a compounding factor within the county variable implemented in the ANOVAs (Tables 1 and 2).

The higher algal abundance at Sandbanks may also be explained by the rock surface heterogeneity. Most seaweeds attach to the rock through a combination of etching, gripping or glue, and carbonate rocks tend to support such attachment more than other rock types [68,69], and the physical and chemical substrate properties are known to affect organism adhesion and persistence [70]. The Sandbanks groynes are made of limestone, while the groynes at Sheringham and Sidmouth are made of syenite and granite, respectively (Table SI1). However, there was no significant difference in algal richness or abundance between structures in Devon, although there was a significantly higher abundance of sessile animals on the Sidmouth groynes. The relatively low algal abundance at Sheringham and Sidmouth may be due to the simpler structure of rock material. We paired sample sites based on proximity rather than rock type in order to address questions pertaining to the spread of geographic distributions and compare nearby biological assemblages. Therefore, we are unable to separate possible effects of structure type and rock type; however, our results (Tables 1 and 2, Figures 2 and 3) do suggest that features associated with the structure type may have a stronger influence than rock type alone. While geographic proximity was a key feature of our analysis, the distance between structures in Devon was approximately 100 km. Stokes Beach is publicly-accessible and not characterised by steep cliff faces (as is typical in the region), and represented the closest suitable, safe and accessible rocky shore, rather than necessarily the closest in Euclidean space. Future research should aim to investigate the role of type in the biological assemblages of groynes, as well the conservation value of ensuring that rock type of artificial structures mimics nearby rocky shores (as a form of ecological engineering).

The higher abundance of sessile animals on rock-rubble groynes has important conservation implications. The dog whelk *Nucella lapillus* was recently included in the OSPAR List of threatened and/or declining species and habitats. We found that this species was more prevalent and abundant on the groynes where there was a higher abundance of sessile animals (e.g., mussels and barnacles), which make up a substantial part of *N. lapillus'* diet [71]. The simple structure of groynes provides refuge for many generalist species, which in turn provide trophic support to the wider ecosystem, resulting in clear communities of dog whelks and barnacles (Figure 4). We also found *Pomatoschistus minutus* on both natural shores and groynes. While the total abundance was higher on natural shores (Table 3), the presence of this species on groynes suggests that they have the potential to mimic natural shores and rock pools, yet currently do not support equivalent abundances. Interestingly, the DCA identified *P. minutus* within communities of Sea Oak *Halidrys siliquosa* and *Corallina officinalis* (Figure 4). *H. siliquosa* supports a range of invertebrate epifauna and epifloral (including red algae such as *C. officinalis*), suggesting that the trophic interactions among these species are vital for the presence of rare species. Moreover, the role of ecosystem engineers [72] in generating microhabitats that provide refuge by reducing the impact of abiotic stresses in coastal ecosystems is important for intertidal communities. The incorporation of biotic interactions in statistical models is a key research frontier in spatial ecology [73,74], and our results (Figure 4) suggest such interactions are key to determining representative communities on artificial structures. However, more research is needed to link the abiotic and biotic factors responsible for intertidal communities on artificial structures and highlight key ecosystem engineering designs that could be introduced to increase biodiversity during construction.

Austrominius modestus was the only species identified in our study that is not native to Great Britain. As such, we cannot sensibly comment on the proposition that artificial structures support a higher richness of non-indigenous taxa, beyond stating that this species was located on both structure types in Devon, but only on the natural shore in Norfolk—contradicting the proposition. We did find higher abundances on the groynes in Devon than on the nearby rocky shores (Table 4), which supports other studies that have found higher densities of *A. modestus* on artificial structures in the UK [75]. However, the indigenous barnacle species were also more abundant on groynes than natural rocky shores, so the non-indigenous species may not differ from the natives in this respect. Possible explanations for the increase in barnacle abundance on groynes could be a function of larval supply, availability of space (i.e., lack of microflora), rock type and post-settlement mortality [76,77], meaning future studies should aim to quantify the possible risk associated with such factors in quantifying non-indigenous taxa spread via artificial structures.

Furthermore, we only observed two native British species beyond their previously-recorded ranges: the lined top shell *Phorcus lineatus* at Sandbanks and the small periwinkle *Melarhaphe neritoides* at Sheringham. According to Crothers [78], *P. lineatus* has an Eastern limit around Kimmeridge bay (Figure 1), due to a combination of unsuitable winter temperature and, importantly, lack of suitable habitat to the East. According to Pizzolla [55], *M. neritoides* is absent between Ipswich (approx. 140 km Southeast of Sheringham) and Scarborough (approx. 400km North of Sheringham). Both species are considered highly-mobile [78,79], and their establishment on the groynes suggests that these artificial structures are providing alternative habitat for species that are limited by a lack of habitat rather than poor dispersal capacity. In sum, while rock-rubble groynes may provide non-indigenous species (whether defined nationally or internationally) with suitable habitat (along with indigenous species), we found no evidence that they enhance the dispersal of these species; more likely is that habitat availability is one of the most important factors limiting intertidal organism distributions [80].

We opted to delineate tidal heights based on the spring tidal levels rather than organism distribution limits for the aforementioned reasons. However, the double high tide at Sandbanks complicates this categorization. A continuous variable measuring distance from a specified point (e.g., the high spring tide mark) may have allowed for more flexibility in investigating the influence of tidal level on the ecological indicators; however, a simple distance metric would not account for variation in the distance and speed of tidal ranges. Discussion on the optimal method of measuring the seaward extent and tidal levels across different study areas remains a priority for studies investigating intertidal ecology. Similarly, we did not measure vertical height. While vertical height influences desiccation rates of species, the range of vertical elevation across tidal levels was quite consistent across study sites, and as such, we are confident that any possible artefacts introduced through the chosen methodology have not biased our results. Similarly, we controlled for aspect through surveying the 'sheltered' side of the structure; while aspect in the sense of compass direction can also influence photosynthetic rates of algae [6,81,82], the predominant aspects of all groynes were East- and West-facing, minimising the influence of this in our study.

Finally, it is important to note that the age of an artificial structure will influence the ecological communities found. All groynes surveyed were of a similar age (Table SI1), so therefore, we are confident that cross-site error was minimised. However, this is an important consideration when discussing the ecosystem services provided by groynes, and particularly comparing with natural rocky shores, and when comparing groynes across different studies. Our results suggest different patterns to those identified by Pinn et al. [21] and Firth et al. [23]. The methodological differences among all three studies (e.g., deconstruction of the ecological indicators, direct statistical testing, unpaired sites) may be sufficient to explain contradictory results; however, it is important to consider the dynamic nature of ecosystems, and the unlikeliness of a balanced equilibrium. More developed groynes (e.g., older) may explain the higher species richness identified in pool habitats on artificial structures by Firth et al. [23], suggesting that these structures may develop complexity with time and support more diverse ecosystems. Similarly, older groynes offer greater opportunity for invasion than younger groynes simply by being around for longer. Therefore, investigating spatiotemporal trends in biodiversity on artificial structures in urban environments should now be a primary avenue of research, to better evaluate their potential as a conservation opportunity, with the data from this study available in Supplementary Information 2 to encourage such studies.

5. Conclusions

Rock-rubble groynes in urban environs have provided researchers and practitioners with a conservation dilemma. On the one hand, they can provide much-needed habitat for intertidal species that are decreasing in abundance, while on the other hand, by providing such habitat they may help non-indigenous species expand their ranges. Our results suggest that the rock surfaces of rock-rubble groynes do provide potential benefit for species for which this habitat is suitable, including species that are of conservation concern. We found similar diversity on rock surface habitats between structure types and found higher abundances of sessile animals on groynes. The rock pool habitat represents a key difference between the groynes and the natural rocky shores, with rock pools on groynes being smaller, fewer and less complex than on natural rocky shores. We identified clear communities of algae and endangered fish species (Figure 4) in pools, suggesting that rock pool creation can result in similar ecological communities on groynes, albeit at lower abundances. We also found very few non-native species on groynes, and those species were also found on corresponding natural rocky shores, offering no support for the notion that groynes pose a threat in terms of biological invasions (at least in the UK). While structure age and equilibrium in the ecosystem must be considered, the results of this study suggest that rock-rubble groynes have the potential to support under-pressure intertidal macro-organisms, particularly in urban environments, where the trend to replace natural shoreline with artificial structures is continuing.

Supplementary Materials: The following are available online at http://www.mdpi.com/2073-445X/9/5/131/s1: Supplementary Information 1: Details of the study sites. Supplementary Information 2: R Code. Supplementary Information 3: Species Data. Supplementary Information 4: Species name and code. Supplementary Information 5: Extra Analysis.

Author Contributions: Conceptualization, P.H. and R.F.; methodology, P.H. and R.F.; data curation, P.H.; formal analysis, P.H. and R.F.; writing—original draft preparation, P.H. and R.F. All authors have read and agreed to the published version of the manuscript.

Funding: This research received no external funding.

Acknowledgments: We would like to thank the reviewers and editors for their comments and suggestions. We would also like to thank our field assistants, Julia, Eddie, and Michael.

Conflicts of Interest: The authors declare no conflict of interest.

References

1. Walther, G.-R.; Post, E.; Convey, P.; Menzel, A.; Parmesan, C.; Beebee, T.J.C.; Fromentin, J.-M.; Hoegh-Guldberg, O.; Bairlein, F. Ecological responses to recent climate change. *Nature* **2002**, *416*, 389–395. [CrossRef]

2. Chen, I.-C.; Hill, J.K.; Ohlemüller, R.; Roy, D.B.; Thomas, C.D. Rapid Range Shifts of Species Associated with High Levels of Climate Warming. *Science* **2011**, *333*, 1024–1026. [CrossRef]

3. Pacifici, M.; Foden, W.B.; Visconti, P.; Watson, J.E.; Butchart, S.H.; Kovacs, K.M.; Scheffers, B.R.; Hole, D.G.; Martin, T.G.; Akçakaya, H.R.; et al. Assessing species vulnerability to climate change. *Nat. Clim. Chang.* **2015**, *5*, 215–224. [CrossRef]

4. Holloway, P.; Miller, J.A.; Gillings, S. Incorporating movement in species distribution models: How do simulations of dispersal affect the accuracy and uncertainty of projections? *Int. J. Geogr. Inf. Sci.* **2016**, *30*, 1–25. [CrossRef]

5. Crowe, T.; Thompson, R.C.; Bray, S.; Hawkins, S. Impacts of anthropogenic stress on rocky intertidal communities. *J. Aquat. Ecosyst. Stress Recover.* **2000**, *7*, 273–297. [CrossRef]

6. Harley, C.D.G.; Helmuth, B. Local- and regional-scale effects of wave exposure, thermal stress, and absolute versus effective shore level on patterns of intertidal zonation. *Limnol. Oceanogr.* **2003**, *48*, 1498–1508. [CrossRef]

7. Lees, F.; Baillie, M.; Gettinby, G.; Revie, C.W. The Efficacy of Emamectin Benzoate against infestations of Lepeoptheirus salmonis on Farmed Atlantic Salmon (Salmo salar L.) in Scotland, 2002–2006. *PLoS ONE* **2008**, *3*, e1549. [CrossRef] [PubMed]

8. Spencer, M.; Birchenough, S.N.R.; Mieszkowska, N.; Robinson, L.A.; Simpson, S.; Burrows, M.T.; Capasso, E.; Cleall-Harding, P.; Crummy, J.; Duck, C.; et al. Temporal change in UK marine communities: Trends or regime shifts? *Mar. Ecol.* **2011**, *32*, 10–24. [CrossRef]

9. Sorte, C.J.B.; Davidson, V.E.; Franklin, M.C.; Benes, K.M.; Doellman, M.M.; Etter, R.J.; Hannigan, R.; Lubchenco, J.; Menge, B.A. Long-term declines in an intertidal foundation species parallel shifts in community composition. *Glob. Chang. Biol.* **2016**, *23*, 341–352. [CrossRef] [PubMed]

10. Obst, M.; Vicario, S.; Lundin, K.; Berggren, M.; Karlsson, A.; Haines, R.; Williams, A.; Goble, C.; Mathew, C.; Güntsch, A. Marine long-term biodiversity assessment suggests loss of rare species in the Skagerrak and Kattegat region. *Mar. Biodivers.* **2017**, *48*, 2165–2176. [CrossRef]

11. Hillebrand, H.; Brey, T.; Gutt, J.; Hagen, W.; Metfies, K.; Meyer, B.; Lewandowska, A.M. Climate Change: Warming Impacts on Marine Biodiversity. In *Handbook on Marine Environment Protection*; Springer Science and Business Media LLC: Berlin, Germany, 2017; pp. 353–373.

12. Smale, D.A.; Wernberg, T.; Oliver, E.C.J.; Thomsen, M.; Harvey, B.P.; Straub, S.C.; Burrows, M.T.; Alexander, L.V.; Benthuysen, J.A.; Donat, M.G.; et al. Marine heatwaves threaten global biodiversity and the provision of ecosystem services. *Nat. Clim. Chang.* **2019**, *9*, 306–312. [CrossRef]

13. Neumann, B.; Vafeidis, A.; Zimmermann, J.; Nicholls, R. Future Coastal Population Growth and Exposure to Sea-Level Rise and Coastal Flooding—A Global Assessment. *PLoS ONE* **2015**, *10*, e0118571. [CrossRef] [PubMed]

14. Seto, K.C.; Fragkias, M.; Güneralp, B.; Reilly, M.K. A Meta-Analysis of Global Urban Land Expansion. *PLoS ONE* **2011**, *6*, e23777. [CrossRef] [PubMed]

15. Browne, M.; Chapman, M. Mitigating against the loss of species by adding artificial intertidal pools to existing seawalls. *Mar. Ecol. Prog. Ser.* **2014**, *497*, 119–129. [CrossRef]

16. Bulleri, F.; Chapman, M.G. The introduction of coastal infrastructure as a driver of change in marine environments. *J. Appl. Ecol.* **2010**, *47*, 26–35. [CrossRef]

17. Nordstrom, K.F. Living with shore protection structures: A review. *Estuar. Coast. Shelf Sci.* **2014**, *150*, 11–23. [CrossRef]

18. Perkins, M.J.; Ng, T.P.; Dudgeon, D.; Bonebrake, T.C.; Leung, K.M. Conserving intertidal habitats: What is the potential of ecological engineering to mitigate impacts of coastal structures? *Estuar. Coast. Shelf Sci.* **2015**, *167*, 504–515. [CrossRef]

19. Chapman, M.; Bulleri, F. Intertidal seawalls—New features of landscape in intertidal environments. *Landsc. Urban Plan.* **2003**, *62*, 159–172. [CrossRef]

20. Moschella, P.; Abbiati, M.; Aberg, P.; Airoldi, L.; Anderson, J.; Bacchiocchi, F.; Bulleri, F.; Dinesen, G.; Frost, M.; Gacia, E.; et al. Low-crested coastal defence structures as artificial habitats for marine life: Using ecological criteria in design. *Coast. Eng.* **2005**, *52*, 1053–1071. [CrossRef]

21. Pinn, E.H.; Mitchell, K.; Corkill, J. The assemblages of groynes in relation to substratum age, aspect and microhabitat. *Estuar. Coast. Shelf Sci.* **2005**, *62*, 271–282. [CrossRef]

22. Lam, N.W.; Huang, R.; Chan, B.K. Variations in Intertidal assemblages and zonation patterns between vertical artificial seawalls and natural rocky shores: A case study from Victoria Harbour, Hong Kong. *Zool. Stud.* **2009**, *48*, 184–195.

23. Firth, L.B.; Thompson, R.C.; White, F.J.; Schofield, M.; Skov, M.W.; Hoggart, S.P.G.; Jackson, J.; Knights, A.M.; Hawkins, S.J. The importance of water-retaining features for biodiversity on artificial intertidal coastal defence structures. *Divers. Distrib.* **2013**, *19*, 1275–1283. [CrossRef]

24. Firth, L.B.; Thompson, R.C.; Bohn, K.; Abbiati, M.; Airoldi, L.; Bouma, T.; Bozzeda, F.; Ceccherelli, V.; Colangelo, M.; Evans, A.; et al. Between a rock and a hard place: Environmental and engineering considerations when designing coastal defence structures. *Coast. Eng.* **2014**, *87*, 122–135. [CrossRef]

25. Aguilera, M.A.; Broitman, B.R.; Thiel, M. Spatial variability in community composition on a granite breakwater versus natural rocky shores: Lack of microhabitats suppresses intertidal biodiversity. *Mar. Pollut. Bull.* **2014**, *87*, 257–268. [CrossRef] [PubMed]

26. Sanabria-Fernandez, J.A.; Lazzari, N.; Riera, R.; Becerro, M.A. Building up marine biodiversity loss: Artificial substrates hold lower number and abundance of low occupancy benthic and sessile species. *Mar. Environ. Res.* **2018**, *140*, 190–199. [CrossRef] [PubMed]

27. Bulleri, F.; Chapman, M.G.; Underwood, A.J. Intertidal assemblages on seawalls and vertical rocky shores in Sydney Harbour, Australia. *Austral. Ecol.* **2005**, *30*, 655–667. [CrossRef]

28. Bacchiocchi, F.; Airoldi, L. Distribution and dynamics of epibiota on hard structures for coastal protection. *Estuar. Coast. Shelf Sci.* **2003**, *56*, 1157–1166. [CrossRef]

29. Pister, B. Urban marine ecology in southern California: The ability of riprap structures to serve as rocky intertidal habitat. *Mar. Biol.* **2009**, *156*, 861–873. [CrossRef]

30. Chapman, M.G. Paucity of mobile species on constructed sea walls: Effects of urbanisation on biodiversity. *Mar. Ecol. Prog. Ser.* **2003**, *264*, 21–29. [CrossRef]

31. Clynick, B.G. Assemblages of fish associated with coastal marinas in north-western Italy. *J. Mar. Biol. Assoc. U. K.* **2006**, *86*, 847–852. [CrossRef]

32. Pinn, E.H.; Rodgers, M. The influence of visitors on intertidal biodiversity. *J. Mar. Biol. Assoc. U. K.* **2005**, *85*, 263–268. [CrossRef]

33. Bishop, M.J.; Mayer-Pinto, M.; Airoldi, L.; Firth, L.B.; Morris, R.L.; Loke, L.H.L.; Hawkins, S.; Naylor, L.; Coleman, R.A.; Chee, S.Y.; et al. Effects of ocean sprawl on ecological connectivity: Impacts and solutions. *J. Exp. Mar. Biol. Ecol.* **2017**, *492*, 7–30. [CrossRef]

34. Bulleri, F.; Airoldi, L. Artificial marine structures facilitate the spread of a non-indigenous green alga, Codium fragile spp tomentosoides, in the north Adriatic Sea. *J. Appl. Ecol.* **2005**, *42*, 1063–1072. [CrossRef]

35. Airoldi, L.; Turon, X.; Perkol-Finkel, S.; Rius, M. Corridors for aliens but not for natives: Effects of marine urban sprawl at a regional scale. *Divers. Distrib.* **2015**, *21*, 755–768. [CrossRef]
36. Holloway, M.G.; Keough, M.J. An introduced polychaete affects recruitment and larval abundance of sessile invertebrates. *Ecol. Appl.* **2002**, *12*, 1803–1823. [CrossRef]
37. Bulleri, F.; Abbiati, M.; Airoldi, L. The colonization of artificial human-made structures by the invasive alga Codium fragile ssp. tomentosoides in the north Adriatic Sea (NE Mediterranean). *Hydrobiologia* **2006**, *555*, 263–269. [CrossRef]
38. Glasby, T.M.; Connell, S.D.; Holloway, M.G.; Hewitt, C. Nonindigenous biota on artificial structures: Could habitat creation facilitate biological invasions? *Mar. Biol.* **2006**, *151*, 887–895. [CrossRef]
39. Airoldi, L.; Bulleri, F. Anthropogenic Disturbance Can Determine the Magnitude of Opportunistic Species Responses on Marine Urban Infrastructures. *PLoS ONE* **2011**, *6*, e22985. [CrossRef]
40. Vaselli, S.; Bulleri, F.; Benedetti-Cecchi, L. Hard coastal-defence structures as habitats for native and exotic rocky-bottom species. *Mar. Environ. Res.* **2008**, *66*, 395–403. [CrossRef]
41. Halpern, B.S.; Walbridge, S.; Selkoe, K.A.; Kappel, C.V.; Micheli, F.; D'Agrosa, C.; Bruno, J.F.; Casey, K.; Ebert, C.; Fox, H.; et al. A Global Map of Human Impact on Marine Ecosystems. *Science* **2008**, *319*, 948–952. [CrossRef]
42. Hawkins, S.J.; Jones, H.D. *Marine Field Course Guide: Rocky Shores*; IMMEL Publishing: London, UK, 1992.
43. Lohrer, A.M.; Fukui, Y.; Wada, K.; Whitlatch, R.B. Structural complexity and vertical zonation of intertidal crabs, with focus on habitat requirements of the invasive Asian shore crab, Hemigrapsus sanguineus (de Haan). *J. Exp. Mar. Biol. Ecol.* **2000**, *244*, 203–217. [CrossRef]
44. Walker, S.J.; Schlacher, T.A.; Thompson, L.M. Habitat modification in a dynamic environment: The influence of a small artificial groyne on macrofaunal assemblages of a sandy beach. *Estuar. Coast. Shelf Sci.* **2008**, *79*, 24–34. [CrossRef]
45. Dyderski, M.K.; Jagodziński, A.M. Low impact of disturbance on ecological success of invasive tree and shrub species in temperate forests. *Plant Ecol.* **2018**, *219*, 1369–1380. [CrossRef]
46. Oksanen, J.; Blanchet, F.G.; Friendly, M.; Kindt, R.; Legendre, P.; McGlinn, D.; Minchin, P.R.; O'Hara, R.B.; Simpson, G.L.; Solymos, P.; et al. *Vegan: Community Ecology Package*; R Package version 2.5-5; R Core Team: Vienna, Austria, 2019. Available online: https://CRAN.R-project.org/package=vegan (accessed on 30 June 2019).
47. Ter Braak, C.; Smilauer, P. Topics in constrained and unconstrained ordination. *Plant Ecol.* **2014**, *216*, 683–696. [CrossRef]
48. Palmer, M.W. Putting Things in Even Better Order: The Advantages of Canonical Correspondence Analysis. *Ecology* **1993**, *74*, 2215–2230. [CrossRef]
49. R Development Core Team. *R: A Language and Environment for Statistical Computing*; R Foundation for Statistical Computing: Vienna, Austria, 2011; ISBN 3-900051-07-0. Available online: http://www.R-project.org/ (accessed on 19 August 2011).
50. Gibson, R.; Hextall, B.; Rogers, A. *Photographic Guide to the Sea and Shore Life of Britain and North-West Europe*; Oxford University Press: New York, NY, USA, 2001.
51. MarLIN (Marine Life Information Network). *Marine Life Information Network*; Marine Biological Association of the United Kingdom: Plymouth, UK, 2016. Available online: www.marlin.ac.uk (accessed on 1 January 2009).
52. Connor, D.W.; Allen, J.H.; Golding, N.; Howell, K.L.; Lieberknecht, L.M.; Northern, K.O.; Reker, J.B. *The Marine Habitat Classification for Britain and Ireland*; Version 04.05; JNCC: Peterborough, UK, 2004.
53. Werner, A. Lichen Growth Rates for the Northwest Coast of Spitsbergen, Svalbard. *Arct. Alp. Res.* **1990**, *22*, 129. [CrossRef]
54. Mieszkowska, N. *Osilinus lineatus. Thick Top Shell. Marine Life Information Network: Biology and Sensitivity Key Information Sub-Programme*; Marine Biological Association of the United Kingdom: Plymouth, UK, 2008. Available online: http://www.marlin.ac.uk/speciesfullreview.php?speciesID=3990 (accessed on 30 March 2012).
55. Pizzolla, P. *Melarhaphe neritoides. Small periwinkle. Marine Life Information Network: Biology and Sensitivity Key Information Sub-Programme*; Marine Biological Association of the United Kingdom: Plymouth, UK, 2007. Available online: http://www.marlin.ac.uk/speciesinformation.php?speciesID=3785 (accessed on 30 March 2012).

56. Avant, P. *Elminius modestus. An Acorn Barnacle. Marine Life Information Network: Biology and Sensitivity Key Information Sub-Programme*; Marine Biological Association of the United Kingdom: Plymouth, UK, 2007. Available online: http://www.marlin.ac.uk/speciesinformation.php?speciesID=3252 (accessed on 30 March 2012).

57. Ji, Y.; Tanaka, J. Effect of desiccation on the photosynthesis of seaweeds from the intertidal zone in Honshu, Japan. *Phycol. Res.* **2002**, *50*, 145–153. [CrossRef]

58. Baugh, T.M.; Yarish, C.; Kirkman, H. Seaweeds: Their Environment, Biogeography, and Ecophysiology. Revised Translation of (Meersbotanik: Verbreitung, ökophysiologie und Nutzung der Marinen Makroalgen), K. Lüning (1985). *Estuaries* **1992**, *15*, 255. [CrossRef]

59. Moreira, J.; Chapman, M.; Underwood, A. Maintenance of chitons on seawalls using crevices on sandstone blocks as habitat in Sydney Harbour, Australia. *J. Exp. Mar. Biol. Ecol.* **2007**, *347*, 134–143. [CrossRef]

60. Chapman, M.G.; Blockley, D.J. Engineering novel habitats on urban infrastructure to increase intertidal biodiversity. *Oecologia* **2009**, *161*, 625–635. [CrossRef]

61. Firth, L.B.; Browne, K.A.; Knights, A.M.; Hawkins, S.J.; Nash, R. Eco-engineered rock pools: A concrete solution to biodiversity loss and urban sprawl in the marine environment. *Environ. Res. Lett.* **2016**, *11*, 94015. [CrossRef]

62. Strain, E.M.A.; Olabarria, C.; Mayer-Pinto, M.; Cumbo, V.; Morris, R.L.; Bugnot, A.; Dafforn, K.; Heery, E.; Firth, L.B.; Brooks, P.; et al. Eco-engineering urban infrastructure for marine and coastal biodiversity: Which interventions have the greatest ecological benefit? *J. Appl. Ecol.* **2017**, *55*, 426–441. [CrossRef]

63. Evans, S.; Finnie, M.; Manica, A. Shoaling preferences in decapod crustacea. *Anim. Behav.* **2007**, *74*, 1691–1696. [CrossRef]

64. Martins, G.M.; Hawkins, S.; Thompson, R.C.; Jenkins, S.R. Community structure and functioning in intertidal rock pools: Effects of pool size and shore height at different successional stages. *Mar. Ecol. Prog. Ser.* **2007**, *329*, 43–55. [CrossRef]

65. Moksnes, P.O.; Pihl, L.; van Montfrans, J. Predation on postlarvae and juveniles of the shorecrab Carcinus maenus: Importance of shelter, size, and cannibalism. *Mar. Ecol. Prog. Ser.* **1998**, *166*, 211–225. [CrossRef]

66. Pugh, D.T. *Changing Sea Levels: Effects of Tide, Weather, and Climate*; Cambridge University Press: New York, NY, USA, 2004.

67. Humphreys, J. Salinity and Tides in Poole Harbour: Estuary or Lagoon? In *The Ecology of Poole Harbour*; Humphreys, J., May, V., Eds.; Elsevier: Oxford, UK, 2005.

68. Bromley, R.G.; Heinberg, C. Attachment strategies of organisms on hard substrates: A palaeontological view. *Palaeogeogr. Palaeoclim. Palaeoecol.* **2006**, *232*, 429–453. [CrossRef]

69. Coombes, M.A.; La Marca, E.C.; Naylor, L.A.; Thompson, R.C. Getting into the goove: Opportunities to enhance the ecological value of hard coastal infrastructure using fine-scale surface textures. *Ecol. Eng.* **2015**, *77*, 314–323. [CrossRef]

70. Fletcher, R.L.; Callow, M.E. The settlement, attachment and establishment of marine algal spores. *Br. Phycol. J.* **1992**, *27*, 303–329. [CrossRef]

71. Underwood, A.; Chapman, M.; Crowe, T. Identifying and understanding ecological preferences for habitat or prey. *J. Exp. Mar. Biol. Ecol.* **2004**, *300*, 161–187. [CrossRef]

72. Jones, C.; Lawton, J.H.; Shachak, M. Organisms as Ecosystem Engineers. *Oikos* **1994**, *69*, 373. [CrossRef]

73. Peterson, A.T.; Soberón, J.; Ramsey, J.; Osorio-Olvera, L. Co-occurrence Networks do not Support Identification of Biotic Interactions. *Biodivers. Inform.* **2020**, *15*, 1–10. [CrossRef]

74. Dormann, C.F.; Bobrowski, M.; Dehling, D.M.; Harris, D.J.; Hartig, F.; Lischke, H.; Moretti, M.D.; Pagel, J.; Pinkert, S.; Schleuning, M.; et al. Biotic interactions in species distribution modelling: 10 questions to guide interpretation and avoid false conclusions. *Glob. Ecol. Biogeogr.* **2018**, *27*, 1004–1016. [CrossRef]

75. Bracewell, S.A.; Robinson, L.A.; Firth, L.B.; Knights, A.M. Predicting Free-Space Occupancy on Novel Artificial Structures by an Invasive Intertidal Barnacle Using a Removal Experiment. *PLoS ONE* **2013**, *8*, e74457. [CrossRef] [PubMed]

76. Maki, J.S.; Ding, L.; Stokes, J.; Kavouras, J.H.; Rittschof, D. Substratum/bacterial interactions and larval attachment: Films and exopolysaccharides ofHalomonas marina (ATCC 25374) and their effect on barnacle cyprid larvae, Balanus amphitriteDarwin. *Biofouling* **2000**, *16*, 159–170. [CrossRef]

77. Miron, G.; Boudrea, B.; Bourget, E. Intertidal barnacle distribution:a case study using multiple working hypotheses. *Mar. Ecol. Prog. Ser.* **1999**, *189*, 205–219. [CrossRef]

78. Crothers, J.H. Common topshells: An introduction to the biology of *Osilinus lineatus* with notes on other species in the genus. *Field Stud.* **2001**, *10*, 115–160.

79. Riel Patrick, V.; Breugelmans, K.; De Wolf, H.; Mikhailova, N.; Backeljau, T. Analysis of mitochondrial DNA variation via PCR-SSCP revels micro- and macrogeographic genetic heterogeneity in the planktonic developing periwinkle, Melaraphe neriotoides (Caenogastropoda, Littorinidae). *Vlis Spec. Publ.* **2004**, *17*, 76.

80. Wang, W.; Wang, J.; Choi, F.M.P.; Ding, P.; Li, X.; Han, G.; Ding, M.; Guo, M.; Huang, X.; Duan, W.; et al. Global warming and artificial shorelines reshape seashore biogeography. *Glob. Ecol. Biogeogr.* **2019**, *29*, 220–231. [CrossRef]

81. Chu, F.J.; Seaward, M.R.D.; Hodgkiss, I.J. Effects of Wave Exposure and Aspect on Hong Kong Supralittoral Lichens. *Lichenologist* **2000**, *32*, 155–170. [CrossRef]

82. Firth, L.B.; White, F.J.; Schofield, M.; Hanley, M.E.; Burrows, M.T.; Thompson, R.C.; Skov, M.W.; Evans, A.; Moore, P.J.; Hawkins, S.J. Facing the future: The importance of substratum features for ecological engineering of artificial habitats in the rocky intertidal. *Mar. Freshw. Res.* **2016**, *67*, 131. [CrossRef]

 *land*

Article

Revisiting the Proximity Principle with Stakeholder Input: Investigating Property Values and Distance to Urban Green Space in Potchefstroom

Zene Combrinck, Elizelle Juanee Cilliers *, Louis Lategan and Sarel Cilliers

Unit for Environmental Sciences and Management, North-West University, Potchefstroom 2520, South Africa; 27135349@student.g.nwu.ac.za (Z.C.); 21441480@nwu.ac.za (L.L.); sarel.cilliers@nwu.ac.za (S.C.)
* Correspondence: juanee.cilliers@nwu.ac.za

Received: 5 June 2020; Accepted: 16 July 2020; Published: 20 July 2020

Abstract: Nature is essential to urban quality of life, yet green spaces are under pressure. In an attempt to strengthen the case for urban greening and to reclaim nature into cities, this research considered green spaces from an economic spatial perspective. The proximity principle, as part of hedonic price analysis, is employed to determine the impact of green spaces on property value in specifically selected residential areas within Potchefstroom, South Africa. Our statistical analysis indicated a rejection of the proximity principle in some areas, contradicting internationally accepted theory. To investigate local trends and possible reasons for the rejection, supporting quantitative data was gathered through structured questionnaires disseminated to local residents of Potchefstroom and Professional Planners in South Africa. Challenges pertaining to the planning of green spaces were emphasised, despite residents' willingness to pay more for such green spaces in close proximity to residential areas, according to the cross-tabulations conducted. The research results contributed to the discourse on the economic benefits of green spaces and presented the trends of such benefits within the local context of Potchefstroom. The results emphasised the need to rethink the planning of green spaces within the local context, and provided recommendations on how to reclaim nature into cities from a spatial planning perspective.

Keywords: green spaces; ecosystem services; ecosystem disservices; economic benefits; proximity principle; hedonic pricing analysis

---

1. Introduction

Urban development often occurs at the expense of urban green spaces (UGSs), separating urbanites from nature [1] and depriving them of the various social and environmental benefits, also known as ecosystem services, that UGSs provide [2–4]. Importantly, UGSs also deliver certain economic benefits. These benefits are often much more difficult to measure, generally leading to a preference for land uses with an explicit monetary value, including commercial and residential development, above UGSs in property development decisions [5]. Various studies have been conducted using different techniques and methods to quantify the potential economic benefits of UGSs with the majority of these studies focusing on the contexts of the global North [6–10]. Among these, the proximity principle has been widely applied [8,10]. The proximity principle suggests that residential property value increases as the distance to UGSs decreases [11], thus providing potential reciprocal increases in property tax for authorities [6]. The majority of studies in the global North confirmed the proximity principle [8,10], whereas in the global South results have been less conclusive. In 2015, Cilliers and Cilliers disproved the proximity principle in a South African case study, testing the principle in the local context of Potchefstroom. [12]. The current paper returned to the 2015 case study to verify, update and refine the work conducted by Cilliers and Cilliers [12] as a preliminary study to direct future research. The paper followed a quantitative approach investigating municipal property valuations for three zones differentiated based on distance from a green space in five residential areas bordering a UGS in Potchefstroom. Additional data were also gathered by means of a survey on the social, environmental and economic values attributed to green spaces by a sample of Potchefstroom residents; and a second survey of South African urban planners concerning perceptions on planning for green spaces. The paper is initiated with a review of core concepts in a South African context to provide orientation and reference existing studies in the field. This is followed by the quantitative investigation and a discussion of the results, finally arriving at our conclusions and ultimate recommendations.

1.1. Green Spaces in the South African Context

Urban green spaces (UGSs) include various types and land uses that are located within the urban boundary, primarily covered by permeable surfaces, soil or plant species including grass, shrubs or trees [13]. Zoning classifications may include recreational, institutional, residential, commercial or agricultural categories to accommodate land uses such as sports fields, private gardens, street trees, playgrounds, greenways, urban farms and urban forests [13,14]. UGSs may also include more informal, residual and seemingly abandoned areas, similar to vacant lots, spaces along and under freeways, railways and side streets, sidewalks and derelict properties. Green land uses in urban areas have been studied extensively for their contributions to urban quality of life [1,2,5,15], as part of urban green infrastructure [15].

The majority of studies have been conducted in the contexts of the global North, with scholarship in the global South still being relatively underrepresented [15], with the exception of South Africa. Like most countries, South Africa is experiencing rapid urbanisation as the most urbanised sub-region in Africa [16]. Sub-Saharan Africa along with Asia, is expected to accommodate 90% of the total increase in population over the next four decades [15]. Urban expansion is driven by development in both formal and informal areas, placing green spaces on the expanding urban periphery and within the existing urban envelope in danger of land use conversion [13,17–19]. According to Lubbe et al. [19], the socio-economic status of an urban area determines what resources are available to change the environment, having a tremendous impact on the vegetation types, plant diversity and green space availability [20]. Moreover, South African standards call for a minimum 10% of development area to be allocated to green space, but as in other global areas, UGSs continuously compete with other land uses that may present more explicit monetary values and returns [21,22]. Green spaces also face additional threats from illegal dumping, pollution, invasive species and other social ails that may damage natural ecosystems and diminish biological diversity [23]. To defend existing and promote the development of UGSs, a compelling argument for the social, environmental and, importantly, the economic benefits UGSs can provide, must be made. However, such arguments must also make balanced reference to the potential disadvantages often related to these green spaces, especially in the contexts of the global South. The following section provides a succinct discussion of these benefits, as ecosystem services, as well as potential disadvantages, as ecosystem disservices.

1.2. Ecosystem Services and Disservices

Ecosystem services refer to the benefits people derive from any form of ecosystem functions delivered by UGSs (urban green spaces) that have a positive impact on human wellbeing [2]. Ecosystem services may be classified according to two broad categories, distinguishing between direct and indirect benefits [1,5]. Direct benefits include environmental and social benefits or services, whereas indirect benefits refer to economic benefits (Table 1).

Table 1. Summary of green space benefits and related ecosystem services.

UGS Benefit Categories	Type of Ecosystem Service or Benefits	Examples of Benefits and Services
Environmental (indirect benefits)	Regulating services	Improved air and water quality, reduction of urban heat island effect, regulating urban temperature, moderating extreme events such as flooding, waste water treatment, erosion control, pollination, biological control
	Supporting services	Enhancing urban biodiversity (presence of suitable urban habitats), conservation of natural ecosystems
	Provisioning services	Protection and restoration of natural resources to provide water, food and medicine
Social (indirect benefits)	Cultural services	Improves mental and physical health, aesthetic value of dense urban centres, encourages social cohesion, promotes sense of place, strengthens historical and cultural values
Economic (direct benefits)	Economic benefits	Replacement of expensive conventional and technical environmental management systems (e.g., storm water management), water retention, microclimate regulation through green spaces and areas of permeable surfaces, pollution reduction, physical and physiological health benefits, increase in property value ("willingness to pay" of property buyers), increase in property tax returns of municipalities, increases the marketability of a city

Sources. [3,5,6,12,15,24–27].

It is imperative that the different categories of UGS and types of ecosystem services and benefits green spaces provide, are well understood and communicated in aid of UGS protection and advancement [1]. A case study in Johannesburg, one of the fastest growing cities in Africa, determined that cities and towns are in desperate need of green planning strategies which are only possible when ecosystem services are recognised in city budgeting and accounting systems and when the "ecological economy" is integrated in urban planning processes [16]. Many of the indirect environmental and social benefits potentially delivered by UGSs may be more readily accepted and anticipated by the public and authorities [28]. The direct economic benefits delivered by UGSs are often less thoroughly understood and accepted, ascribed to the complexities of determining and expressing such values [5]. De Wit et al. [29] determined the monetary value of specific ecosystem services in Cape Town, South Africa, and concluded that these services generated economically valuable services in the context of the metro's economy and should, thus, be considered in its budgeting processes. Cognisance of such economic contributions may incentivise investment in and spending on maintenance and protection of UGSs [6] and may justify expenditure to address possible associated disadvantages (ecosystem disservices). Ecosystem disservices EDS) refer to the same ecosystem functions that provide benefits in social, environmental and economic terms, but recognise related actual or perceived negative impacts on human wellbeing [3,30]. Ecosystem disservices are an especially important consideration in the South African context as [30,31]. This paper highlights the ecosystem disservices that are present in towns and cities in South Africa, similar to Potchefstroom, that pose a threat to human wellbeing and require financial resources and private investment to address these disservices in spatial planning, as seen in the third column of Table 2.

The valuation of the economic benefits delivered by UGSs can, thus, be influenced by both ecosystem services and the ecosystem disservices derived from the same spaces. Various approaches and methods have been developed to quantify the economic benefits of UGSs in this regard [6–9,27,32]. A comprehensive review of all existing methods falls beyond the scope of this paper. As this research is based on the work completed by Cilliers and Cilliers [12], only the approach used in that study is recapitulated in Section 1.3 of this paper.

Table 2. Summary of ecosystem disservices.

Category	General Example	South African Example
Ecological	Invasive species that outcompete indigenous species, change in species interactions and populations, change in abiotic species variables, decrease in air quality due to production on volatile organic compounds (VOC's)	Similar to general examples
Social (negative impact on human wellbeing)	Security concerns, negative health impact (allergic reaction to VOC's and pollen), high levels of noise, unpleasant due to exposure to the elements (excessive winds), safety hazard (tree falls), poisonous plants and pests	Unsafe (frequent criminal activity and drug trade), aesthetically unpleasing due to a lack of maintenance and littering, unappealing and nuisance, initial plan for land use not followed through (play area for children)
Economic and financial	Damage to infrastructure (tree roots), maintenance costs, promoting accessibility to green space	Sensitive approaches to green space planning in low-income areas to address ecosystem disservices, fragmented town and cities due to apartheid era resulted in lower-income groups being located far away from green spaces with high transportation cost to access green spaces and child friendly areas.

Sources. [28,30,31].

1.3. The Proximity Principle as A Hedonic Price Analysis Method

The proximity principle, also called the proximate principle, suggests that the value of an amenity, like a green space, is determined by the property values of adjacent residential properties [11,12,33]. The proximity principle reflects on the concept of "willingness to pay", as residents are willing to pay more for properties adjacent to or enjoying a view of such an amenity, or green space [33]. As a result, residents pay more property tax leading to higher tax returns available to local municipalities to direct towards planning, developing and maintaining quality UGSs [6]. Hedonic Price Analysis is generally the most commonly employed method to quantify the value of green spaces [8]. Hedonic Price Analysis is used to identify and quantify the different factors and characteristics that influence the value of property by using several regression models and considering property value as a function of measures of proximity to green space [8,33–35]. Thus, the Hedonic Price Analysis method considers that residential properties are not homogeneous, but reflect various factors that influence property value including property size, the physical condition of the property and accessibility [8]. Several studies [6,22] have determined the positive impact of green spaces on property value, thus, testing the proximity principle using Hedonic Price Analysis. Cilliers and Cilliers [12] invoked the proximity principle as a hedonic pricing method in the local context of Potchefstroom, South Africa, to compare local case study findings with case study results proving the proximity principle in international literature. In the Potchefstroom case study, Cilliers and Cilliers [12] rejected the proximity principle, as proximity to nearest UGS exerted a negative impact on residential property value, as elaborated on in Section 2. These contrasting findings underscore the potential impact of ecosystem services and disservices within the parameters of socio-cultural and physical context in South Africa in realising certain economic values derived from UGSs.

1.4. The Importance of Context and Planning at Community Level

As referenced in Section 1.2 (ecosystem disservices in SA context) and the contrasting findings in the previous section, context is key in eliciting ecosystem services and downplaying potential disservices in the realisation of economic benefits. Context specificities must be acknowledged in planning for UGSs, local level research, data collection and spatial planning [36]. As part of this process, public participation and stakeholder engagement can render valuable inputs regarding perceptions, experiences and acknowledged benefits and disadvantages related to UGSs at local level. Not all inputs from community members can be incorporated in planning and decision making [36], but such processes at least provide nuanced information related to citizen experience and may highlight previously unknown cultural, biodiversity, recreational values, community needs and preferences [36] that may not be captured in spreadsheets dealing with economic data. A study was conducted in South Africa investigating the inequality of public green space, finding that poorer suburbs are endowed with less public green space area than provided in more affluent areas, resulting in poor access to such spaces for the poor [15]. The South African context's severe socio-economic disparities, which are also expressed spatially and in access to amenities, provides fertile ground for community-based participatory planning. Stakeholder engagement may thus consider the interests and opinions of urban residents in the planning and management of public green spaces [15] and provide nuanced understanding of the impacts of such spaces in varied contexts in recognition of generally acknowledged challenges. As mentioned in Section 1.2, UGSs in South Africa are often under-maintained, unattractive and perceived as crime hotspots [37], leading to underutilisation and further dereliction. Such conditions demand continuous efforts to maintain and improve UGSs, akin to Dempsey and Burton's [38] conceptualisation of 'place-keeping' as a collaborative process that surpasses a focus on mere physical upkeep to include ongoing inputs from public and private stakeholders, as well as communities. Thus, calling for community stewardship and the use of local knowledge and community networks to increase the quality of UGSs and by extension, their potential economic benefits.

2. Methodology

This section provides a discussion of the methodologies followed in data acquisition for the two components of quantitative research completed. The results are presented for each component in Section 3. As previously mentioned, this paper presents a refinement of and elaboration on the study conducted by Cilliers and Cilliers (2015) [12] investigating the proximity principle in the case study of Potchefstroom using updated municipal valuations in its analysis (See Section 2.1). The paper tested the proximity principle in the local context in 2019, and compared 2019 findings with 2015 findings, to identify if the proximity principle would still be rejected, as in the 2015 study. The 2019 study was further substantiated with supporting quantitative data collected through two structured questionnaires capturing perceptions amongst a sample of Potchefstroom residents pertaining to green space values; and a questionnaire distributed amongst Professional Planners in South Africa regarding green space planning in the local context (See Section 2.2). Although these questionnaires were limited in sample size, and should be expanded on in future research endeavours, results provide preliminary insights into community and professional perspectives on UGSs in support of the importance of context and community input discussed in Section 1.4.

2.1. Data Acquisition Part A: Proximity Principle in Potchefstroom

The data were collected in Potchefstroom (26.7145° S, 27.0970° E), situated in the North-West province of South Africa. Potchefstroom was chosen as the local case study area as international literature were divergent from the results of a study conducted in Potchefstroom by Cilliers and Cilliers in 2015 [12]. This paper refined the previous study by Cilliers and Cilliers, employing the same methods of data collection and data analysis to verify the results study four years later. As such, the same residential areas and respective properties within each residential area included in the 2015 study by Cilliers and Cilliers [12] were reselected in 2019 to determine the impact of green spaces on property value, using 2019 municipal property valuations. These residential areas were originally selected based on their proximity and accessibility to UGSs, and included the following, as seen in Figure 1:

- **Area A**: Grimbeek Park, bordering the Potchefstroom Country Club
- **Area B**: Van der Hoff Park, situated next to an equestrian open space and wetland.
- **Area C**: Potchefstroom Dam Area, situated next to Potch Dam and the surrounding open space.
- **Area D**: Heilige Akker, situated adjacent to the Fanie du Toit sports grounds.
- **Area E**: Oewersig, situated adjacent to the Mooi River and surrounding open space.

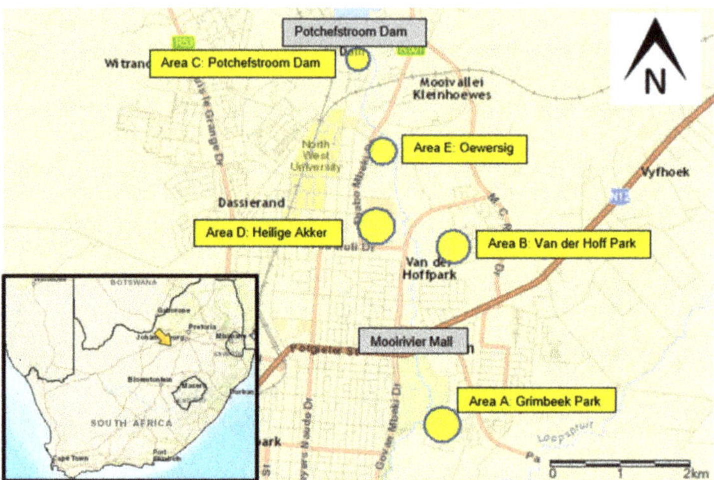

Figure 1. The five selected residential areas in Potchefstroom.

Within each residential area, the selected properties were divided into three zones to test the proximity principle, as illustrated in Figure 2. Zone 1 bordered a UGS, whereas zone 3 was located furthest away from the green the same UGS. Firstly, the property price per square meter was determined by dividing the municipal property values of June 2019 with the total area (in square meters). Thereafter, the mean square meter value per zone in each selected residential area was calculated. Therefore, in each selected residential area, zone 1, zone 2 and zone 3 were assigned a mean ZAR per square meter value that was used to statistically analyse the data. As verified in the international literature, the proximity principle suggests, that zone 1 should present the highest property value [12]. The data obtained from the 2019 Potchefstroom case study were accordingly analysed using three analytical methods including the Analysis of Variance (ANOVA), the Kruskal–Wallis analysis and the Dependent T-test.

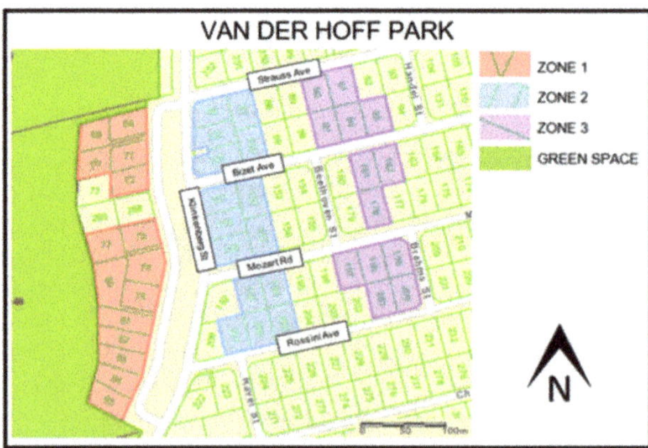

Figure 2. Zone 1 to 3 within area B, Van der Hoff Park.

2.2. Data Acquisition Part B: Community Survey and Professional Planner Survey

As a refinement on the study conducted by Cilliers and Cilliers [12] and gain further insights on the findings derived from the quantitative investigation, two additional surveys were conducted. The data were collected through two structured questionnaires employing a Likert scale (1 = Fully agree,

2 = Agree, 3 = Not sure, 4 = Disagree and 5 = Fully disagree). The first questionnaire, referred to as the Potchefstroom resident survey, focussed on a sample of Potchefstroom residents and their perceptions and appreciation of green space benefits in Potchefstroom. A total of 74 residents completed the survey. The aim was to understand how communities perceive open spaces in Potchefstroom, as the primary users of these spaces. Data were statistically analysed using Cramer's V test and cross-tabulations of selected questions. The second questionnaire focussed on the perspectives of a sample of Professional Planners relating to UGS planning in the context of South Africa. The Professional Planner questionnaire focused on green space aspects including the available financial resources (local budgeting) in green space management, community engagement in green space planning, environmental considerations in practice and green space typologies. A total of 26 planners completed the survey where 17 were Professional Planners, 8 were Candidate Planners and 1 was not registered as a planner, but did work in planning practice. Both questionnaires were distributed electronically, and consent was granted by virtue of completion of a questionnaire.

3. Data Analysis and Results

3.1. Data Results Part A: Proximity Principle in Potchefstroom

A verdict on the rejection or acceptance of the proximity principle can be made solely based on the mean ZAR per square meter values obtained for the 3 zones within each residential area; however, the statistical evidence provides credence. In observing the 2019 mean ZAR per square metre for each zone in each area the following could be identified. The mean ZAR per square meter values in four of the five residential areas (Grimbeek Park, Van der Hoff Park, Potchefstroom Dam Area and Heilige Akker) indicated that zone 1 represented the lowest value in ZAR per square meter compared to zone 2 and zone 3. Thus, zone 1 that is located closest to the UGS presented the lowest mean ZAR per square meter value instead of the highest value as suggested by the proximity principle. Zone 3 displayed the highest mean ZAR per square meter value in three of the five residential areas (Grimbeek Park, Van der Hoff Park and Potchefstroom Dam area), the zone furthest away from the green space.

By using the municipal property values of 2019, the mean ZAR per square meter values of each zone within each selected residential area were calculated as mentioned in the previous section. The latter values were used to complete the ANOVA (mean1 − mean 2/max SD, also referred to as the "standardized difference between the means" and Kruskal–Wallis analyses (Z/sqrt(N). The effect sizes presented by the ANOVA and Kruskal–Wallis analyses were used to verify the significance of the results, as the sample sizes did not provide enough power to test for normality, requiring both parametric and the non-parametric tests. Where results differ, the non-parametric test was preferred.

The effect sizes obtained determined whether a practically significant difference was present between zone 1 and zone 2, as well as between zone 1 and zone 3 within each residential area. The effect sizes obtained from analyses were interpreted. A small effect size of 0.2 indicated no practically significant difference, whereas a medium effect size of 0.5 indicated a practically visible difference. A large effect size of 0.8 indicated a practically significant difference [39].

The results showed that the comparison between zone 1 and zone 2 in the respective residential areas delivered an overall medium effect size, thus, a practically visible difference between the mean ZAR per square meter values (≈0.5), as seen in Table 3. Four of the five residential areas indicated a practically visible difference (≈0.5) (Van der Hoff Park, Potchefstroom Dam Area, Heilige Akker and Oewersig), while only one residential area indicated a practically significant difference (≈0.8) (Grimbeek Park). The effect sizes produced by comparing zone 1 and zone 3 in the respective residential areas delivered an overall visible (≈0.5) to a significant difference (≈0.8) (Table 3). Two residential areas presented a practically significant difference (≈0.8) (Grimbeek Park and Van der Hoff Park) and another two residential areas delivered a practically visible difference (≈0.5) (Potchefstroom Dam Area and Oewersig). Only one residential area presented no practically significant difference between zone 1 and zone 3 (≈0.2) (Heilige Akker).

Table 3. ANOVA and Kruskal–Wallis statistical analysis (2019).

Area	Zone	N (188)	R/m^2	ANOVA Effect Sizes a ≈ 0.2 small (No practically sig. diff.) b ≈ 0.5 medium (Practically visible diff.) c ≈ 0.8 large (Practically sig. diff.)		Kruskal-Wallis Effect Sizes a ≈ 0.1 small (No practically sig. diff.) b ≈ 0.3 medium (Practically visible diff.) c ≈ 0.5 large (Practically sig. diff.)		ANOVA p-Values Statistically significant difference between the means ($p < 0.05$)	Kruskal-Wallis p-Values Statistically significant difference between the groups ($p < 0.05$)
				1 with …	2 with …	1 with …	2 with …		
Grimbeek Park	1	14	1260.70					0.0002	0.0006
	2	14	1611.67	1.19 c		0.54 c			
	3	13	1699.25	1.63 c	0.30 a	0.68 c	0.23 b		
Van der Hoff Park	1	15	1290.59					0.015	0.005
	2	15	1472.43	0.53 b		0.38 b			
	3	13	1624.30	0.98 c	0.54 b	0.58 c	0.27 b		
Potchefstroom Dam Area	1	9	1116.44					0.396	0.237
	2	9	1303.45	0.53 b		0.43 c			
	3	9	1448.64	0.44 b	0.19 a	0.26 b	0.09 a		
Heilige Akker	1	10	1751.96					0.641	0.640
	2	12	1904.15	0.42 b		0.17 a			
	3	14	1850.28	0.24 a	0.13 a	0.07 a	0.15 a		
Oewersig	1	14	1668.44					0.105	0.052
	2	14	1852.15	0.44 b		0.26 b			
	3	13	1549.20	0.36 b	0.72 c	0.23 b	0.45 c		

The effect sizes indicated an overall medium to large effect size ($\approx 0.5/\approx 0.8$), indicating a visible difference to practical significant difference between the mean ZAR per square meter values of zone 1 and the zones further away from the green space. Zone 1 also presented values lower than the zones further away from the green space, refuting much of the existing research on the positive impact of UGS on property value. The proximity principle was thus rejected, with the statistical analyses supporting the data showing a medium to large differences between values. Thus, in the case study a more proximate location to a UGS did not indicate an increase in property value, but showed an inverse effect.

The data from Cilliers and Cilliers (2015) [12] were statistically compared to the data obtained in the 2019 Potchefstroom case study, as captured in Table 4. The effect sizes obtained from the Dependent T-test, once again, indicated whether a practically significant difference existed between the data. The effect sizes presented a practically significant difference between the mean ZAR per square meter values (effect size of $\approx 0.8/0.5$). The latter was clear by observing the old and new mean ZAR per square meter values in Table 4. Property values increased between 2015 and 2019 by 68% in Grimbeek Park, 35% in Van der Hoff Park, 55% in the Potchefstroom Dam area, 46% in Heilige Akker and 45% in Oewersig, thus presenting an aggregate increase on average of almost 50%. Various potential factors could have led to the increase in property values identified from 2015 to 2019. Inflation was a potential factor as the Consumer Price Index (CPI) increased from 5.78% in 2013 to 6.59% in 2016 and decreased, thereafter, to 4.25% in 2019 [40]. Other factors influencing property value in South Africa, according to the Absa residential property market database, include migration trends, security issues, income levels, employment, monetary and fiscal policies, investment returns, the condition of the property and foreign property buying of South African properties [41].

Table 4. Dependent T-test statistical analysis.

Area	Zone	N (188)	R/m^4 Old	R/m^2 New	% Increase between Old and New	Effect Sizes $a \approx 0.2$ small $b \approx 0.5$ medium $c \approx 0.8$ large	Dependent T-Test Statistically significant difference between Old and New ($p < 0.05$)
Area A:	1	14	798.20	1260.70		1.13	0.006
Grimbeek	2	14	953.12	1611.67	68%	1.64	0.001
Park	3	13	974.76	1699.25		0.69	0.011
Area B:	1	15	938.29	1290.59		1.95	0.0001
Van der Hoff	2	15	1105.07	1472.43	35%	2.23	0.0001
Park	3	13	1202.56	1624.30		2.69	0.0001
Area C:	1	9	718.97	1116.44		1.88	0.0001
Potchefstroom	2	9	843.41	1303.45	55%	1.36	0.0001
Dam Area	3	9	925.29	1448.64		1.47	0.0001
Area D:	1	10	1114.23	1751.96		1.75	0.0001
Heilige Akker	2	12	1413.52	1904.15	46%	1.33	0.0001
	3	14	1238.36	1850.28		1.39	0.0001
Area E:	1	14	1079.50	1668.44		1.03	0.0001
Oewersig	2	14	1292.09	1852.15	45%	1.54	0.0001
	3	13	1120.30	1549.20		1.51	0.0001

3.2. Data Results Part B: Community Survey and Professional Planner Survey

3.2.1. Potchefstroom Resident Survey

The results obtained from the survey indicated that the sample of Potchefstroom residents recognise the social, environmental and economic value of green spaces; however, fewer residents recognise the economic value of green spaces. Question 4 focused on the residents' perceptions of green spaces in Potchefstroom, referring to safety matters. A total of 52% of respondents agreed that green spaces in Potchefstroom are perceived as crime hotspots, thus, contributing to unsafe neighbourhoods and indicating a related ecosystem disservice (cross-reference to Section 2.2). A total of 60% of respondents agreed that they would pay more for a property that is located next to a green space in Question 5; however, many residents were unsure or disagreed. Interestingly, some residents who perceived green spaces as crime hotspots in Question 4, still agreed that they would be willing to pay more for a property that is located next to green space (Figure 3).

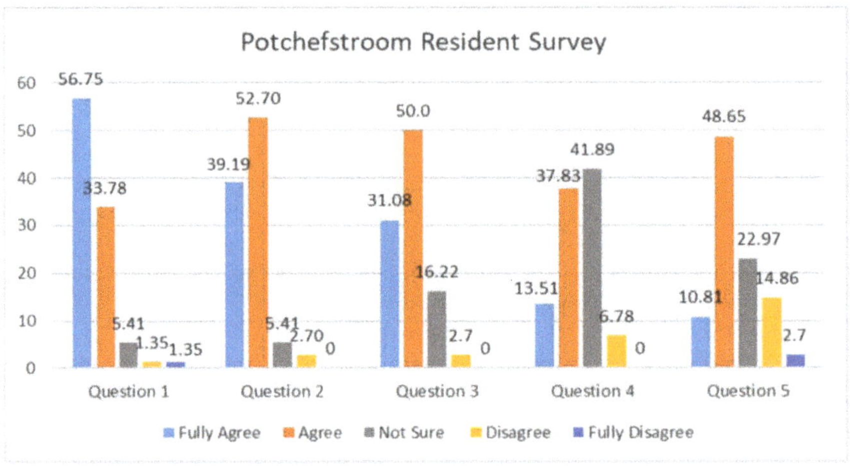

Figure 3. Potchefstroom resident survey results. Question 1: Urban green spaces have environmental value; Question 2: Urban green spaces have social value; Question 3: Urban green spaces have economic value; Question 4: Urban green spaces are perceived as crime hot spots in Potchefstroom; Question 5: I would pay more for a property because it is located next to an urban green space.

As a result of the answers to Question 4 and 5, in recognition of the ecosystem disservices linked to the South African and Potchefstroom contexts, a cross-tabulation were conducted as part of the analysis, in an attempt to further clarify findings: Question 4 (Urban green spaces are perceived as crime hot spots) was cross-tabulated with Question 5 (I would pay more for a property because it is located next to an urban green space). The cross-tabulation was completed with 74 valid cases and a medium practical significant difference as the Cramer's V test value was V = 0.287 (V≈0.3) was presented. Thus, findings supported the observation that although certain residents perceived UGSs as crime hotspots (ecosystem disservices), they residents are still willing to pay more for a property is located next to an UGS.

3.2.2. Professional Planner Survey

The results of Question 1 and Question 2, as shown in Figure 4, both indicated that 88% of the planners agreed that unattractive green spaces are due to a lack of maintenance by local authorities and a lack of community engagement. Question 3 delivered interesting results as 50% of respondent planners agreed that environmental considerations are not prioritised in the planning process; however, the other 50% were either unsure or disagreed that environmental considerations are not prioritised in the planning process Question 4 of the survey focussed on local budgeting for green space planning and indicated that only 62% of respondents agreed with the statement that green spaces are not prioritised in local budgeting. A total of 38% of the planners were either unsure or disagreed that green spaces are not prioritised in local budgeting. The majority of planner respondents reported being familiar with green space typologies (92%).

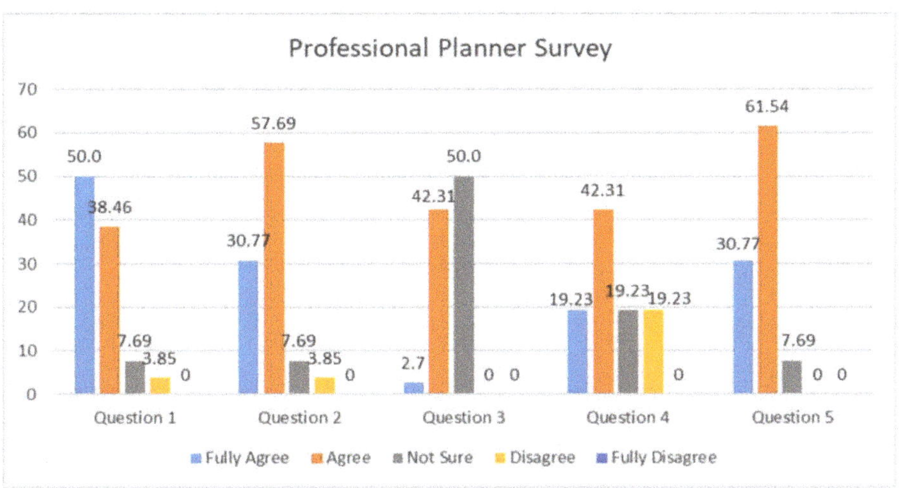

Figure 4. Professional Planner survey results. Question 1: Unattractive urban green spaces are the result of a lack of maintenance by local authorities; Question 2: Unattractive urban green spaces are the result of a lack of community engagement; Question 3: Environmental considerations are not prioritised in the planning process; Question 4: Environmental considerations are not prioritised in local budgeting; Question 5: I am familiar with green space typologies.

4. Conclusions

The challenges pertaining to green spaces within urban environments, as captured in the literature review, were also evident from the empirical investigation and local case study. Despite the theoretically recognised values and benefits linked to green spaces, the economic value thereof is still underestimated to a large extent in the South African context [16].

According to international case studies, as discussed in Section 1.3, UGSs can have a positive influence on proximate property value [6]. The proximity principle may hold true in the global North planning context, but was rejected in the Potchefstroom case study in 2015 [12] and 2019. In the Potchefstroom case most properties located further away from a UGS in the purposefully selected residential areas indicated a higher value in ZAR (South African Rand) per square meter in comparison to properties located adjacent to said UGS. Both the 2015 and 2019 case studies thus delivered contrasting results compared to the findings in international literature. From these findings it was evident that the proximity principle could not be applied to all contexts, indicating the need to consider context in the planning of UGSs.

The Professional Planner survey investigated perceptions of green spaces, from the perspective of a sample of South African Professional Planners, in attempt to gain insight into planning practice in the South African context. This survey indicated that a lack of maintenance, community engagement and efficient local budgeting are the core challenges inhibiting UGS planning approaches. Half of the planners included in the survey stated that environmental considerations are not prioritised in the planning process, even though Environmental Management is considered a crucial component of local Urban Planning approaches, policy and legislative frameworks. The latter findings questioned the extent to which the environment is prioritised in mainstream Urban Planning and emphasised the opportunities for more comprehensive, trans-disciplinary planning approaches going forward.

Even though the proximity principle was rejected for the Potchefstroom case study, and that Professional Planners indicated that environmental considerations are under-prioritized from a planning perspective, the local resident survey revealed a certain level of recognition of the importance of green spaces.

The Potchefstroom resident survey investigated local perceptions towards green spaces, in an attempt to further understand the challenges and needs pertaining to green space planning in Potchefstroom for a participatory perspective. The survey indicated that the majority of respondents recognise the environmental and social values of green spaces, and to a lesser extent the economic value of green spaces. The majority of residents are willing to pay more for a property that is located next to green space, indicating that the residents perceive green spaces benefits as valuable (willingness to pay), regardless of the ecosystem disservices associated with UGSs (poorly maintained green spaces and crime hotspots). Thus, from a community perspective, green space value should be prioritised. Preliminary results indicate a discrepancy between community perspectives, the provision of green spaces within the local context, and economic valuation to support such initiatives.

5. Recommendations

5.1. Future In-Depth Research Based on This Paper's Preliminary Findings

As the 2019 data confirmed the trends established in 2015 regarding the rejection of the proximity principle in the Potchefstroom case study, it is recommended that comparable research be conducted in various locations around South Africa to provide further clarification on the generalisability of the research premise. Such investigations could consider the influence of specific UGS characteristics and differentiation in socio-economic gradients [19] on the proximity principle. The results of the Potchefstroom resident survey, in addition to the emphasis placed on stakeholder engagement in the literature, further call for future research on the proximity principle to include community surveys to deepen understanding. Such approaches could include qualitative investigations in the form of interviews and focus group discussions with proximate communities to triangulate findings. Participatory planning strategies can be crucial in ensuring long-term returns on investments in UGSs and should thus be included in both research and practice.

5.2. Municipal Valuations Should be Reviewed

If residents are indeed willing to pay more for properties located next to UGS as the preliminary survey data suggests, local authorities should consider investing in UGSs (following stakeholder engagement processes) and revise municipal valuation of proximate residential properties in accordance. Such adjustments could result in increases in property tax revenue to reinvest in UGSs to enhance indirect benefits derived from ecosystem services and address ecosystem disservices.

The latter may build a compelling argument to convince local municipalities, private investors and the community to invest in UGSs. This may influence decision-making processes regarding policy formulation to promote environmental protection, green space planning and environmental management within cities.

5.3. Enhance Both the Quality and Quantity of Green Spaces

The quality of green spaces could be enhanced by implementing green planning initiatives as part of broader green infrastructure planning approaches [35]. As dense urban centres do not have adequate space available for development, revitalisation may be considered. The quantity and quality of green spaces within dense urban centres may be increased and improved by compact city and mixed-use planning approaches where green spaces are prioritised as a land use requirement. Local budgeting, maintenance and public participation are also aspects that may ensure that the quality of green spaces is enhanced as the issues regarding ecosystem disservices of green spaces will be addressed. An increase in the quality of green spaces may result in more economic benefits. The latter aspects will be discussed in the next section. Further research may be conducted that investigate the quality and economic benefits of private open spaces, such as gardens in contrast with the quality and economic benefits of public open spaces. The approaches will include "willingness to pay" for amenities, services, green space benefits and properties close to a UGS (stated preference approach) and the impact of the UGS on property value (revealed preference approach). The results will determine whether functional green spaces with high quality have more economic benefits.

5.4. Prioritise Environmental Considerations in Mainstream Urban Planning

Green space planning should be emphasised in the local context and Professional Planners should collaborate with local authorities to prioritise green space planning in local budgeting. Planning practice should consider participatory planning approaches that include the active involvement of the public specifically in UGS planning. This calls for trans-disciplinary planning approaches. Ecosystem services and disservices should play a more prominent part within urban planning, in an attempt to enhance green space quality and increase the economic benefits. Land Use Schemes (LUS) of local municipal areas could be amended to accommodate, not only the minimum green space provision requirements in applications but also mixed-use zonings. Mixed-use developments may increase the quantity and enhance the quality of green spaces. Mixed-use developments ensure that various land uses similar to residential, business and green spaces are included in a single development [42]. Currently, local authorities follow an approach to development where green spaces are perceived as potential areas for development (cross-reference to Professional Planner survey). Local planning should follow an approach to development that encourages the protection of green spaces and natural systems, rather than the development of these areas.

5.5. A Broader Spatial Planning Approach

It is crucial to determine and provide evidence of the economic benefits of green spaces to ensure that green space planning is prioritised in the local context. Therefore, it should form part of a broader spatial planning approach. The Spatial Development Frameworks (SDF) of each local municipality should include a section that discusses green spaces in economic terms and what is aimed to be achieved in terms of environmental management and green space planning in the future and such should be supported by the Spatial Planning and Land Use Management Act (Act 16 of 2013) [43]. To measure the quality of green spaces, the stated preference and revealed preference approach may be considered. The stated preference approach includes "willingness to pay" that determines what residents are willing to pay for amenities, services, green space benefits and for properties that are located adjacent to green space. The revealed preference approach includes the proximity principle as a hedonic price analysis method, as employed in the 2019 Potchefstroom case study of this research paper (cross-reference to Section 2). The proximity principle may be used as an indicator of the economic benefits of green spaces by determining how green spaces influence property value in different urban areas including public open spaces, private open spaces and neighbourhoods of different social status (the socio-economic gradient). Areas that indicate a green space with a higher economic value (approved proximity principle) may indicate what functions and characteristics should be prioritised to increase the economic benefits of green spaces in other areas. When attempting to valuate green spaces in economic terms, six important aspects should be considered including market value (green space impact on property value), enhancement value (green space influence on adjacent land), production value (contribution to production referring to resources), natural systems value (urban biodiversity and ecosystems), direct and indirect value (social, environmental and economic benefits) and intangible value (how people perceive green spaces referring to ethics, knowledge and opinions) [44].

5.6. Focus on Context-Based Planning

Further research may be conducted on the impact of green spaces on property values along a socio-economic gradient as this may influence residents' perceptions of the value of urban green areas, especially where environmental inequity is a reality [19]. Contrasting results may indicate the importance to plan in context. The current research mentioned the difference in planning contexts of Global North and Global South countries, as the proximity principle holds true in the Global North context, but is rejected in the local context of Potchefstroom, South Africa (cross-reference to Section 2).

To plan according to context, a shift in planning approaches will be required. Local communities should participate in planning on a high and empowering level. For local municipalities to shift towards context-/community-based planning, legislative transformation will be required that will support planners in active community engagement and encourage planners to consider context-based planning. Nature and the environment, basic human rights, should be prioritised in urban planning approaches, in an attempt to reclaim nature in cities.

Author Contributions: All authors have read and agree to the published version of the manuscript. Supervision, E.J.C., L.L. and S.C.; Writing—original draft, Z.C.

Funding: This research was funded by National Research Foundation South Africa, grant number 116243.

Acknowledgments: This research is supported in part by the National Research Foundation of South Africa (Grant Numbers: 116243).

Conflicts of Interest: The authors declare no conflict of interest.

References

1. Cilliers, E.J.; Timmermans, W. An Integrative Approach to Value-Added Planning: From Community Needs to Local Authority Revenue. *Growth Chang.* **2015**, *46*, 675–687. [CrossRef]
2. Bolund, P.; Hunhammar, S. Ecosystem services in urban areas. *Ecol. Econ.* **1999**, *29*, 293–301. [CrossRef]
3. Gómez-Baggethun, E.; Gren, A.; Barton, D.N.; Langemeyer, J.; McPhearson, T.; O'Farrell, P.; Andersson, E.; Hamstead, Z.; Kremer, P. *Urbanization, Biodiversity and Ecosystem Services: Challenges and Opportunities*; Springer: London, UK, 2013; pp. 175–251.
4. O'Brien, E. Social housing and green space: A case study in Inner London. *Forestry* **2006**, *79*, 535. [CrossRef]
5. Cilliers, J.; Timmermans, W. Approaching value added planning in the green environment. *J. Place Manag. Dev.* **2013**, *6*, 144–154. [CrossRef]
6. Cilliers, E.J. Rethinking Sustainable Development: The Economic Value of Green Spaces. PhD. Thesis, North-West University, Potchefstroom, South Africa, 2010.
7. Vandermeulen, V.; Verspecht, A.; Vermeire, B.; Van Huylenbroeck, G.; Gellynck, X. The use of economic valuation to create public support for green infrastructure investments in urban areas. *Landsc. Urban Plan.* **2011**, *103*, 198–206. [CrossRef]
8. Melichar, J.; Vojáček, O.; Rieger, P.; Jedlička, K. Measuring the value of urban forest using the Hedonic price approach. *Reg. Stud.* **2009**, *2*, 13–20.
9. Klimas, C.; Williams, A.; Hoff, M.; Lawrence, B.; Thompson, J.; Montgomery, J. Valuing ecosystem services and disservices across heterogeneous green spaces. *Sustainability* **2016**, *8*, 853. [CrossRef]
10. Nicholls, S.; Crompton, J.L. The impact of greenways on property values: Evidence from Austin, Texas. *J. Leis. Res.* **2005**, *37*, 321–341. [CrossRef]
11. Czembrowski, P.; Laszkiewicz, E.; Kronenberg, J.; Engström, G.; Andersson, E. Valuing individual characteristics and the multifunctionality of urban green spaces: The integration of sociotope mapping and hedonic pricing. *PLoS ONE* **2019**, *14*, 1–16. [CrossRef]
12. Cilliers, S.; Cilliers, J. From green to gold: A South African example of valuing urban green spaces in some residential areas in Potchefstroom. *Town Reg. Plan.* **2015**, *67*, 1–12.
13. Girma, Y.; Terefe, H.; Pauleit, S.; Kindu, M. Urban green spaces supply in rapidly urbanizing countries: The case of Sebeta Town, Ethiopia. *Remote Sens. Appl. Soc. Environ.* **2019**, *13*, 138–149. [CrossRef]
14. Wolch, J.R.; Byrne, J.; Newell, J.P. Urban green space, public health, and environmental justice: The challenge of making cities 'just green enough'. *Landsc. Urban Plan.* **2014**, *125*, 234–244. [CrossRef]
15. du Toit, M.J.; Cilliers, S.S.; Dallimer, M.; Goddard, M.; Guenat, S.; Cornelius, S.F. Urban green infrastructure and ecosystem services in sub-Saharan Africa. *Landsc. Urban Plan.* **2018**, *180*, 249–261. [CrossRef]
16. Schäffler, A.; Swilling, M. Valuing green infrastructure in an urban environment under pressure—The Johannesburg case. *Ecol. Econ.* **2013**, *86*, 246–257. [CrossRef]
17. Cilliers, J. Economic value of green spaces: South Africa in contrast with Europe. *Acta Acad.* **2013**, *45*, 1–27.
18. Lategan, L.G.; Cilliers, E.J. The value of public green spaces and the effects of South Africa's informal backyard rental sector. *WIT Trans. Ecol. Environ.* **2014**, *191*, 427–438.

19. Lubbe, C.S.; Siebert, S.J.; Cilliers, S.S. Political legacy of South Africa affects the plant diversity patterns of urban domestic gardens along a socioeconomic gradient. *Sci. Res. Essays* **2010**, *5*, 2900–2910.

20. Kabisch, N.; Korn, H.; Stadler, J.; Bonn, A. *Nature-based Solutions to Climate Change Adaptation in Urban Areas: Theory and Practice of Urban Sustainability Transitions*; Springer: Cham, Switzerland, 2017; p. 215.

21. Cilliers, E.J. Urban green compensation. *Int. J. Green Econ.* **2012**, *6*, 346–356. [CrossRef]

22. Biao, Z.; Gaodi, X.; Bin, X.; Canqiang, Z. The effects of public green spaces on residential property value in Beijing. *J. Resour. Ecol.* **2012**, *3*, 243–253. [CrossRef]

23. Funmilayo, O.A.; Adegboyega, S.A.; Orimoogunje, O.O.; Banjo, O.O. Population growth: Implications for environmental sustainability. *IFE Psychol. Int. J.* **2011**, *19*, 56–69.

24. Simpeh, K.; Smallwood, J. Analysis of the benefits of green building in South Africa. *J. Constr. Proj. Manag. Innov.* **2018**, *8*, 1829–1851.

25. Elmqvist, T.; Setälä, H.; Handel, S.N.; Van der Ploeg, S.; Aronson, J.; Blignaut, J.N.; Gómez-Baggethun, E.; Nowak, D.J.; Kronenberg, J.; De Groot, R. Benefits of restoring ecosystem services in urban areas. *Curr. Opin. Environ. Sustain.* **2015**, *14*, 101–108. [CrossRef]

26. Silvennoinen, S.; Setälä, H.; Taka, M.; Koivusalo, H.; Yli-Elkonen, V.; Ollikainen, M. Monetary value of urban green space as an ecosystem service provider: A case study of urban runoff management in Finland. *Ecosyst. Serv.* **2017**, *28*, 17–27. [CrossRef]

27. Tyrväinen, L.; Väänänen, H. The economic value of urban forest amenities: An application of the contingent valuation method. *Landsc. Urban Plan.* **1998**, *43*, 105–118. [CrossRef]

28. Ives, C.D.; Oke, C.; Hehir, A.; Gordon, A.; Wang, Y.; Bekessy, S.A. Capturing residents' values for urban green space: Mapping, analysis and guidance for practice. *Landsc. Urban Plan.* **2017**, *161*, 32–43. [CrossRef]

29. de Wit, M.; van Zyl, H.; Crookes, D.; Blignaut, J.; Jayiya, T.; Goiset, V.; Mahumani, B. Including the economic value of well-functioning urban ecosystems in financial decisions: Evidence from a process in Cape Town. *Ecosyst. Serv.* **2012**, *2*, 38–44. [CrossRef]

30. Lategan, L.; Cilliers, J. Considering urban green space and informal backyard rentals in South Africa: Disproving the compensation hypothesis. *Town Reg. Plan.* **2016**, *69*, 1–6. [CrossRef]

31. Cilliers, S.; Cilliers, J.; Lubbe, R.; Siebert, S. Ecosystem services of urban green spaces in African countries—perspectives and challenges. *Urban Ecosyst.* **2013**, *16*, 681–702. [CrossRef]

32. Van Leeuwen, E.; Nijkamp, P.; de Noronha Vaz, T. The multifunctional use of urban greenspace. *Int. J. Agric. Sustain.* **2010**, *8*, 20–25. [CrossRef]

33. Panduro, T.E.; Veie, K.L. Classification and valuation of urban green spaces—A hedonic house price valuation. *Landsc. Urban Plan.* **2013**, *120*, 119–128. [CrossRef]

34. Daams, M.N.; Sijtsma, F.J.; van der Vlist, A.J. The effect of natural space on nearby property prices: Accounting for perceived attractiveness. *Land Econ.* **2016**, *92*, 389–410. [CrossRef]

35. Cilliers, E.J.; Timmermans, W.; Van den Goorbergh, F.; Slijkhuis, J. Green Place-making in Practice: From Temporary Spaces to Permanent Places. *J. Urban Des.* **2015**, *20*, 349–366. [CrossRef]

36. Haaland, C.; Van den Bosch, C.K. Review: Challenges and strategies for urban green-space planning in cities undergoing densification: A review. *Urban For. Urban Green.* **2015**, *14*, 760–771. [CrossRef]

37. Adegun, O.B. Residents' relationship with green infrastructure in Cosmo City, Johannesburg. *J. Urban* **2018**, *11*, 329–346. [CrossRef]

38. Dempsey, N.; Burton, M. Defining place-keeping: The long-term management of public spaces. *UrbanForestry Urban Green.* **2012**, *11*, 11–20. [CrossRef]

39. Ellis, S.M.; Steyn, H.S. Practical significance (effect sizes) versus or in combination with statistical significance (p-values). *Manag. Dyn.* **2003**, *12*, 51–53.

40. Inflation.eu. Available online: https://www.inflation.eu/inflation-rates/south-africa/historic-inflation/cpi-inflation-south-africa-2013.aspx (accessed on 29 August 2019).

41. Luüs, C. The Absa Residential Property Market Database for South Africa - Key Data Trends and Implications. Available online: https://www.bis.org/publ/bppdf/bispap211.pdf (accessed on 5 June 2020).

42. Hill Country News. Available online: http://hillcountrynews.com/stories/new-mixed-use-development-planned-in-cedar-park,78057 (accessed on 10 October 2019).

43. South Africa. *Spatial Planning and Land Use Management Act*; Act 16 of 2013, Gazette 36730 South Africa; South African Government: Cape Town, South Africa, 2013.
44. Tekel, A.; Akbarishahabi, L. Determination of Open-green Space's Effect on Around House Prices by Means of Hedonic Price Model; in Example of Ankara/Botanik Park. *Gazi Univ. J. Sci.* **2013**, *26*, 347–361.

 *land*

Article

A Theoretical Framework for Bolstering Human-Nature Connections and Urban Resilience via Green Infrastructure

Jackie Parker [1,2,*] and Greg D. Simpson [1]

[1] Environmental and Conservation Sciences, College of Science, Health, Engineering and Education, Murdoch University, South Street, Perth, WA 6150, Australia; g.simpson@murdoch.edu.au

[2] Manager of Parks, Leisure and Environment, City of Belmont, 215 Wright Street, Perth, WA 6105, Australia

* Correspondence: Jacqueline.Parker@murdoch.edu.au

Received: 9 July 2020; Accepted: 28 July 2020; Published: 29 July 2020

Abstract: Demand for resources and changing structures of human settlements arising from population growth are impacting via the twin crises of anthropogenic climate change and declining human health. Informed by documentary research, this article explores how Urban Resilience Theory (URT) and Human-Nature Connection Theory (HNCT) can inform urban development that leverages urban green infrastructure (UGI) to mitigate and meditate these two crises. The findings of this article are that UGI can be the foundation for action to reduce the severity and impact of those crises and progress inclusive and sustainable community planning and urban development. In summary, the URT promotes improvement in policy and planning frameworks, risk reduction techniques, adaptation strategies, disaster recovery mechanisms, environmentally sustainable alternatives to fossil fuel energy, the building of social capital, and integration of ecologically sustainable UGI. Further, the HNCT advocates pro-environmental behaviors to increase the amount and accessibility of quality remnant and restored UGI to realize the human health benefits provided by nature, while simultaneously enhancing the ecological diversity and health of indigenous ecosystems. The synthesis of this article postulates that realizing the combined potential of URT and HNCT is essential to deliver healthy urban settlements that accommodate projected urban population growth towards the end of the 21st-century.

Keywords: climate change; green infrastructure; human health, human-nature connection theory; urbanization; urban resilience theory

---

1. Introduction

Over the past two centuries, urbanization has changed the relationship that urban dwellers share with the surrounding environment [1,2]. Resultant are adverse effects, with some proving to be devastating to urban communities and indigenous ecosystems [3–10]. Urban centers are now gripped by two crises, exacerbated by a cascade of factors related to the short term doubling of the global human population, rapid urbanization of humanity, unsustainable lifestyle choices, short-term economically focused development, and the resultant changes in the structure of human settlements [5–10]. Those twin crises are anthropogenic climate change (hereafter referred to simply as climate change) and declining human health in urbanized populations [11–13]. Following a short exploration of those crises in the urban context, this article postulates that the Urban Resilience Theory (URT) and the Human-Nature Connection Theory (HNCT) provide complementary opportunities to mitigate land mediate the drivers and impacts of those two crises.

Informed by the reviews of Parker [14–18] and others [19,20] and the key research over the past decade delivered by Lovell and Taylor [5], Tzoulas et al. [7], Mathey et al. [8], Norton et al. [9], Meerow,

Newell and Stults [10], and Burley [12]; a gap is apparent in the existing literature with respect to applying a complementary combination approach to the implementation of the URT and HNCT in urban centers. The novel approach of integrating the application of those two theories posited in this article addresses this gap in the literary discourse and positively contributes to the refocusing and amendment of the unsustainable development many urban centers are currently pursuing. As such, the conceptual model proffered later in this article postulates that the integration of URT and HNCT to inform the provision of accessible, quality UGI will produce urban centers that can alleviate climate change impacts and declining human health for the betterment of current and future generations.

Explicitly defining the URT and the HNCT is problematic given the breadth of disciplines those theories transect, meaning that definitions are highly contested in current literature [19–22]. Both theories move beyond *Homo sapiens* merely surviving within urban centers, to thriving as humans, communities, and cities in coherence with the surrounds [11]. Recently redefined by Romero-Lankao et al. [11] (p. 2), a combined approach to URT and HNCT seeks to: "develop strategies for environmental protection, economic prosperity, inclusivity, and community wellbeing, while increasing their cities' resilience to both chronic and acute physical, social, and economic challenges."

In practice, urban resilience relates to the ability of an urban center to withstand and recover in the event of a shock such as a natural disaster, terrorist attack, economic failure, or pandemic [19]. With the ability to recover from a city-scale shock being one measure for the level of urban resilience, Hobor [19] reports that urban resilience was closely linked to the economic history of a city, as recovery and adaption requires significant financial and infrastructure resources. More recently, urban resilience, in the context of community planning and urban development, is becoming increasingly integrated with higher considerations of human health and wellbeing [23]. Economic factors, social factors, and the intersecting socio-economic factors, play a larger role in recovery efforts, and therefore the resilience of urban centers.

Human-nature connection is a cognitive, emotional, spiritual, and biophysical attachment or affinity that humans feel for natural places [20,21]. Humans can connect to nature at local, regional, national, or international scales, or connections may be location non-specific (i.e., a connection to water or nature at large). Two grains exist for human-nature connection reported within the literature; fine and coarse. Fine grain connections arise from personal nature experiences, interactions with features within a natural setting, and/or direct interactions with the land, (i.e., gardening or farming) [20]. Coarse grain connections come from cultural significance, cultural landscapes, and broader place attachments [20,22]. Human-nature connections have changed over the past few decades [2,21,24]. The change in this relationship, within the context of declining human mental and physical health of urbanized human populations, has sparked the interest of researchers across many disciplines [14,25,26].

This high-level synthesis opens dialogue around the overlapping contribution that application of those two theories make to urban planning and building urban communities as humanity looks towards the second half of the 21st-century. To that end, the specific purpose of this article is to:

1. Summarize the challenges that the twin crises of climate change and declining health pose for urbanized human populations.
2. Highlight the identified gap in research regarding a combined approach to the implementation of the URT and the HNCT.
3. Set an agenda for future URT and HNCT grounded research that explores how UGI can contribute to inclusive community planning and sustainable urban development as humanity realigns and refocuses on the 21st-century and beyond.

2. Methods and Application

The synthesis presented in this article is informed and inspired by the extensive systemic literature reviews and the UGI research that is reported in several earlier articles by the authors [2,14–18,22,26]. In addition, 32 globally focused review articles and 20 geographically specific UGI research articles are directly cited to support the synthesis presented in this article (Tables 1 and 2 below). Further, a summary of the geographic scope and UGI research focus of the case study, empirical, and documentary research reported in those publications is presented in Appendix A (Tables A1–A4).

As such, this article is not a systematic. Rather, it is aligned to the "narrative meta-review" of Nieuwenhuijsen [27] (p. 2) that was constructed "around a number of cutting edge and visionary studies on urban . . . planning and health reported in the literature over the past few years". However, this article goes beyond providing a meta-review to postulate a new conceptual model for the combined application of URT and HCNT in relation to the provision of accessible, high-quality UGI. As highlighted in the introduction, it is the premise of this article that utilization of that combined model of URT and HCNT can provide a UGI-based response to the crises of climate change and declining human health in the face of the rapid urban population growth.

Cognizant of the multifaceted approach that Rist [28] championed for social research undertaken to influence the policy process, the documentary research [29–35] that underpins this article builds on the knowledge reported in the aforementioned literature to scope the key peer reviewed URT and HNCT literature. On the basis of that research, the URT and the HNCT have not previously been connected in other peer reviewed literature.

In addition to summarizing the breadth of research that underpins the proposition of this article that an integrated model of the URT and the HNCT can help urban centers to respond to the challenges of climate change and declining human health, Tables A1–A3 also provide evidence that supports the global applicability of that proposition. From the outset, this project was structured to avoid a Euro or Western centric filter, which was aided by the scholarship regarding urban centers reporting the globalization of Western approaches to urban development for creating cities to accommodate increasingly urbanized human populations [34–46], (Tables A1–A3).

3. Challenges for Urban Centers

3.1. The Climate Change Crisis

With most global energy and steel production requiring burning of fossil carbon stores/fuels, the increase in the human population is driving an increase in greenhouse gas emissions [46,47]. As greenhouse gases exceed the natural stable state operating volumes, documented devastating effects of climate change are becoming ever more apparent [46,47]. Tipping-point events currently observable include the accelerated melting of the Arctic Tundra, Greenland Ice Sheet, and Polar Icecaps creating feedback loops that further exacerbate global warming [46,47].

While severity in impact is geographically dependent, the rapidly warming climate is: changing global climate patterns of normal and extreme weather events (i.e., rainfall, wind patterns, and storm frequency and tracks); affecting the biophysical and ecological function of agricultural and natural systems; and is causing micro and macros scale displacement and extinction of flora, fauna, and human populations [48,49].

3.2. The Crisis for Declining Health in Urban Populations

Despite predictions from theorists such as Thomas Malthus in 1798 that humanity would double every 25 years [48], this only occurred during a short window in the late 20th and early 21st centuries [48]. Otherwise, human population growth has not been uniform in either a temporal or a geographical context, nor has it always trended in positive direction. World War One (1914–1918), World War Two and the associated Holocaust (1939–1945), and The Great Famine (1959–1961) are examples of periods when the global human population measurably declined [49–52]. The human population reached 800 million during the Industrial Revolution, but took until the early 1800's to reach 100 million and was 1.6 billion at the start of the 20th-century, had grown to 6 billion at the dawn of the 21st century, currently sits at (pre-COVID-19 pandemic projections) 7.8 billion, and is projected to reach 9.5 billion by 2050 [49–52]. Reflecting the slowing trend of global population growth that has been evident since the 1950s, humanity is predicted to only grow by 1 billion people between 2050 and 2100 to start the 22nd Century at 11.2 billion on pre-COVID-19 pandemic projections [49–52].

While populations of some long-established cities are shrinking because of the demographic transition and/or economic migration [49–52], humanity became an increasingly urbanized species during the 20th-century, and that trend is predicted to continue until 2100. Approximately one quarter of the human population (1.95 billion people) lived in urban centers at the start of the 20th-century [51–54]. Further, urban populations are predicted to more than double by the end of the century, with the current level just over 4 billion people (55% of all humans) growing almost 6.5 billion in 2050 (68% of humanity) to reach to 9.5 billion people (85% of the global human population) by 2100 [51–56].

That rapid change in lifestyle has resulted in human disconnection from nature, with little opportunity for adaptation to those changed circumstances [2,14,56–58]. As a result, significant adverse health impacts, both physiologically and psychologically, are beginning to emerge [14,17,18]. Physiologically: cardiovascular health has decreased; diabetes has increased; obesity has increased; and biological intolerances have increased [14,15]. Psychologically: depression and anxiety have increased; cognition recovery ability has decreased; and stress related conditions have increased [6, 7,12–14]. Human-nature disconnection is likely to have further effects that are not yet apparent, the subject of which is an area of growing interest for researchers and practitioners [14–16].

4. Framing of the Urban Resilience Theory

4.1. Origins and Perspectives of Urban Resilience Theory

Urban Resilience Theory research over the past two decades has been driven by growing environmental, social, and political uncertainty, combined with increased prevalence and severity of risks to urban centers [10]. Tensions arising from the ambiguity and disparity in defining the concept of resilience may go some way in explaining the limited application of the URT principles and teachings by decision makers [10]. To progress the definitional debate and improve wide adoption of the URT propositions, Meerow & Newell [10] (p. 315) proposed the following definition of urban resilience that is appropriate in the context of this article:

> "Urban resilience refers to the ability of an urban system-and all its constituent socio-ecological and socio-technical networks across temporal and spatial scales-to maintain or rapidly return to desired functions in the face of a disturbance, to adapt to change, and to quickly transform systems that limit current or future adaptive capacity ".

The URT emerged from the fields of ecology, engineering, and psychology [59–61]. Summaries in the applications of the term resilience in each of those fields appear below. Each perspective provides a unique contribution to the current understanding of URT. Given the multidisciplinary nature of URT, agreement on an explicit definition remains elusive, although all the published formulations share the notion of the ability of a system or urban center to bounce-back from external pressures, stresses, or shocks [11].

Ecological resilience has been defined as the amount of disturbance or pressure that an ecosystem is able to withstand without permanently changing self-organizing processes and structures [59]. The greater the capacity of an ecosystem to recover and adapt to stochastic changes in circumstance, the higher the resilience. An allied ecological theory with relevance to the crises reported in this article is the Alternative Stable State Theory [59–62]. This theory holds that a tipping point may be reached whereby the associated feedback that may not allow the system to recover/return to previous state.

With respect to engineering, the concept of urban resilience arose from the need to respond to new threats within modern society. Concentrating on infrastructure and networks, Bozza et al. [60] defined resilience to be the ability to recover, absorb, and restore equilibrium after a perturbation. From the engineering perspective, the original formulation for resilience was based on the idea that systems and networks need to realize a post-shock equilibrium position [60]. That perspective has evolved, and the current usage of the term resilience has moved towards a performance-based assessment of recovery and is thus seeking those assets and systems perform in the same capacity as prior to the disturbance or shock [60]. In this context, resilience relates to the recovery of complex systems, which are usually composed of physical subsystems occurring in an urban context [60].

Psychological resilience is the ability to adapt to stress, significant challenges, or adversities, which may include challenging life events, acute trauma, and/or chronic adversity [61]. Those experiences have the potential to substantially impact on brain function and brain structure [61]. A lack of resilience becomes visible through the development of psychological responses such as anxiety, depression, and/or post-traumatic stress disorder [61]. This perspective strongly adheres to the notion of adaptation in pursuit of ongoing health.

4.2. Converting Theory to Practice

High density urban centers provide access and diversity of services, however also overcrowding, sensory overload, and increased levels of stress [23,62]. Samuelsson et al. [23] posit that greater consideration of resilience principles can improve planning and formation of policy for design and development of urban centers, with respect to densities and provision of services, by assessing opportunities and threats at a city scale.

Building on work by the Organization of Economic and Community Development and other stakeholders and researchers, the 2014 researcher of Kim et al. [63] proposed a new Green Growth model to assess the development and growth of urban centers. Kim et al. [63] recommended twelve indicators of sustainable urban development that include measures such as greenhouse gas emissions, energy use, energy sources, water asset usage, portion of land covered by forest, public transportation opportunities, and more [63].

The Green Growth Model is not dissimilar to the Sustainable Development Goals (SDG's) when looking forward to urban development principles and practices. The recommendations and conclusions within this research show the contributions of URT and HNCT making some way to support the SGD's for achievement by 2030. Examples of this are SDG 3 Good Health and Well-being, SDG 11 Sustainability Development and Communities, and SDG 13 Climate Action.

In line with the Green Growth Model and the SDG's, this new model lends itself to the approach of urban development that provides a socio-ecological focus, in contrast to traditional socio-economic approaches. In that context, URT proposes that greater concentration be placed on [64]:

1. Building and supporting the robustness of cities, systems, and networks;
2. Increasing the efficiency with which the city, system, or network can bounce-back or bounce-forward; and
3. Increasing the ability of practitioners to decentralize, predict opportunities, discontinue redundant practices, and provide transparent and authentic feedback to community and industry stakeholders.

However, potential problems with implementation of URT that are identified in current literature [62] include the:

1. Observed disconnect between researchers and practitioners;
2. Ambiguity surrounding the specific issues being addressed through the URT and the associated ambiguity in the plan to overcome those issues;
3. Broad scope of URT allowing both research and practitioners to use the concept as a buzzword, rather than applying the theory in genuine progression of urban resilience; and
4. Lack in the thorough understanding of resilience characteristics and therefore how to bolster the principles into new policy, strategy, planning, and implementation.

5. Framing of the Human-Nature Connection Theory

5.1. Origins and Perspectives of the Human-Nature Connection Theory

Many models and theories attempt to explain the relationship between humans and nature and the resulting impacts or benefits for human health. Models include the Environment of Health, the Mandala of Health, the Wheel of Foundational Health Need, the Healthy Communities, the One Health approach, and more [63]. Each of those models attempt to describe the balance and interactions of the biological, social, and spatial influences of human-nature connections.

Through the milestones of urbanization, from ancient city states, through to the 20th-centruy, the connection of humans and nature has been the focus of origin stories and philosophical and scientific writings. Given the ongoing interest in HNCT and similarly the URT, a number of multidisciplinary perspectives inform our current understanding of how exposure to nature influences the health and welfare of human populations. Key perspectives that inform the HNCT are summarized below.

The environmentalism perspective suggests that humans traditionally had a relationship with nature being one of power and dominance, which is embedded in the Judeo-Christian belief systems that shaped Western civilizations [63–65]. That relationship has weakened over the past few decades, anticipated to reflect a natural balancing out between humans and the environment [63].

From an evolutionary biology perspective, culture-genetic interactions affect our lifestyle choices and thus our health. One example of such an interaction is that humans predominantly gain nutrition derived from food and farming processes [63]. Traditionally, humanity sourced food from farming the land and/or preparing food from natural ingredients that contained varied microflora [63]. For modern human populations living in urban centers, processed and packaged alternatives are now readily available and provide a larger proportion of food intake. Food tolerances and intolerances are suggested to be based largely on food choices, with reduced exposure to microbial activity increasing susceptibility to allergies [63]. An increased incidence of more extreme food-induced allergic reactions is one of the factors contributing to the crisis of declining health in urbanized human populations [63].

The evolutionary psychology perspective is founded on the apparent preference of humans for scenes dominated by nature and natural elements. Emerging only within the past few decades, this perspective suggests that human psychological characteristics are shaped and adapted based on the prevailing ecological and environmental conditions [24,25]. This perspective has delivered concepts that incorporate the inclusions of nature in self, deep ecology, extinction of nature experience, connectedness to nature, and the Biophilic Hypothesis [21,24].

5.2. Converting Theory to Practice

Embracing the HNCT to inform community planning and urban development practices can provide diverse opportunities to improve the security and quality of urban life.

These opportunities include [20–24]:

1. Incorporating nature into design principles and material pallets for commercial buildings, homes, public spaces, and other elements of the built environment [66];
2. Integrating plants, nature, and natural elements into everyday lives, work places/stations, and institutional settings such as hospitals and prisons;
3. Providing opportunities for renaturing and re-connecting with nature through implementation and improvement of the public UGI [66];
4. Restoring and revegetating degraded land and environmental assets such as lakes and riverine zones to assist urban dwellers to engage with nature;
5. Reclaiming abandoned and baron land for the purposes of public enjoyment and recreation as well as increasing biodiversity and other environmental functions; and
6. Advocating for improvements in public accessibility and opportunity for diverse nature experiences via public UGI, nature trails, green walls, green roofs, and more [66].

6. Setting the Scene for Green Infrastructure

The term green infrastructure appeared in the peer-reviewed literature and language of urban planning and management practices during the 1980s [16]. While initially highly contested, the term is now widely accepted in published literature and has generated significant research interest over the past decade [14,16]. The components of UGI encompass a broad range of assets, including green POS, urban trees, urban stormwater management, green roofs and green-walls, and many other green assets [14,16,67]. A key aspect that distinguishes UGI from other forms of infrastructure (i.e., blue or grey) is the simultaneous delivery of social and environmental services and relief from the hard forms of the built environment [14,16,67].

Urban land managers already rely upon UGI assets, generally in the form of green POS, to deliver opportunities to urban dwellers to connect to nature and mediate the impacts of climate change [14,26]. Arguably the most prominent and widespread occurrence of UGI, green POS, is heavily relied upon to provide public health benefits [14,26]. There is a growing body of published research in which green POS users report and/or have demonstrated improved physiological and psychological health, a better outlook on life, enhanced cognitive recovery, because of the opportunities that quality green POS provides for exercise, recreation, relaxation, and reflection [14,15].

In summary, the opportunities offered by UGI are becoming increasingly significant assets that support healthy communities and enhance liveability in urban centers. In addition, UGI has relevance in migrating and mediating the twin crises of climate change (Table 1) and declining human health (Table 2).

Table 1. Examples of urban green infrastructure offerings that could mitigate and/or meditate impacts arising from the crisis of climate change on urban centers.

Impact	Green Infrastructure Offering	Example References
Ambient temperatures warming faster than urban communities can adapt.	Urban trees, green POS elements, and other porous UGI surfaces reduce reflective and embedded heat, and provide evaporative cooling, which reduces ambient temperatures.	Lovell & Taylor, 2013 [5]; Li et al. 2018 [68]; Roe & Mell, 2013 [69]; Norton et al. 2015 [9]; Mathey et al. 2015 [8].
Increased frequency of extreme weather events	Urban trees act as barriers to extreme events such as wind, hail, and rain. Green POS elements and other porous UGI surfaces absorb rainfall and slow the wind.	Lovell & Taylor, 2013 [5]; Li et al. 2018 [68]; Roe & Mell, 2013 [69]; Norton et al. 2015 [9]; Mathey et al. 2015 [8].
Increased severity and unseasonal timing of storms	Urban tree canopy assists in capturing rain and reducing the velocity of rainfall, which reduces the impact. Green POS elements and other porous UGI surfaces absorb rainfall and slow the wind.	Lovell & Taylor, 2013 [5]; Li et al. 2018 [68]; Roe & Mell, 2013 [69]; Norton et al. 2015 [9]; Mathey et al. 2015 [8].
Flora displacement	Quality green POS and vegetated biofiltration systems can conserve and protect remnant and restored indigenous vegetation.	Lovell & Taylor, 2013 [5]; Cameron et al. 2012 [6]; Tzoulas et al. 2007 [7]; Mathey et al. 2015 [8]; Norton et al. 2015 [9].
Fauna displacement	Urban trees, green POS, green walls, and/or green roofs of remnant/restored indigenous vegetation provide fauna with habitat, food, and refuges.	Cameron et al. 2012 [6]; Tzoulas et al. 2007 [7]; Mathey et al. 2015 [8]; Norton et al. 2015 [9]; Meerow & Newell, 2017 [10].

Table 2. Examples of urban green infrastructure offerings that could mitigate and/or meditate impacts arising from the crisis of declining human health in urban centers.

Declining Health Impacts	Green Infrastructure Offerings	Example References
Increase in anxiety disorders.	Contact with and experiences within quality remnant and restored nature spaces shown to reduce anxiety.	Burley, 2018 [12]; Suppakittpaisarn et al. 2017 [13]; Parker & Simpson 2018 [14]; Parker, 2017 [15]; Cameron et al. 2012 [6]; Mekala et al. 2014 [70]; Tzoulas et al. 2007 [7]; Mathey et al. 2015 [8]; Heckert & Rosan, 2018 [71]
Increase in depression.	Contact with and experiences within quality nature spaces shown to reduce depression. Quality public open spaces (POS) also provide opportunities for recreation and socialization, which is known to reduce depression.	Burley, 2018 [12]; Suppakittpaisarn et al. 2017 [32]; Parker & Simpson 2018 [14]; Parker, 2017 [15]; Cameron et al. 2012 [6]; Mekala et al. 2014 [70]; Tzoulas et al. 2007 [7]; Mathey et al. 2015 [8]; Heckert & Rosan, 2018 [71]
Increase in stress related illness	Contact with and experiences within quality remnant and restored nature spaces shown to reduce stress and therefore reduce stress related illness. Quality POS also provide opportunities for recreation and socialization, which is known to reduce stress.	Suppakittpaisarn et al. 2017 [12]; Cameron et al. 2012 [6]; Tzoulas et al. 2007 [7]; Sammuelsson et al. 2019 [23];
Decrease in cardiovascular health	Quality POS provide opportunities for formal and informal recreation and exercise, which is known to improve cardiovascular health.	Suppakittpaisarn et al. 2017 [13]; Cameron et al. 2012 [6]; Mekala et al. 2014 [70]; Tzoulas et al. 2007 [7]; Mathey et al. 2015 [8]; Heckert & Rosan, 2018 [71].
Increase in diabetes	Quality POS provide opportunities for formal and informal recreation and exercise, which is known to assist in the avoidance and/or management of diabetes.	Urakami, 2017 [72]; Fang, 2018 [73].
Increase in heat related hospitalizations	Urban trees, green POS elements, and other porous UGI surfaces reduce reflective and embedded heat, and provide evaporative cooling, which reduces ambient temperatures.	Knowlton et al. 2009 [74]; Sun et al. 2019 [75]; Heaviside et al. 2016 [76].

7. Concurrent Application of URT and HNCT

This section consolidates the information presented above to summarize how the equitable provision of easily accessible quality UGI aligns to both the Urban Resilience and Human-Nature Connection theories. This section provides responses that help mediate and mitigate the impacts from the twin crises of climate change and declining levels of human health within urban centers. Building from this content, Section 9 presents a conceptual model (Figure 1) that shows how the unified application of URT and HCNT can inform community planning and urban development that is preadapted to mitigate and mediate impacts from both climate change and declining human health in this century and beyond.

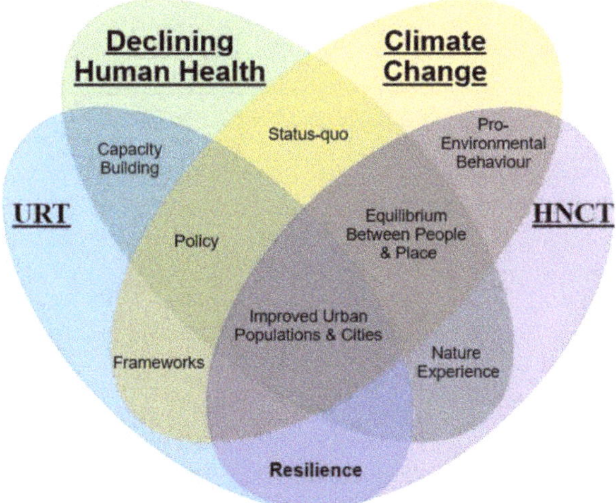

Figure 1. Conceptual model of the interrelationships between the twin crises of climate change and declining human health in urban centers and mitigation and mediation measures offered by the Urban Resilience Theory (URT) and the Human-Nature Connection Theory (HNCT).

7.1. Urban Resilience Theory

As previously mentioned, URT can contribute to the mediation and mitigation of climate change [77]. In summary, URT provides teachings and direction for practitioners and decision makers that can: improve traditional policy and planning frameworks, stimulate new systems and measures that aim to reduce greenhouse gas emissions, provide tools for climate change adaptation programs, promote capacity building through inter-governmental cooperation, and direct recovery planning for identified crises and emerging challenges.

Over the past three decades, natural disasters that impact urban centers have quadrupled [77]. Further, climate change place many urban centers at risk and increases the susceptibility of cities also impacted by natural disasters [77]. Cities were once considered a place of refuge for inhabitants. Urban centers are, however, increasingly seen as hotspots of climate hazards and climate risks [77]. Climate hazards comprise floods, windstorms, droughts, fires, large temperature fluctuations, sea level rise, and landslides [77]. Climate risks represent the likelihood of climate hazards occurring, as well as the likelihood of adverse impacts to human health, green and grey infrastructure assets, and environmental and urban services or even loss of human life [77]. Implementing climate change mediation and mitigation measures informed by the URT, can avoid (mediate) climate hazards or reduce (mitigate) climate risks to: reduce the susceptibility of the affected locations and increase the ability of those locations to withstand imminent hazards, and improve in post-disaster response and post-disaster recovery. Specifically, URT contributes to overcoming climate change hazards and climate change risks in the following ways [77]:

- Informs risk reduction programs at a city (and wider) scale;
- Informs measures to modify current work practices to reduce the likelihood of passively (unintentionally) increasing risk;
- Informs modifications to current management practices, policy, legislation, working structures, and tools to reduce risk and increase the ability for adaptation;

- Informs the modifications of internal organizational level policies to reduce risk for individual organizations;
- Promotes cooperation between government agencies and the public to optimize risk reduction, disaster response, and disaster recovery; and
- Supports a conceptual shift in the philosophy that drives professional and public education.

For the above strategies to be effective, industry change makers, policy advisors, practitioners, and decision makers need to ensure the following [76]:

- Actions, changes, and programs do not unintentionally increase the risk to urban centers;
- Institutionalize the idea of risk reduction for the implementation of all public works and policy;
- Ensure high level commitment to disaster recovery and climate change to aid acceptance;
- Cooperate with government agencies, industry partners, and competitors to create multi-level systems to manage risk; and
- Promote and support professional development and education on risk reduction in support of urban resilience principles.

The application of the URT for mediating and mitigating climate change is strengthened by the scope of the theory and strategic planning of policy creation, infrastructure investment, building and construction, resource extraction and utilization, and environmental asset management.

Declining human health in urbanized populations manifests as physiological conditions such as declining cardiovascular health and/or increasing diabetes and obesity and through psychological afflictions such as increases in stress related illness and the associated increase in anxiety and depression [12–15]. The teachings and direction of URT offers three key responses to declining human health among urban dwellers and supports several additional opportunities to stabilize and improve human health within urbanized human populations.

Firstly, in addition to other measures, URT advocates and provides frameworks to ensure that urban dwellers have fair and equitable access to the environmental services and benefits provided by UGI assets. In this frame of reference, the UGI may constitute or be incorporated into urban POS that supports exercise and recreation, safe urban spaces suitable for engagement and connection with other urban dwellers, frameworks for engagement and deeper involvement of individuals with their local community in order to build social capital and social resilience, and the fair and equitable access to health related services that support and provide care for individuals whose health is compromised [10,65].

Secondly, URT supports and advocates for changes in common practice and approach to the consumption of resources that support modern urban life [10,65]. This is particularly evident in the space of energy production. Application of the teachings and direction of URT in this context supports the implementation of renewable energy sources in conjunction with new technology that proves successful in reducing emissions. Carbon emissions, a large proportion of the waste from traditional energy production, negatively impacts human health in urban centers in directly through the effects of climate change and directly through exposure to atmospheric contaminants [44,45]. Regarding the direct exposure to atmospheric contaminants, energy related emissions reducing air quality precipitates an estimated 3.4 million pre-mature deaths, globally, each year [78] from disease mechanisms linked to adverse respiratory health conditions and cancers [46,47]. Adoption of renewable energy sources and other new technologies aligned to increased urban resilience can significantly reduce emissions and therefore reduce the health risk among urban populations. Further, several policy and legislative changes that can, and already do, contribute to improving public health are aligned to URT principles [78].

Thirdly, briefly returning to the climate change crisis, URT offers frameworks and strategies to support human populations in adapting to the threats and challenges of climate change. This may be in the form of employable technology (i.e., improved heating/cooling, off the grid power back-ups), behavior change (i.e., reduction in waste, reduction in electricity dependence), UGI support (i.e., urban trees, urban POS, revegetation) and implementation, and building and construction changes (i.e., solar passive design, new materials) [10,65]. Those adaptations, informed by the URT, can reduce negative health outcomes, which can manifest in premature death, caused by climate hazards such as extended heatwaves and/or flooding and landslides associated with more frequent extreme weather events [10,65].

7.2. Human-Nature Connection Theory

The HNCT provides multiple practical opportunities for modern cities to respond to climate change by cultivating and increasing pro-environmental behaviors, advocating for the creation and implementation more UGI assets, improving practices and approaches to current community planning and urban development regimes, and advocating for the renaturing of urban communities [20–23]. The level of human-nature connection is seen to be a reliable predictor for pro-environmental behavior [20]. Pro-environmental behavior is defined as individuals that display behaviors that contribute to environmental sustainability, such behaviors including limiting waste, limiting energy consumption, improving recycling habits, and more [77]. Pro-environmental behavior represents ground level action in reducing the anthropogenic drivers of climate change. Therefore, the higher the level of human-nature connection, the higher the level of pro-environmental behaviors and the resultant decline in behaviors that contribute to climate change [20–25]. Human-nature connections, in this context, show human-influenced measures of environmental protection and conservation.

Urban green infrastructure is at the core of HNCT (and Biophilic Design) principles as it contributes to mediating and mitigating climate change. As highlighted above, the HNCT advocates for increased rates of implementation for UGI assets such as green POS, green walls, green roofs, urban trees, and more. These UGI assets significantly increase the area of spongy surfaces in the built environment. Those spongy surfaces reduce the urban heat island effect by absorbing and reducing reflective heat, providing shade that allows for human and fauna refuge, absorbing rainfall and stormwater to help reduce local-scale flooding and erosion, and increasing the amount of biodiversity, all of which increases urban resilience in relation to climate hazards [20–25].

Community planning and urban development strategies play a significant role in shaping how cities are designed and constructed. Improvements to community planning and urban development policies and frameworks greatly influence how humans cohabitate with the environment and each other in urban centers. Planning and development designed around people, and their deep seeded desire to be within and around nature and natural elements can also provide answers to climate change. Urban centers can contribute to the mediation and mitigation of climate through means such as solar passive heating and cooling design principles, providing focus on natural light and natural air flow to reduce the reliance on electricity, employing sustainable building materials to reduce the need for resource extraction, and more [20–25]. Therefore, approaches that align with the HNCT and Biophilic Design principles prove to be essential in addressing the climate change crisis.

Similarly, HNCT offers some unique contributions to address the crisis of declining human health amongst urban populations. As highlighted by this article, these contributions include the teachings from the ecological health perspective, the direct physiological and psychological benefits available from engaging with nature, the contributions that the presence of quality urban nature make to communities and social capital that supports of public health, and a reduction in vulnerability to adverse health conditions [20–25,46,47].

For the past 50 years, as humans have become an ever more urbanized species, the interrelationships between people and the environment around them, and how these relationships affect human health, have been a growing focus for research [14,78,79]. Dating back to the 1980s, the Mandala of Health model was advanced to explain the complexities and holistic influences, systems, and relationships of human interactions with nature. Similarly, Wilson [24] proposed his Biophilic Hypothesis in the 1990s. These anthropocentric models, like others before and after, focus on the benefits that interacting with quality nature spaces provides for mind, body, and spirit of individuals. The Mandala of Health also incorporated the interactions and entanglements at the family level that arise from personal behavior from psycho-social, economic, and environmental factors; from the physical environment, and from human biology [63]. The third layer proposed by Hancock [63] is the human-made environment (community) and fourth an all-encompassing layer of the Mandala is the biosphere (culture). The Mandala of Health proposes that each of these layers affect human health and wellbeing and need cognition to all factors to find the equilibrium required for peak health performance and outcomes. Research over the past few decades has built on ecologic health models, however the notion remains much the same. To achieve good health and wellbeing outcomes, balance must be realized with the built environment and the green infrastructure of urban centers [14,78,79]. A large amount of research now underpins our growing knowledge about the relationship between nature and human physiological and psychological health. A summary of that research is provided below:

- Engaging with nature, whether by exercising, gardening, relaxing, reflecting, volunteering, or other, provides physiological health benefits. Research has shown that physical activity within a nature setting results in improvements to physical fitness, cardiovascular health, immune response systems, reduces stress hormones, and increases the operation of the parasympathetic nervous system [15,79–82].
- Engaging with nature has been shown to improve an individuals' outlook on life, reduce stress, improve both cognitive performance and cognitive recovery, reduce depression and anxiety, and facilitate personal reflection that aids spiritual health [21,79,83–86]. In addition, engaging with nature provides opportunities for socialization that build social capital and community connections. Feeling part of a community and social network has been found essential for mental health.
- Functioning urban ecosystems, biophilic principles, and UGI assets have further been shown to reduce crime rates, increase hospital patient recovery, and reduce the need and reliance on pain medication in hospital patients [21,87]. Equitable access to quality UGI that aligns with the HNCT can also reduce the vulnerability of populations in respect to health impacts brought about by heatwaves and extreme weather events [21].

8. Realigning and Refocusing Towards 2100 and Beyond

Due to the dominant impact of humans on the systems of planet Earth, the new geological series of the Anthropocene, also sometimes referred to as the Human Dominated Geological Epoch, has been recognized [80–88]. While somewhat contested, this epoch is proposed to have commenced in the 1950s, the Anthropocene is characterized by trends of urbanism and the depletion and/or contamination of natural resources [1,89]. Despite these negative trends, surveys show that urban dwellers, in the main, consider urban city life as largely positive [14,90]. However, concern is growing about the equity of access and opportunity, a correct power balance, and the current social and environmental crises, which includes the impact of climate change and declining human health addressed in this article. Current approaches to urban development has been seen to be lucrative for a number of industries, which can perpetuate challenges such as poor planning practices and development for financial gain, lack of availability and demand for alterative sustainable practices, lack of demand for different outcomes and community planning values, a perception of a cost premium for the adoption and implementation of alternative practices and measures, and low support and incentives from regulatory bodies [90].

The current dominant paradigm suggests that the primary purpose of nature is the provision of raw materials and environmental services for the benefit and enjoyment of humans. Under this paradigm, humans are not part of nature, but rather nature exists to be conquered for gain on the basis that short-term human growth and progression will provide benefits in the future that are able to compensate for abstraction and destruction now [91]. Also known as the Technocratic Paradigm, this approach is prefaced on most of humanity perceiving that technological advancements will be able to overcome current and future threats to human populations [11,91]. The dominant paradigm suggests that economic growth needs be continuous for communities, cities, and societies to advance [9,11].

Evolving a sustainable dominant paradigm that is focused on equity, opportunity, risk reduction, and resilience is greatly needed. The barriers to adopting a sustainable paradigm need to be investigated, debunked, and worked through to achieve change.

9. The Promise of Urban Green Infrastructure

The URT and HNCT both show that there are suitable alternatives to the current dominant paradigm, many of which can be realized through the employment and implementation of UGI. While this article only examined the contributions of these theories in the context of climate change and declining human health, it is thought that many other crises may be similarly mediated and mitigated by the application of the URT and HNCT to conserve, protect and reintroduce UGI through inclusive community planning and sustainable development.

As reported above, substantial mitigation and mediation of the twin crises can be achieved through the conservation, reintroduction, enhancement, and protection of UGI as a response that addresses both climate change and declining human health, as advocated by the URT and HNCT. Informed by the documentary analysis presented in this article, the four-set Venn diagram conceptual model presented in Figure 1 demonstrates the complex and highly interrelated nexus between the twin crises examined and the combined application of the URT and HNCT. Key points are provided for each interrelationship to provide a starting context to the relationship. Further, the existing literature of this field has a gap with respect to advocating for that combined approach to the implementation of the URT and HNCT in urban centers. The novel approach of integrating the application of those two theories via the equitable provision of easily accessed quality UGI can mediate and mitigate both climate change and declining human health and can positively contribute to the refocusing and reframing of unsustainable development many urban centers are currently pursuing.

Adopting the evidence-based approach, advocated in this article and illustrated in the conceptual model provided in Figure 1, is essential for delivering the inclusive community planning and sustainable urban development as humanity recalibrates our focus towards the end of the 21st-century and beyond.

10. Conclusions

This article highlights that the URT and the HNCT can both make multifaceted contributions to mitigating and mediating the drivers and impacts of the twin crises of climate change and declining health among urban dwellers. With respect to both theories, conserving, protecting, and restoring quality UGI is the foundation for action to reduce the severity and impact of those crises and for progressing inclusive and sustainable community planning and urban development that focuses beyond 2050. While some UGI can be found within the fabric of most modern cities, the URT, the HNCT, and this article advocate for an increase in the amount, the resourcing, and the perceived value of those UGI assets.

Historically, the benefits that UGI provides in terms of mitigating and mediating climate change and providing physiological and psychological health benefits for urban dwellers have been considered as intangibles. As such, UGI has not been valued in terms of the economic return-on-investment those spaces provide with respect to the resources expended to realize the environmental and social value of the ecosystem services that quality UGI delivers for modern urban centers. However, this article highlights how, congruent with the URT and the HNCT, access to quality UGI can assist in changing from "business as usual" to a more sustainable and resilient approach to community planning and urban development in the second half of the 21st-century.

Implementing the combined approach to URT and HNCT advocated in this article is likely to have limitations in the global context. The limitations could include geographical and cultural considerations at the local scale, resources that are available to land managers, community demand and expectations for UGI installations, stability of current political environments, and the quality of governments and government structures. However, given the previously identified gap in the literature regarding the complementary implementation of URT and HNCT through the medium of UGI, the nature and magnitude of such limitations remains unknown. Clarification of such limitations will most likely be facilitated by the anticipated increase in volume of research URT, HNCT, and UGI. Further, the rapid global changes and growing research interest highlighted in this article, will generate additional insights that are not yet apparent.

Further research is therefore required to investigate and understand the contribution that UGI can make in terms of inclusive community planning and sustainable urban development as our increasingly urbanized human population begins to recalibrate and refocus beyond 2100. That research should investigate different forms of UGI to deliver specific and robust findings aligned to the climate change, human health, urban resilience, community planning, and sustainable urban development foci of this article. To that end, two such studies have been completed in support and are under manuscript development:

1. A study that explores and quantifies how economic, environmental, and social factors influence the theoretical carrying capacity and realized planting of urban tree canopy as an UGI asset.
2. A study that provides proof-of-concept for a framework that quantifies and informs the efficacy of POS management with respect to the delivery of quality UGI spaces with limited resources.

Author Contributions: Conceptualization, J.P.; methodology, J.P.; investigation, J.P.; writing—original draft preparation, J.P.; writing—review and editing, J.P. and G.D.S. All authors have read and agreed to the published version of the manuscript.

Funding: This research received no external funding.

Acknowledgments: The authors thank Richard Harper for his supervision of JPs PhD program and allowing us to progress this article without being listed as an author. We also thank our former and current colleagues and management at the City of Belmont and the Towns of Claremont and Mosman Park for their support of the research projects that help to inform this article. The authors also thank the *Land* editors and four reviewers whose insightful feedback enhanced our article.

Conflicts of Interest: The authors declare no conflicts of interest.

Appendix A

The tables reported below (Tables A1–A4) summarize the geographic scope and UGI research focus of the case study, empirical, and documentary research reported in the 87 articles analyzed by Parker and Simpson [14], the 171 articles analyzed by Parker and Zingoni de Baro [16], and the 38 UGI related articles cited as evidence in support of the synthesis of the URT and HNCT presented in this article.

Table A1. Geographic distribution of case studies and sources of data that informed this article.

Continental Region	Parker and Simpson [1] [14]	Parker and Zingoni de Baro [2] [16]	Cited in this Article [3]
Global Reviews	18	31	32
East & SE Asia	16	12	4
Europe	9	49	7
Middle East	0	0	1
North America	6	48	7
Oceania	20	16	7
South America	0	0	0
Sub-Saharan Africa	3	4	0

1. Data extracted from included articles published in Data Descriptor by Simpson and Parker [17]. 2. Data extracted from included articles published in Data Descriptor by Parker and Simpson [19]. 3. The global-scale analysis of GIS data by Richards and Belcher [43] that covered the five most inhabited continents was coded as a Global Review article.

Table A2. Number of countries represented in the case studies and sources of data that informed this article.

Continental Region	Parker and Simpson [1] [14]	Parker and Zingoni de Baro [2] [16]	Cited in this Article [3.]
East & SE Asia	5	3	4
Europe	7	16	8
Middle East	0	0	1
North America	2	2	7
Oceania	2	2	7
South America	0	0	0
Sub-Saharan Africa	3	1	0

1. Data extracted from included articles published in Data Descriptor by Simpson and Parker [17]. 2. Data extracted from included articles published in Data Descriptor by Parker and Simpson [19]. 3. Richards and Belcher [43] conducted a global analysis of GIS data that covered multiple countries on each of the five most inhabited continents.

Table A3. Number of cities reported in the case studies and sources of data that informed this article.

Continental Region	Parker and Simpson [1] [14]	Parker and Zingoni de Baro [2] [16]	Cited in this Article [3.]
East & SE Asia	7	5	6
Europe	10	18	6 [4.]
Middle East	0	0	1
North America	7	21	7
Oceania	4	4	6
South America	0	0	0
Sub-Saharan Africa	3	1	0

1. Data extracted from included articles published in Data Descriptor by Simpson and Parker [17]. 2. Data extracted from included articles published in Data Descriptor by Parker and Simpson [19]. 3. In addition, Richards and Belcher [43] conducted a global review of GIS data that graphically reported UGI data for at least 1000 cities across all five of the most inhabited continents (excluding Antarctica). 4. In addition, Bieganska et al. [37] report a GIS-based country-scale comparative analysis of the UGI provided in peri-urban developments associated with approximately 100 urban centers in Germany, Latvia, and Poland.

Table A4. Focus for case studies of the aspects of urban communities and development that informed this article.

Focus of Case Studies	Parker and Simpson [1] [14]	Parker and Zingoni de Baro [2] [16]	Cited in this Article
Climate	NR	8	23
Economic	45	23	33
Environmental/Ecological	54	41	41
Health/Wellbeing	59	24	37
Liveability/Quality of Life	57	NR	22
Planning/Policy	50	43	40
Greenspace/Public Open Space (POS)	87	8	31
Quality/Performance of Greenspace/POS	59	21	25
Social	80	38	44
Other	NR	5	NA

1. Data extracted from included articles published in Data Descriptor by Simpson and Parker [17]. 2. Data extracted from included articles published in Data Descriptor by Parker and Simpson [19]. NR = Not Reported – NA = Not Assessed.

References

1. Urbanization. Available online: https://ourworldindata.org/urbanization (accessed on 20 March 2020).
2. Simpson, G.; Newsome, D. Environmental history of an urban wetland: From degraded colonial resource to nature conservation area. *Geo Geogr. Environ.* **2017**, *4*, e00030. [CrossRef]
3. Wu, H. Protecting Stream Ecosystem Health in the Face of Rapid Urbanization and Climate Change. *ProQuest Diss. Publ.* **2014**.
4. Ziter, C. The biodiversity–ecosystem Service Relationship in Urban Areas: A Quantitative Review. *Oikos* **2016**, *125*, 761–768. [CrossRef]
5. Lovell, S.T.; Taylor, J.R. Supplying urban ecosystem services through multifunctional green infrastructure in the United States. *Landsc. Ecol.* **2013**, *28*, 1447–1463. [CrossRef]
6. Cameron, R.W.F.; Blanuˇsa, T.; Taylor, J.E.; Salisbury, A.; Halstead, A.J.; Henricot, B.; Thompson, K. The domestic garden—Its contribution to urban green infrastructure. *Urban For. Urban Green.* **2012**, *11*, 129–137. [CrossRef]
7. Tzoulas, K.; Korpela, K.; Venn, S.; Yli-Pelkonen, V.; Kaźmierczak, A.; Niemela, J.; James, P. Promoting ecosystem and human health in urban areas using green infrastructure: A literature review. *Landsc. Urban Plan.* **2017**, *81*, 167–178. [CrossRef]
8. Mathey, J.; Rößler, S.; Banse, J.; Lehmann, I.; Bräuer, A. Brownfields as an element of green infrastructure for implementing ecosystem services into urban areas. *J. Urban Plan. Dev.* **2015**, *141*, 1–13. [CrossRef]
9. Norton, B.A.; Coutts, A.M.; Livesley, S.J.; Harris, R.J.; Hunter, A.M.; Williams, N.S. Planning for cooler cities: A framework to prioritise green infrastructure to mitigate high temperatures in urban landscapes. *Landsc. Urban Plan.* **2015**, *134*, 127–138. [CrossRef]
10. Meerow, S.; Newell, J.P.; Stults, M. Defining urban resilience: A review. *Landsc. Urban Plan.* **2016**, *147*, 38–49. [CrossRef]
11. Romero-Lankao, P.; Gnatz, D.; Wilhelmi, O.; Hayden, M. Urban Sustainability and Resilience: From Theory to Practice. *Sustainability* **2016**, *8*, 1224. [CrossRef]
12. Burley, B.A. Green infrastructure and violence: Do new street trees mitigate violent crime? *Health Place* **2018**, *54*, 43–49. [CrossRef] [PubMed]
13. Suppakittpaisarn, P.; Jiang, X.; Sullivan, W.C. Green infrastructure, green stormwater infrastructure, and human health: A review. *Curr. Landsc. Ecol. Rep.* **2017**, *2*, 96–110. [CrossRef]
14. Parker, J.; Simpson, G.D. Public Green Infrastructure Contributes to City Livability: A systematic Quantitative Review. *Land* **2018**, *7*, 161. [CrossRef]
15. Parker, J. A Survey of Park User Perception in the Context of Green Space and City Liveability: Lake Claremont, Western Australia. Master's Thesis, Murdoch University, Perth, Australia, 2017. Available online: http://researchrepository.murdoch.edu.au/id/eprint/40856/ (accessed on 8 October 2018).
16. Parker, J.; de Baro, M.E.Z. Green infrastructure in the urban environment: A quantitative literature review. *Sustainability* **2019**, *11*, 3182. [CrossRef]
17. Simpson, G.; Parker, J. Data on Peer Reviewed Papers about Green Infrastructure, Urban Nature, and City Liveability. *Data* **2018**, *3*, 51. [CrossRef]
18. Simpson, G.; Parker, J. Data for an Importance-Performance Analysis (IPA) of a public green infrastructure and urban nature space in Perth, Western Australia. *Data* **2018**, *7*, 69–78. [CrossRef]
19. Hobor, G. New Orleans' remarkably (un)predictable recovery: Developing a theory of Urban Resilience. *Am. Behav. Sci.* **2015**, *59*, 1214–1230. [CrossRef]
20. Klaniecki, K.; Leventon, J.; Abson, D.J. Human-nature connectedness as a 'treatment' for pro-environmental behaviour: Making the case for spatial consideration. *Sustain. Sci.* **2018**, *13*, 1375–1388. [CrossRef]
21. Beatley, T. *Biophilic Cities: Integrating Nature into Urban Design and Planning*; Island Press: Washington, DC, USA, 2011.
22. Simpson, G.D.; Patroni, J.; Teo, A.C.K.; Chan, J.K.L.; Newsome, D. Importance-Performance Analysis to Inform Visitor Management at Marine Wildlife Tourism Destinations. *Tour. Futures Early View.* **2019**, *6*, 165–180. [CrossRef]
23. Samuelsson, K.; Colding, J.; Barthel, S. Urban resilience at eye level: Spatial analysis of empirically defined experiential landscapes. *Landsc. Urban Plan.* **2019**, *187*, 70–80. [CrossRef]
24. Wilson, E.O. *Biophilia*; Harvard University Press: Cambridge, MA, USA, 1984; ISBN 978-0-6740744-2-2.

25. Seymour, V. The human-nature relationship and its impact on health: A critical review. *Front. Public Health* **2016**, *4*, 260. [CrossRef] [PubMed]
26. Parker, J.; Simpson, G. Visitor satisfaction with a public green infrastructure and urban nature space in Perth, Western Australia. *Land* **2018**, *7*, 159. [CrossRef]
27. Nieuwenhuijsen, M.J. Urban and transport planning pathways to carbon neutral, liveable and healthy cities; A review of the current evidence. *Environ. Int.* **2020**, *140*, 105661. [CrossRef] [PubMed]
28. Rist, R.C. Influencing the Policy Process with Qualitative Research. In *Collecting and Interpreting Qualitative Materials*, 2nd ed.; Denzin, N.K., Lincoln, Y.S., Eds.; Sage Publications Inc.: Thousand Oaks, CA, USA; pp. 619–644.
29. Sarantakos, S. *Social Research*, 4th ed.; Macmillan International Higher Education: London, UK, 2013.
30. Stake, R.E. Qualitative Case Studies. In *The Sage Handbook of Qualitative Research*, 3rd ed.; Denzin, N.K., Lincoln, Y.S., Eds.; Sage Publications Ltd.: London, UK, 2005; pp. 443–466.
31. Eisenhardt, K.M. Building theories from case study research. In *The Qualitative Researcher's Companion*; Huberman, M., Matthew, B., Miles, M.B., Eds.; Sage Publications Ltd.: London, UK, 2002; pp. 5–36.
32. Cunningham, J.A.; Menter, M.; Young, C. A review of qualitative case methods trends and themes used in technology transfer research. *J. Technol. Transf.* **2017**, *42*, 923–956. [CrossRef]
33. Tight, M. *Documentary Research in the Social Sciences*; SAGE Publications Ltd.: London, UK, 2019; pp. 15–17.
34. Spencer, J.H. Health, human security and the peri-urban transition in the Mekong Delta: Market reform, governance and new analytic frameworks for research in Southeast Asia. *Int. J. Asia Pac. Stud.* **2007**, *3*, 43–64.
35. Biegańska, J.; Środa-Murawska, S.; Kruzmetra, Z.; Swiaczny, F. Peri-urban development as a significant rural development trend. *Quaest. Geogr.* **2018**, *37*, 125–140. [CrossRef]
36. Kantor, P.; Savitch, H.V. How to study comparative urban development politics: A research note. *Int. J. Urban Reg. Res.* **2005**, *29*, 135–151. [CrossRef]
37. Weitz, J.; Moore, T. Development inside urban growth boundaries: Oregon's empirical evidence of contiguous urban form. *J. Am. Plan. Assoc.* **1998**, *64*, 424–440. [CrossRef]
38. Douglass, M.; Huang, L. Globalizing the city in Southeast Asia: Utopia on the urban edge—The case of Phu My Hung, Saigon. *Int. J. Asia Pac. Stud.* **2007**, *3*, 1–42.
39. Cowell, R. Community Planning: Fostering Participation in the Congested State? *Local Gov. Stud.* **2004**, *30*, 497–518. [CrossRef]
40. United Nations, Department of Economic and Social Affairs, Population Division. World Urbanization Prospects: The 2018 Revision, Online Edition. 2018. Available online: https://population.un.org/wup/Download/ (accessed on 20 February 2020).
41. Richards, D.R.; Belcher, R.N. Global Changes in Urban Vegetation Cover. *Remote Sens.* **2020**, *12*, 23. [CrossRef]
42. Sadan, E. Macro Practice, Policy and Advocacy, Populations and Practice Settings. In *Encyclopedia of Social Work*; Franklin, C., Ed.; National Association of Social Workers (NASW Press) and Oxford University Press: Oxford, UK, 2015.
43. Adams, D. *Urban Planning and the Development Process*; Routledge: Abingdon, UK, 1994; pp. 2–9.
44. McBain, B.; Lenzen, M.; Albrecht, G.; Wackernagel, M. Building robust housing sector policy using the ecological footprint. *Resources* **2018**, *7*, 24. [CrossRef]
45. Świąder, M.; Szewrański, S.; Kazak, J.; van Hoof, J.; Lin, D.; Wackernagel, M.; Alves, A. Application of ecological footprint accounting as a part of an integrated assessment of environmental carrying capacity: A case study of the footprint of food of a large city. *Resources* **2018**, *7*, 52. [CrossRef]
46. Intergovernmental Panel on Climate Change (IPCC). *Climate Change 2014: Synthesis Report. Contribution of Working Groups I, II and III to the Fifth Assessment Report of the Intergovernmental Panel on Climate Change*; Core Writing Team, Pachauri, R.K., Meyer, L.A., Eds.; IPCC: Geneva, Switzerland, 2014; p. 151.
47. Shukla, P.R.; Skea, J.; Buendia, E.C.; Masson-Delmotte, V.; Pörtner, H.-O.; Roberts, D.C.; Zhai, P.; Slade, R.; Connors, S.; van Diemen, R.; et al. *Climate Change and Land: An IPCC Special Report on Climate Change, Desertification, Land Degradation, Sustainable Land Management, Food Security, and Greenhouse Gas Fluxes in Terrestrial Ecosystems*; Intergovernmental Panel on Climate Change (IPCC): Geneva, Switzerland, 2019; in press.
48. Bashford, A.; Chaplin, J.E. *The New Worlds of Thomas Robert Malthus: Rereading the Principle of Population*; Princeton University Press: Princeton, NJ, USA, 2016.

49. Baird, V. A Brief History of Population. Available online: https://newint.org/features/2010/01/01/history (accessed on 20 February 2020).

50. Jenson, J.M.; Howard, M.O. Research and Evidence-Based Practice. In *Encyclopedia of Social Work*; Franklin, C., Ed.; National Association of Social Workers (NASW Press) and Oxford University Press: Oxford, UK, 2013.

51. Jedwab, R.; Pereira, D.; Roberts, M. *Cities of Workers, Children, or Seniors? Age Structure and Economic Growth in a Global Cross-Section of Cities*; World Bank: Washington, DC, USA, 2019.

52. Lawler, A. Ancient DNA. How Europe exported the black death. *Science* **2016**, *352*, 501. [CrossRef]

53. Roberts, L. 9 Billion? *Science* **2011**, *333*, 540–543. [CrossRef]

54. Roy, S.; Byrne, J.; Pickering, C. A systematic quantitative review of urban tree benefits, costs, and assessment methods across cities in different climate zones. *Urban For. Urban Green.* **2012**, *11*, 351–363. [CrossRef]

55. Frumkin, H.; Frank, L.D.; Jackson, R. *Urban Sprawl and Public Health: Designing, Planning, and Building for Healthy Communities*; Island Press: Washington, DC, USA, 2004.

56. Reid, E.; Meakins, G.; Hamidi, S.; Nelson, A.C. Relationship between Urban Sprawl and Physical Activity, Obesity, and Morbidity-Update and Refinement. *Health Place* **2014**, *26*, 118–126. [CrossRef]

57. Pradhan, B. (Ed.) *Spatial Modeling and Assessment of Urban Form: Analysis of Urban Growth: From Sprawl to Compact Using Geospatial Data*, 1st ed.; Springer International Publishing: Cham, Switzerland, 2017.

58. Miller, J.R. Biodiversity conservation and the extinction of experience. *Trends Ecol. Evol.* **2005**, *20*, 430–434. [CrossRef]

59. Gunderson, L.H. Ecological resilience—In theory and application. *Annu. Rev. Ecol. Syst.* **2000**, *31*, 425–439. [CrossRef]

60. Bozza, A.; Aspron, D.; Fabbrocino, F. Urban resilience: A civil engineering perspective. *Sustainability* **2017**, *9*, 103. [CrossRef]

61. Wu, G.; Feder, A.; Cohen, H.; Kim, J.J.; Calderon, S.; Charney, D.S.; Mathé, A.A. Understanding resilience. *Front. Behav. Neurosci.* **2013**, *7*, 1–15. [CrossRef] [PubMed]

62. Gårdmark, A.; Casini, M.; Huss, M.; van Leeuwen, A.; Hjelm, J.; Persson, L.; de Roos, A.M. Regime Shifts in Exploited Marine Food Webs: Detecting Mechanisms Underlying Alternative Stable States using Size-Structured Community Dynamics Theory. *Biol. Sci.* **2015**, *370*, 1–10. [CrossRef]

63. Hancock, T. The mandala of health: A model of the human ecosystem. *Fam. Community Health* **1985**, *8*, 1–10. [CrossRef] [PubMed]

64. Kim, S.E.; Kim, H.; Chae, Y. A new approach to measuring green growth: Application to OECD and Korea. *Futures* **2014**, *63*, 37–48. [CrossRef]

65. Meerow, S.; Stults, M. Comparing conceptualizations or urban climate resilience in theory and practice. *Sustainability* **2016**, *8*, 701–717. [CrossRef]

66. Blaut, J.M. Environmentalism and Eurocentrism. *Geogr. Rev.* **1999**, *89*, 391–408. [CrossRef]

67. Naumann, S.; McKenna, D.; Kaphengst, T.; Pieterse, M.; Rayment, M. Design, Implementation and Cost Elements of Green Infrastructure Projects: Final Report to the European Commission. 2011 Ecologic Institute and GHK Consulting. 2019. Available online: http://ec.europa.eu/environment/enveco/biodiversity/pdf/GI_DICE_FinalReport.pdf (accessed on 20 February 2019).

68. Li, C.; Peng, C.; Chiang, P.; Cai, Y.; Wang, X.; Yang, Z. Mechanisms and applications of green infrastructure practices for stormwater control: A review. *J. Hydrol.* **2018**, *568*, 626–637. [CrossRef]

69. Roe, M.; Mell, I. Negotiating value and priorities: Evaluating the demands of green infrastructure development. *J. Environ. Plan. Manag.* **2013**, *56*, 650–673. [CrossRef]

70. Mekala, G.D.; Jones, R.N.; MacDonald, D.H. Valuing the benefits of creek rehabilitation: Building a business case for public investments in urban green infrastructure. *J. Environ. Manag.* **2014**, *55*, 1354–1365. [CrossRef]

71. Heckert, M.; Rosan, C.D. Creating GIS-based planning tools to promote equity through green Infrastructure. *Front. Built Environ.* **2018**, *4*, 1–5. [CrossRef]

72. Urakami, T. Increased Trend in the Incidence of Diabetes among Youths in the USA during 2002–2012. *J. Diabetes Investig.* **2017**, *8*, 748–749. [CrossRef] [PubMed]

73. Fang, M. Trends in the Prevalence of Diabetes among U.S. Adults: 1999–2016. *Am. J. Prev. Med.* **2018**, *55*, 497–505. [CrossRef] [PubMed]

74. Knowlton, K.; Rotkin-Ellman, M.; King, G.; Margolis, H.G.; Smith, D.; Solomon, G.; Trent, R.; English, P. The 2006 California Heat Wave: Impacts on Hospitalizations and Emergency Department Visits. *Environ. Health Perspect.* **2006**, *117*, 61–67. [CrossRef]

75. Sun, S.; Cao, W.; Mason, T.G.; Ran, J.; Qiu, H.; Li, J.; Yang, Y.; Lin, H.; Tian, L. Increased Susceptibility to Heat for Respiratory Hospitalizations in Hong Kong. *Sci. Total Environ.* **2019**, *666*, 197–204. [CrossRef] [PubMed]

76. Heaviside, C.; Tsangari, H.; Paschalidou, A.; Vardoulakis, S.; Kassomenos, P.; Georgiou, K.E.; Yamasaki, E.N. Heat-Related Mortality in Cyprus for Current and Future Climate Scenarios. *Sci. Total Environ.* **2016**, *569*, 627–633. [CrossRef] [PubMed]

77. Wamsler, C.; Brink, E.; Rivera, C. Planning for climate change in urban areas: From theory to practice. *J. Clean. Prod.* **2013**, *50*, 68–81. [CrossRef]

78. Ritchie, H.; Roser, M. Outdoor Air Pollution. 2020. Available online: https://ourworldindata.org/outdoor-air-pollution (accessed on 3 March 2020).

79. Mesmer-Magnus, J.; Viswesvaran, C.; Wiernik, B.M. *The Role of Commitment in Bridging the Gap between Organizational Sustainability, and Environmental Sustainability. Managing Human Resources for Environmental Sustainability*; Jossey-Bass/Wiley: San Francisco, CA, USA, 2012; pp. 155–186.

80. Shanahan, D.F.; Fuller, R.A.; Bush, R.; Lin, B.B.; Gaston, K.J. The health benefits of urban nature: How much do we need? *BioScience* **2015**, *65*, 476–485. [CrossRef]

81. Keniger, L.E.; Gaston, K.J.; Irvine, K.N.; Fuller, R.A. What are the benefits of interacting with nature? *Int. J. Environ. Res. Public Health* **2013**, *10*, 913–935. [CrossRef]

82. Gladwell, V.F.; Brown, D.K.; Wood, C.; Sandercock, G.R.; Barton, L.J. The great outdoors: How a green exercise environment can benefit all. *Extrem. Physiol. Med.* **2013**, *2*, 3. [CrossRef]

83. Li, Q.T.; Kobayashi, O.M.; Wakayama, Y.; Inagaki, H.; Katsumata, M.; Hirata, Y.; Li, Y.; Hirata, K.; Shimizu, T.; Suzuki, H.; et al. Acute effects of walking in forest environments on cardiovascular and metabolic parameters. *Eur. J. Appl. Physiol.* **2011**, *111*, 2845–2853. [CrossRef]

84. Cracknell, D.; White, M.P.; Pahl, S.; Nichols, W.J.; Depledge, M.H. Marine biota and psychological well-being: A preliminary examination of dose-response effects in an aquarium setting. *Environ. Behav.* **2016**, *48*, 1242–1269. [CrossRef]

85. Dallimer, M.; Irvine, K.N.; Skinner, A.; Davies, Z.G.; Rouquette, J.R.; Maltby, L.L.; Warren, P.H.; Armsworth, P.R.; Gaston, K.J. Biodiversity and the feel-good factor: Understanding associations between self-reported human well-being and species richness. *BioScience* **2012**, *62*, 47–55. [CrossRef]

86. Bratman, G.N.; Hamilton, P.; Daily, G.C. The impacts of nature experience on human cognitive function and mental health. *N. Y. Acad. Sci.* **2012**, *1249*, 118–136. [CrossRef] [PubMed]

87. Totaforti, S. Applying the benefits of biophilic theory to hospital design. *City Territ. Archit.* **2018**, *5*, 1. [CrossRef]

88. Heymans, A.; Breadsell, J.; Morrison, G.M.; Byrne, J.J.; Eon, C. Ecological urban planning and design: A systematic literature review. *Sustainability* **2019**, *11*, 3723. [CrossRef]

89. Anthropocene Working Group. Results of binding vote by Anthropocene Working Group. Available online: http://quaternary.stratigraphy.org/working-groups/anthropocene/ (accessed on 3 March 2020).

90. Boschetti, F.; Gaffier, C.; Price, J.; Moglia, M.; Walker, I. Myths of the city. *Sustain. Sci.* **2017**, *12*, 611–620. [CrossRef]

91. Vos, R.O. Perspective defining sustainability: A conceptual orientation. *J. Chem. Technol. Biotechnol.* **2017**, *82*, 334–339. [CrossRef]

 *land* 

Article

Ways Forward for Advancing Ecosystem Services in Municipal Planning—Experiences from Stockholm County

Sara Khoshkar [1],*, Monica Hammer [2], Sara Borgström [1] and Berit Balfors [1]

[1] Department of Sustainable Development, Environmental Science and Engineering, KTH Royal Institute of Technology, 100 44 Stockholm, Sweden; sara.borgstrom@abe.kth.se (S.B.); balfors@kth.se (B.B.)

[2] School of Natural Sciences, Technology and Environmental Studies, Södertörn University, 141 89 Huddinge, Sweden; monica.hammer@sh.se

* Correspondence: khoshkar@kth.se

Received: 4 August 2020; Accepted: 25 August 2020; Published: 26 August 2020

Abstract: This case study from Stockholm County, Sweden, explores practitioners' experiences of barriers and bridges in municipal planning practices to support actions for ecosystem services. This qualitative study is based on information gathered from a focus group, workshops, and semi-structured interviews, which aided in identifying key factors for integrating ecosystem services in municipal planning. We identified 10 key factors divided into three themes: (i) regulatory framework and political support, (ii) local organizational capacity, and (iii) local adaptation of tools and practices. In particular, the practitioners pointed to the need for the development of legal support and regulations for ecosystem services on the national and EU policy levels. Furthermore, the need for local capacity building and understanding of ecosystem services as well as increased regional support to enhance local knowledge exchange and learning was emphasized. Also, in a decentralized local governance system such as in Sweden, to fully implement ecosystem services in urban planning for sustainable development, locally adapted practical tools and monitoring procedures were considered important.

Keywords: capacity building; ecosystem services; municipal planning practice; urban governance

---

1. Introduction

Globally, multiple human drivers, degrading biodiversity and ecosystem functions and services [1] have significantly altered ecosystems. To prevent further decline and restore ecosystems, there is a need for transformation and change for socioeconomic development to strengthen the delivery of ecosystem services (ES) [2]. ES are defined generally as the functions performed by nature that directly or indirectly benefit humankind and contribute to well-being (e.g., water regulation, air regulation, recreation) [1]. Agenda 2030 states that ES and biodiversity are relevant to all sustainable development goals (SDGs) and emphasizes the need for action [3].

Urban areas are rapidly growing in Europe, for example, Berlin, Malaga, and Stockholm [4]. The increasing number and expansion of urban areas underline the importance of integrating ES in spatial planning [5,6]. Strategic decisions on land use allocations made during the planning process can have significant impacts on urban green spaces and their capacity to support the provision of ES. The integration of ES into urban planning processes is considered to provide a more comprehensive understanding of the values at stake and of the synergies and trade-offs that might arise from land-use decisions [7].

Although the knowledge and awareness of the relevance of ES has increased, there remains a gap in the mainstreaming and implementation of ES in practical planning and decision-making [8,9].

Spatial planning in Europe varies amongst the different countries, influenced by the governance arrangements and institutional settings [10,11]. In Sweden, municipalities have the main responsibility for spatial planning [12] and local management practices play a key role in sustaining urban ES [13]. Decisions that consider ES at the local planning level can enhance the provision of biodiversity and ES and generate benefits that occur at multiple levels from local to global [14,15]. Also, affiliated planning instruments like strategic environmental assessment (SEA) have been recognized as opportunities to enhance decision-making by including the ES approach [16,17].

The importance of governance systems that have the potential to safeguard the long-term delivery and use of ES at different levels of governance is emphasized [6]. Different challenges confronted when implementing ES into practical governance have been identified in the literature related to awareness and interest among practitioners, coordination between planning departments, as well as processes and routines [18,19]. In a study from Berlin, based on qualitative content analysis and semi-structured interviews, several governance challenges for the implementation of ES were identified, including financial constraints, loss of expertise, and insufficient communication about benefits from urban green spaces to key decision-makers [20]. In Rotterdam, the main challenges included a lack of coordination between planning departments and a lack of evaluation of and learning from pilot projects [21]. The governance challenges for the implementation of ES can be different depending on the countries' governance arrangements. However, current ES approaches often do not take existing governance structures and practices as a starting point, and tend to focus on issues such as the spatial extent of ES [22]. Hence, to assimilate the ES framework in urban planning practice rather than in single projects, it is important to gain an understanding of the governance context including institutional frameworks, existing policies, and planning systems [23]. Furthermore, the views of practitioners and decision-makers involved in the planning process are important for the implementation of ES, particularly on the potentials and added value of using the concept and what tools could be used in practical integration [24]. In light of the transition towards decentralized governance in many European countries [25] and the uniquely strong position of Swedish municipalities in planning [26], the Stockholm case provides an interesting context to examine and contribute to the knowledge on ES implementation in local planning practice.

In our study, drawing on empirical experiences of practitioners in Swedish municipal planning, we explored and analyzed local spatial planning practices to identify key factors for advancing ES implementation in planning processes focusing on municipalities in Stockholm County as a case study. The following research questions formed the basis of this paper:

i.   What were the practitioner's experiences and views on challenges and needs for integrating ecosystem services in municipal planning practice?
ii.   What key factors could be identified for supporting the integration of ecosystem services in municipal planning practices?

2. Theoretical Framework

Urban Governance and Spatial Planning

Governance can be defined as the institutions, structures, and processes that determine who make decisions, how and for whom decisions are made, whether, how, and what actions are taken and by whom [27]. In urban regions, spatial planning is a key component of governance efforts to guide development processes [28]. To achieve changes for planning [29] emphasizes that the use of knowledge is a central element. More specifically, [30] highlights the role of expert knowledge in order to take planning and development systematically in directions for contributing to the achievement of defined goals. The authors emphasize that it is important to note that various forms of knowledge interplay in planning processes, with actors from different sectors and levels involved. Therefore, even with the existence of the relevant expert knowledge, it may be overruled by the knowledge of actors involved. In the context of planning for ES, new knowledge about ES does not necessarily influence decisions, because complex interactions within and across governance systems may have implications for the actual implementation in policies and plans [31,32]. [33] also discuss that environmental knowledge alone does not shift priorities from unsustainable practices to sustainable development since priority setting is influenced by competing interests and political agendas, power relations and modes of governance, regulatory frameworks, and property rights.

Furthermore, multiple governmental and non-governmental actors and institutions are involved in planning for ecosystem services and decision-making, which can lead a heterogeneity of practices dependent on the governance context [34]. This involves that the management of ecosystems and landscapes requires an integration of various sources of information and knowledge from various levels and sectors of society [22,35]. Moreover, coordination between involved actors is required to create coherent action and strengthen the role of ES in spatial planning. However, changing land-use patterns, diverse interests, and values of actors on different scales poses challenges for the governance and planning of ES when negotiating the trade-offs in the provision of such ES [36,37].

Lastly, many methods, approaches, and tools have been tested to support decision-making in applying the ES concept [38]. Actions and tools are said to constitute the core of a plan [39], and the probability of success of an action depends on the type of tools used [40]. However, [41] emphasizing that tools and methods alone are not enough to support ES, there remains a need to encourage better ES management. For example, there is a need to provide access to the tools and foster their use through knowledge exchange and application in practice [38–42].

Overall, as highlighted by [43] more attention should be given on how to facilitate change that moves the concept of ES from an "ideal into reality". According to [31], empirical evidence is needed on the governance of ecosystem services, including issues of decision-making and policy implementation. Our paper contributes to the above research gap by providing insight on practices and experiences of ES implementation from a decentralized governance setting and presenting key factors for advancing ES in planning practice.

3. Swedish Governance Context

In Sweden, ES was introduced in the 2010 Swedish Environmental Policy (Government Bill 2009/10:155) and was further elaborated in 2012 [44]. More specifically, two milestone targets related to the implementation of ES were added in 2012 [44]. Furthermore, in 2013, the Swedish government assigned an inquiry (SOU:2013:68) to analyze actions and propose methods and measures to better evaluate ES and integrate the importance of biodiversity and the value of ES into economic plans and other decisions in society [45]. As a result, the ES concept has been implemented to varying degrees in the different parts and at different levels of Swedish society. In 2018, two additional targets regarding ES were added which called for municipalities to (i) have access to a developed method for utilizing and integrating urban greenery and ecosystem services in urban environments, in planning, construction and management by 2020, and (ii) shall utilize and integrate urban greenery and ecosystem services in urban environments in planning, construction and management by 2025 [46]. Hence, at a national level, there is a strong political will to implement ES approaches as a strategy for urban sustainable development in Sweden.

The local level of governance, consisting of 290 municipalities, has an important role in meeting the sustainability targets related to ES. More specifically, the municipalities hold a planning monopoly, where regional and national authorities can intervene when national regulations are at risk of being violated [47]. Spatial planning in Sweden is regulated through the Planning and Building Act (PBA) (National Board of Housing, Building, and Planning) and the Environmental Code (Swedish Environmental Protection Agency) (SEPA) [48]. The PBA mainly regulates exploitation, while the Environmental Code mainly addresses the conservation/protection of land or water [49]. Municipal comprehensive plans (MCP) and detailed development (local) plans are the main statutory planning instruments in Swedish spatial planning [50]. In accordance with the PBA, municipalities have an important role in coordinating their comprehensive plans with plans and policies at other levels of government and demonstrating how the international, national, regional, and local goals will be met. A recent analysis of the integration of ES in the MCPs revealed that these plans are increasingly addressing concrete strategies or measures for the provision and/or conservation of ES [51].

4. Materials and Methods

4.1. Case Study Area

The geographic scope for this study was set as Stockholm County, which is Sweden's largest urbanized region, consisting of 26 municipalities [52]. The planning practices and challenges presented in this study are gathered from six municipalities within Stockholm County: Huddinge, Haninge, Nacka, Stockholm, Täby, and Upplands Väsby (Figure 1). The selection of municipalities represents a mix of rapidly growing urban and peri-urban areas with extensive ongoing urban densification. Moreover, these municipalities have initiated different activities to strengthen ES in their planning practice by, for example, developing approaches and tools for ES mapping, applying for external funding for ES projects, and developing ES strategies. Table 1 provides a brief description of the municipalities studied including population size, projected population size by 2030, percentage of population growth between 2018 and 2030, population density, total land area, and percentage of developed land.

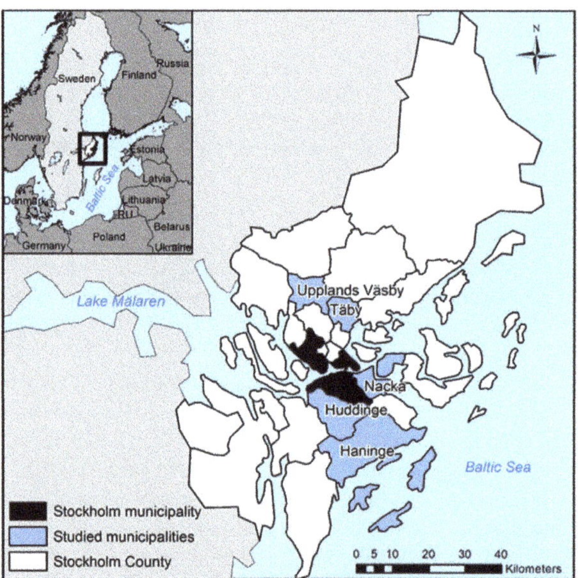

Figure 1. Map of Sweden, zoomed to Stockholm County and municipalities comprised in the study.

Table 1. Overview of municipality statistics (Source: Region Stockholm, 2019; SCB, 2019).

Municipality	Population Size (2018)	Projected Population Size by 2030	% of Population Growth (2018–2030)	Population Density (Inhabitants/km^2)	Total Land Area (km^2) (2015)	Built-up Land (%) (2015)
Haninge	89,989	121,160	35	42	455	12
Huddinge	111,722	145,698	30	790	131	28
Nacka	103,656	119,205	15	800	94	37

Table 1. *Cont.*

Municipality	Population Size (2018)	Projected Population Size by 2030	% of Population Growth (2018–2030)	Population Density (Inhabitants/km^2)	Total Land Area (km^2) (2015)	Built-up Land (%) (2015)
Stockholm	962,154	1,050,660	9	5200	187	55
Täby	71,397	81,783	14	1000	60	36
Upplands Väsby	45,543	58,374	28	2653	75	18

4.2. Research Design

The empirical basis of the paper is a qualitative research study, including a mixed set of methods: one focus group, three workshops, and five semi-structured interviews between June 2018 and September 2019. Gathering data and information through different methods provides divergent perspectives, which according to [53] creates a more complex understanding of the phenomena studied. The study began with a focus group, which allowed for in-depth exploration and discussion on ES practice in spatial planning, and aided in the development of the discussions for the workshops in accordance with [54]. With support from the theoretical framework, themes for improved ES practices were derived and refined through an iterative process based on the information and discussions in the focus group, workshops and semi-structured interviews. Throughout the research process, at least two researchers documented discussions as written notes, which were compiled and cross-checked.

4.3. Focus Group Discussion

An exploratory focus group discussion was organized in accordance with [54] with three experienced municipal practitioners from three municipalities in June 2018, with the aim to discuss future directions for ES in municipal planning. The size of focus groups can vary, however, according to [55], a group size of between 3 and 5 is easy to manage and provides opportunities for all participants to discuss and share views and experiences. The participants were selected based on purposeful sampling, which is a strategy in which particular settings, persons or activities are selected deliberately to provide information that is of relevance to the research questions [56]. The selected participants were three of the most experienced municipal practitioners working with ES in their respective municipalities. More specifically, in their respective municipalities, they had experience of working closely with several research projects regarding ES implementation, as well as in other municipalities and governance levels from which they shared their experiences. The discussions were centered on two questions formulated by the research team: (i) what are the lessons and experiences of ES implementation from ongoing or completed urban development projects? (ii) What are the future directions for ES in municipal planning practice? The outcome of the focus group provided a basis for the topics of discussion for the following workshops, as well as insight into the municipal practitioner's experiences of working with ES.

4.4. Workshops

Based on the discussion in the focus group, a sequence of three half-day workshops were organized on three separate occasions (April, May, and June 2019), discussing the same topic and involving different participant constellations. Workshops were chosen to gather information because they provide the opportunity to elicit rich information from participants selected through purposeful sampling [57]. Furthermore, workshops provide the opportunity for stakeholders of different organizations to collaborate with one another and share experiences [58]. The participants of the workshops included representatives from local and regional governance, construction sector, consultants, and academia (Table 2). In general, the same participants did not participate in more than one workshop. The mixed group of participants allowed for the gathering of different perspectives and views. The discussions in the sequence of workshops focused on factors to advance ES practice in municipal planning, along with challenges that may hinder the implementation.

Table 2. The constellation of participants attending the workshops in April, May, and June 2019.

Workshop	Number of Participants	Constellation of Participants
April 2019	16	Municipal practitioners, environmental consultants, construction sector, academia.
May 2019	30	Municipal practitioners, environmental consultants, construction sector, Stockholm County Administrative Board (SCAB), National Board of Housing, Building and Planning, Swedish Environmental Protection Agency, municipal politicians, academia
June 2019	18	Municipal practitioners, environmental consultants, construction sector, Stockholm County Administrative Board (SCAB), National Board of Housing, Building and Planning, Swedish Environmental Protection Agency, municipal politicians, academia

4.5. Semi-structured Interviews

Following the focus group and workshops, five semi-structured interviews (60–90 min) between July and September 2019 were conducted with selected municipal practitioners to validate the researcher's understanding of the gathered information. Prior to the interviews, an interview guide was developed following [59], which summarized the challenges, needs, and key factors identified in the preceding research settings. Identified key factors for advancing ES in planning practice were presented to the participants, providing them the opportunity to elaborate on the researcher's findings. The interviewees were selected based on their involvement in the focus group and workshops, as well as their leading role in working with activities involving ES in the respective municipalities.

4.6. Thematic Analysis

A thematic analysis in accordance with [54] was conducted based on the theoretical framework and the views and experiences gathered from the focus group, workshops, and interviews. Three main themes were identified: (i) regulatory framework and political support (ii) organizational capacity building for implementation of ES in municipal planning (iii) tools and practices for ES in the local context. The themes were refined through an iterative process in relation to the workshop and interview series. For each of the themes, challenges, and needs were identified. Next, key factors for strengthening ES implementation were linked to the challenges and needs within the three themes.

5. Results

Below, the identified key factors for implementing ES in local planning in relation to the practitioner's experiences and views are discussed according to the three themes (see also Table 3). In most cases, general patterns are presented rather than individual viewpoints of the different participants.

5.1. Regulatory Framework and Political Support

Within the Swedish context, the planning of ES is regulated through both the Planning and Building Act (PBA) and the Environmental Code. According to many participants in our study, ES is still regarded as an aesthetic question, rather than a technical requirement in the legislation. More specifically, the participants pointed out that in accordance with the legislation, the detailed development plans can control what actions cannot be taken, but the possibility of placing stronger demands on developers to conduct specific actions is limited, including implementation of measures to enhance ES. The limitations of PBA regarding ecosystem services are also highlighted by [60], who argue that, in contrast to, for example, parking spaces, there are a lack of clear standards related to the size and number of blue-green space, which places them at risk in land use decision making processes. The participants emphasized that the need arises to develop voluntary municipal policies related to ES in urban development projects in order to provide targets and measures for the inclusion of ES in planning and decision-making. One practitioner highlighted the importance of integrating knowledge from green space plans into the comprehensive plans because then there is a "possibility to really push ecosystem services into the comprehensive plans". Furthermore, in the municipality of Huddinge, ES has been incorporated into the municipality's overarching goals to become a sustainable municipality by 2030. One goal identified was "balanced ecosystem services" (see also [61]), including several sub-goals related to ES (e.g., reduced climate impact, water), which the municipality was working towards. Aligning ES work with existing regulations and municipal organizational goals has been suggested to be important for motivating the integration of ES thinking [62]. The participants also emphasized that to support ES in local policy, there is a need to strengthen the integration of ES in EU policies and regulations. The importance of EU legislation was emphasized by exemplifying how different issues become more or less prioritized due to how they are framed in a juridical sense. For example, in the studied municipalities, regulating ES related to water quality tends to have stronger support in the legislation because of binding instruments at higher governance levels, such as the EU Water Framework Directive (2000/60/EC), with clear indicators and requirements that all EU member states must achieve in relation to ecological and chemical water quality.

Even though there is a strong national political support to integrate ES into societal decision-making including urban planning processes, the political support at the local municipal level was identified as a key factor for advancing the implementation of ES. Political shifts were described as directly influencing the available resources, as well as prioritizations among planning objectives and approaches. According to the participants in our study, the benefits of working with ES to achieve national and regional visions must be conveyed to the politicians. More specifically, a participant stated that, "decision-makers are on board when you can translate nature into their conceptions (e.g., storm water, economic benefits)." Hence, communicating the ES concept to local decision-makers was another key factor identified. In Stockholm County, a network of eight southern municipalities in the central urban region employed a development coordinator tasked with communicating within and across municipalities, as well as with municipal decision-making bodies (e.g., local politicians, municipal CEOs). More concretely, in one of the workshops, a local politician suggested using more scenarios to illustrate and serve as a communication tool with the potential to gain interest among politicians.

5.2. Organizational Capacity Building for Implementing ES in Municipal Planning

In a decentralized planning system like in Sweden, local capacity building is important. There is a need for local knowledge to maneuver confronted challenges in the specific local contexts. According to the participants, an important success factor is the internal basic knowledge of ES. Furthermore, the level of integration of ES in the detailed development plans could vary depending on the basic knowledge of ES of the persons involved in the planning process. Consequently, it was suggested that when the ES knowledge base is lacking, it becomes important to build capacity and learn from others. There is also a need for the knowledge to be maintained and developed within the municipal organization and their daily practices. One format to facilitate knowledge exchange and learning discussed was the development of meeting platforms where municipal practitioners could share their experiences and examples of working with ES. One such national platform exists regarding climate change adaptation in Sweden (Swedish Portal for Climate Change Adaptation). Also, the new online guidance on ES in urban planning, developed by the National Board of Housing, Building, and Planning has ambitions to provide such a platform [63]. Furthermore, the Swedish Environmental Protection Agency (SEPA) has also developed a guide on how to value ES, with step-by-step directions and examples [64]. However, the practitioners discussed that for those municipalities that have already started developing their own routines for integrating ES into planning, these guidance documents were developed too late to provide the needed support. They expressed a need of advisory support for keeping and extending the use of ES within the planning practice, as well as with capacity building within the organization. More specifically, the practitioners emphasized the importance of strengthening the advisory role on the regional level where the County Administrative Board (CAB), is tasked with reviewing the detailed development plans before an application for a construction permit is submitted. According to the practitioners, a limited number of ES are amongst the criteria that are checked in the review process. For example, ES connected to water quality is checked, which has led to water issues gaining momentum and more resources being allocated to studies for handling storm water and flooding within the municipalities. As emphasized by the practitioners, if legislation is strengthened to enable the regulation of a full range of ES in the detailed development plans, the advisory role of the CAB could have the potential of ensuring the implementation of actions related to ES.

Additionally, within municipal organizations, there is a frequent change of employees, also confirmed by [65]. This could present challenges, especially if key personnel leading ES initiatives leave a project, process, or the organization. The next person to take over the position may or may not have the same level of knowledge, experience, or motivation to continue with the previous initiatives. Thus, developing processes and routines for transferring knowledge efficiently and institutionalizing ES knowledge and procedures within municipal organizations is a key factor. Also, the development of systematic monitoring plans with relevant indicators can aid in identifying if ES was integrated in the planning process and actually resulted in implementation as well as contributing to capacity building. Examples provided by the practitioners from the case study municipalities included the municipality of Huddinge, which since 2017, has an annual follow up of their 89 environmental goals in order to monitor the status of implemented actions, initiated measures, delayed actions, and goal completion. The conservation of green areas and biodiversity are amongst these goals. In the municipality of Täby, a practitioner shared the experience of how to structure a follow-up phase. This has been "incorporated into the planning process, which takes place after the adoption of the detailed development plan, and before submission of an application for a construction permit." The follow-up aims to ensure that the quality level of the planned development follows in accordance with what had been agreed by the municipality and developers during the planning process.

5.3. Tools and Practices for ES Implementation in the Local Context

In Sweden, the 290 municipalities differ in land use, population, and financial resources affecting the possibilities of implementation of ES in local planning. When working with ES in local municipal planning, the need for incorporating ES into municipal routines was discussed by the participants, with many emphasizing the need to find established ways of working, evaluating practice in the different departments, identifying any gaps and who should fill those gaps. The participants pointed out that several of the progressive municipalities in the Stockholm region have developed their own planning instruments on a voluntary basis as a complement to required municipal comprehensive plans, in order to support planning processes that recognize specific issues such as ES. One example is Upplands Väsby municipality, that has developed extensive planning instruments on ES including "Mapping of Ecosystem Services" and "Development Plan for Ecosystem Services". More recently, the municipality has developed an action plan to support the implementation of the established strategies for ES in municipal activities. Parallel to the voluntary instruments, the participants emphasized the importance of driven individuals to lead the learning process on how to use these planning instruments within the organization for successful implementation.

Also, the participants highlighted the importance of specific roles and clear responsibilities for implementing the ES approach. The planning process was described as being long and extensive and the individuals involved in setting the strategic vision are often not involved in the implementation phase. Consequently, the vision set at the start regarding specific ES (e.g., tree planting) may not be implemented. As one of the participants stated, "when developing the detailed development plan in particular, there should be at least one person involved with the competence of ES." However, depending on the resources available in the municipalities, this is not always the case. Furthermore, the material concerning ES was described as being extensive but fragmented. There is an overall lack of clarity regarding who should be responsible for consolidating ES at the municipal level and there are different approaches by different municipalities, which also impact how it is implemented [66].

In the studied municipalities, two potential tools for ES in the planning of new developments were highlighted, Point System 2.0 and the green space factor (GSF). Point System 2.0 was an adaptation of the original Point System to include ES and was developed in Upplands Väsby. The GSF has been implemented in several of the municipalities in Stockholm County. The aim of the Point System is to facilitate collaboration and encourage construction companies to work with sustainability measures at an early stage. Through the application of this tool, the developers will receive points for implementing sustainability measures, which, in turn, generate discounts on the price of the land. In Upplands Väsby, a points system was applied to a large-scale project, Fyrklövern, which has been described as the largest residential building project in the municipality in 30 years, involving 14 different construction companies (see also [67]). As a basis for the points system, a series of dialogue meetings with the citizens and different actors involved in the planning and development process were organized to identify novel ideas for the new development project. The experiences gained from the implementation of the points system were being used to develop an updated version, which includes measures for ES. According to the practitioner, this version has the potential to be a "tool for discussion to keep the ES alive, providing more room to discuss what we want in a particular place". The GSF is a tool for increasing green spaces within development projects while minimizing the extent of sealed or paved surface designs that have been applied in many cities around the world (e.g., Malmö, Seattle, Berlin) [68,69]. In Stockholm, it has also been used in several of the municipalities and was appreciated by the practitioners as a way to facilitate dialogue between municipalities and the developers. GSF enabled concrete measures to increase the proportion of greenery on the land that had been set for development.

As a means of building capacity in terms of internal learning and establishing routines, Haninge municipality organized an educational workshop about GSF for the different departments to enhance the usage in detailed development planning processes.

Furthermore, the experience from Upplands Väsby was that "maps to identify ES were the starting point to bring ES into the planning process and permitted knowledge to be gained on what exists, what is needed and what means exist for fulfilling the needs." However, a question was raised regarding how long these maps would be relevant and the need to continuously update them was emphasized, although there was often a lack of resources to do so.

Table 3. Summary of challenges and key factors for strengthening ES implementation.

Themes	Challenges and Needs	Keys Factors for Strengthening ES Implementation
Regulatory framework and political support	ES still regarded as an aesthetic question, rather than a technical requirement in legislation Lack of clear standards related to size and number of blue/green space, placing them at risk in land-use decision-making processes Lack of political support	Strengthen EU policies and regulations National legislation and regulations Communication and local support
Organizational capacity building for implementation of ES in municipal planning	Different levels of knowledge of ES in the municipal organization Frequent change of employees in the municipalities Need for structured monitoring and learning from experiences	Increase local knowledge on ES Platforms (local and regional) for knowledge exchange and learning Strengthen support and advisory role of the regional (county level) Develop routines for monitoring and evaluation of ES in the planning process
Tools and practices for ES implementation in the local context	Long and extensive planning process with different expertise involvedLack of clear roles and responsibilitiesNeed for tailored tools and routines to implement ES in local development projects	Develop action plans on how to implement ES strategies Established individual roles and responsibilities Develop, test and adapt existing tools including maps/GIS, point systems to the local context

6. Discussion

Our study explored and analyzed practitioner's views and experiences of local spatial planning practices in municipalities in Stockholm County in order to identify key factors for supporting the integration of ES in municipal planning practice. We found a heterogeneity of practices and experiences amongst municipalities studied. This is expected in light of the strong local planning monopoly and the responsibility of each individual municipality to derive strategies and measures to reach the national and regional goals. Consequently, it may lead to the consideration of relevant goals for ES to be applied differently across municipalities [49]. Also, the diverse development patterns within the municipalities put forth different prerequisites and challenges that each municipality must work with [51]. Similarly, it has been observed in European countries, that the differences in methods, policy mandates and funding mechanisms for ES implementation, results in heterogeneous practices and needs [70].

As discussed in the preceding sections, urban governance systems have the potential to facilitate the use of ES at different levels of governance [6]. However, many factors can influence the uptake of ES, such as political priorities, available knowledge base, and the municipal arrangements [30–34]. In the Swedish local planning context, we identified 10 key factors divided into three themes for supporting the integration of ES in the municipal planning practice (See Table 3). The factors are linked to the identified challenges and needs based on the practitioner's views and experiences.

Our findings indicate that regulatory frameworks that set mandatory requirements for a range of ES are important for supporting the implementation of ES in local planning practice. This is in line with [33] emphasis on the importance of integrating ES into EU level regulatory frameworks such as the Thematic Strategy on the Urban Environment and other regulatory frameworks that guide spatial planning and natural resource management. Also, [71] argues that the management of biodiversity has binding instruments, for example, the Habitat Directive (92/43/EEC) for specific species and non-binding instruments such as the Biodiversity Strategy 2020, but there remains a need for a specific EU policy devoted to governing urban ES. In line with supporting EU regulation, as a legally required instrument, Strategic Environmental Assessment has been identified to be a good entry point for integrating ES in spatial planning, as it is already established as a widespread (often mandatory) process to assess effects of policies, programs and plans [6,72]. For example, the SEA of spatial plans can ensure that ES is taken into account when evaluating different planning alternatives [73]. Furthermore, although the municipalities have a local planning monopoly, political support and resources are necessary in order to enable local initiatives to be implemented, such as local municipal action plans for ES or the development of new tools. Therefore, communication with local decision-makers is important, also emphasized by [74], who suggest that the values of ES should be communicated to decision-makers in a transparent and viable way. A development coordinator, similar to the example provided in the eight southern municipalities in Stockholm County, can facilitate communication between the municipalities and local politicians.

Furthermore, developing and maintaining the knowledge base related to ES in municipal organizations and their daily practices is needed to strengthen the implementation of ES. This is especially important with the frequent transition of employees in municipal organizations. A study of ES planning in Australia [75] highlighted that the turnover of government staff combined with factors such as weak regulatory support resulted in stagnation in the development and adoption of ES policies and practices. Measures for organizational capacity building can aid in confronting the challenges associated with the changing knowledge bases. For example, the development of meeting platforms between municipalities can provide a means for knowledge exchange amongst the practitioners working with ES. Also, clearly established roles and responsibilities can aid in maintaining the ES knowledge throughout a long and extensive planning process. Previous studies have also highlighted the importance of individuals who take an active role in promoting the ES approach in municipal organizations [33,62]. Additionally, our findings indicate the important role of regional authorities (CAB) to enhance local knowledge exchange and learning.

Adapting existing tools to different contexts and challenges faced is also important for supporting the implementation of ES as emphasized by the practitioners. In Berlin, a governance challenge identified when implementing ES was how to address certain groups of stakeholders (e.g., private property owners), which have the ability to implement strategic goals [20].

In this context, existing tools such as the points system implemented in Upplands Väsby or the green space factor can be adopted and adapted to enable dialogue with the groups of stakeholders. In the Swedish context, these tools aided in facilitating dialogue with the developers to work with sustainability issues in the development projects. Furthermore, the points system invited the local community to contribute to ideas. However, it is important to have an understanding of the limits of the planning tools. For example, a risk that may occur with the use of tools such as the Green Space Factor at the local project scale is the possibility to lose the overall regional perspective on green infrastructure.

A common finding in previous studies is the need for structured monitoring and learning from experiences [20,21]. Developing routines for monitoring and evaluation of ES in the planning process following the examples of Täby and Huddinge can provide a way forward for strengthening monitoring and facilitating learning. However, it is important to note that incorporating follow-up procedures in the planning process also requires the process to be evaluated in order to learn from the experiences [76]. According to [74], data and information regarding ecosystem services and the impact of development on them should continuously be gathered and integrated with the goal of learning, adapting and better-informing policy. Overall, our study was directed towards practitioners in the municipalities due to their active role and responsibility for implementing ES in their daily practice. Examining the views and experiences of the practitioners involved in the planning and decision making processes for implementing ES in the local planning context allowed us to gain insight on the bridges and barriers for operationalizing ES. However, future studies directed towards citizens and local actor's perceptions of ES could further contribute to the understanding of challenges for ES implementation by practitioners.

7. Conclusions

The case study from Stockholm County provided wide empirical material and experiences of practitioners working under different municipal arrangements in a rapidly growing urban region. Consequently, the findings of this paper can contribute to better understand the factors that can facilitate the integration of ES in local planning practice. Four main conclusions can be drawn from this study. Firstly, there is a need for legal support and regulations for ES on the national and EU policy level. Secondly, local political support is necessary to enable the municipalities to take initiatives. Therefore, there is a need for communication between the municipal practitioners and local politicians so to foster learning concerning ES implementation. Thirdly, knowledge related to ecosystem services and motivation amongst the municipal actors is a key factor for implementing ecosystem services in the municipal planning practice. Capacity building initiatives are necessary within the municipalities to maintain and develop the ES knowledge in the municipal organization. Fourthly, monitoring is important in order to ensure that the visions set strategically are translated into concrete actions and to learn from experiences. Overall, advancing ES in local planning requires a combination of top-down political support and bottom-up planning initiatives from local actors.

Author Contributions: Conceptualization, S.K.; methodology, S.K.; writing—original draft preparation, S.K.; writing—review and editing, M.H., S.B. and B.B.; supervision, B.B. and M.H.; funding acquisition, B.B. All authors have read and agreed to the published version of the manuscript.

Funding: This research was funded by FORMAS Research Council, grant number 2015-00133.

Acknowledgments: We thank Mona Petersson for help with the layout of the map.

References

1. Reid, W.V.; Mooney, H.A.; Cropper, A.; Capistrano, D. *Ecosystems and Human Well-Being: Synthesis. A Report of Millenium Ecosystem Assessment*; Island Press: Washington, DC, USA, 2005.
2. IPBES. *IPBES Global Assessment on Biodiversity and Ecosystem Services. Second order draft*; IPBES: Bonn, Germany, 2018.
3. United Nations. *Transforming Our World—The 2030 Agenda for Sustainable Development*; United Nations: New York, NY, USA, 2015.

4.   Eurostat. *Urban Europe—Statistics on Cities, Towns and Suburbs*; Publications office of the European Union: Luxembourg, 2016.

5.   Geneletti, D.; Cortinovis, C.; Zardo, L.; Esmail, B.A. *Planning for Ecosystem Services in Cities*; Springer: Cham, Switzerland, 2019.

6.   Ronchi, S. *Ecosystem Services for Spatial Planning-Innovative Approaches and Challenges for Practical Applications*; Springer: Cham, Switzerland, 2018.

7.   De Groot, R.; Alkemade, R.; Braat, L.; Hein, L.; Willemen, L. Challenges in integrating the concept of ecosystem services and values in landscape planning, management and decision making. *Ecol. Complex.* **2010**, *7*, 260–272. [CrossRef]

8.   Albert, C.; Aronson, J.; Fürst, C.; Opdam, P. Integrating ecosystem services in landscape planning: Requirements, approaches, and impacts. *Landsc. Ecol.* **2014**, *29*, 1277–1285. [CrossRef]

9.   Lautenbach, S.; Mupepele, A.-C.; Dormann, C.F.; Lee, H.; Schmidt, S.; Scholte, S.S.K.; Seppelt, R.; Van Teeffelen, A.J.A.; Verhagen, W.; Volk, M. Blind spots in ecosystem services research and challenges for implementation. *Reg. Environ. Chang.* **2019**, *19*, 2151–2172. [CrossRef]

10. Davoudi, S.; Evans, E.; Governa, F.; Santangel, M. Territorial governance in the making. Approaches, methodologies, practices. *Bol. Asoc. Geogr. Esp.* **2008**, *46*, 33–52.

11. Oliveira, E.; Hersperger, A.M. Disentangling the governance configurations of strategic spatial plan-making in European urban regions. *Plan. Pr. Res.* **2018**, *34*, 47–61. [CrossRef]

12. Högström, J.; Balfors, B.; Hammer, M. Planning for sustainability in expansive metropolitan regions: Exploring practices and planners' expectations in Stockholm, Sweden. *Eur. Plan. Stud.* **2017**, *26*, 439–457. [CrossRef]

13. Andersson, E.; Barthel, S.; Borgström, S.; Colding, J.; Elmqvist, T.; Folke, C.; Gren, Å. Reconnecting cities to the biosphere: Stewardship of green infrastructure and urban ecosystem services. *Ambio* **2014**, *43*, 445–453. [CrossRef]

14. Fisher, B.; Turner, R.K.; Morling, P. Defining and classifying ecosystem services for decision making. *Ecol. Econ.* **2009**, *68*, 643–653. [CrossRef]

15. Farley, J.; Costanza, R. Payments for ecosystem services: From local to global. *Ecol. Econ.* **2010**, *69*, 2060–2068. [CrossRef]

16. Geneletti, D. A Conceptual Approach to Promote the Integration of Ecosystem Services in Strategic Environmental Assessment. *J. Environ. Assess. Policy Manag.* **2015**, *17*. [CrossRef]

17. Mascarenhas, A.; Ramos, T.B.; Haase, D.; Santos, R. Ecosystem services in spatial planning and strategic environmental assessment—A European and Portuguese profile. *Land Use Policy* **2015**, *48*, 158–169. [CrossRef]

18. Mann, C.; Loft, L.; Hansjürgens, B. Governance of ecosystem services: Lessons learned for sustainable institutions. *Ecosyst. Serv.* **2015**, *16*, 275–281. [CrossRef]

19. Galler, C.; Albert, C.; Von Haaren, C. From regional environmental planning to implementation: Paths and challenges of integrating ecosystem services. *Ecosyst. Serv.* **2016**, *18*, 118–129. [CrossRef]

20. Kabisch, N. Ecosystem service implementation and governance challenges in urban green space planning—The case of Berlin, Germany. *Land Use Policy* **2015**, *42*, 557–567. [CrossRef]

21. Haase, D.; Frantzeskaki, N.; Elmqvist, T. Ecosystem services in urban landscapes: Practical applications and governance implications. *Ambio* **2014**, *43*, 407–412. [CrossRef] [PubMed]

22. Primmer, E.; Furman, E. Operationalising ecosystem service approaches for governance: Do measuring, mapping and valuing integrate sector-specific knowledge systems? *Ecosyst. Serv.* **2012**, *1*, 85–92. [CrossRef]

23. Opdam, P.; Albert, C.; Fürst, C.; Grêt-Regamey, A.; Kleemann, J.; Parker, D.; La Rosa, D.; Schmidt, K.; Villamor, G.B.; Walz, A. Ecosystem services for connecting actors—Lessons from a symposium. *Chang. Adapt. Socio Ecol. Syst.* **2015**, *2*, 1–7. [CrossRef]

24. La Rosa, D. Why is the inclusion of the ecosystem services concept in urban planning so limited? A knowledge implementation and impact analysis of the Italian urban plans. *Socio Ecol. Pr. Res.* **2019**, *1*, 83–91. [CrossRef]

25. Nadin, V.; Maldonado, A.M.F.; Zonneveld, W.; Stead, D.; Dąbrowski, M.; Piskorek, K.; Sarkar, A.; Schmitt, P. *COMPASS—Comparative Analysis of Territorial Governance and Spatial Planning Systems in Europe: Applied Research 2016–2018: Final Report*; ESPON & TU Delft: Luxembourg, 2018.

26. Johnson, G. Regional planning in Sweden. In *Planning and Sustainable Urban Development in Sweden*; Lundström, M.J., Fredriksson, C., Witzell, J., Eds.; Swedish Society for Town & County Planning: Stockholm, Sweden, 2013.

27. Bennett, N.; Satterfield, T. Environmental governance: A practical framework to guide design, evaluation, and analysis. *J. Soc. Conserv. Biol.* **2018**, *11*, e12600. [CrossRef]

28. Albrechts, L.; Healey, P.; Kunzmann, K.R. Strategic Spatial Planning and Regional Governance in Europe. *J. Am. Plan. Assoc.* **2003**, *69*, 113–129. [CrossRef]

29. Rydin, Y. Re-examining the role of knowledge within planning theory. *Plan. Theory* **2007**, *6*, 52–68. [CrossRef]

30. Tennøy, A.; Hansson, L.; Lissandrello, E.; Næss, P. How planners' use and non-use of expert knowledge affect the goal achievement potential of plans: Experiences from strategic land-use and transport planning processes in three Scandinavian cities. *Prog. Plan.* **2016**, *109*, 1–32. [CrossRef]

31. Primmer, E.; Jokinen, P.; Blicharska, M.; Barton, D.N.; Bugter, R.; Potschin, M. Governance of ecosystem services: A framework for empirical analysis. *Ecosyst. Serv.* **2015**, *16*, 158–166. [CrossRef]

32. Ratamäki, O.; Jokinen, P.; Sørensen, P.B.; Breeze, T.; Potts, S. A multilevel analysis on pollination-related policies. *Ecosyst. Serv.* **2015**, *14*, 133–143. [CrossRef]

33. Saarikoski, H.; Primmer, E.; Saarela, S.-R.; Antunes, P.; Aszalós, R.; Baró, F.; Berry, P.; Blanko, G.G.; Goméz-Baggethun, E.; Carvalho, L.; et al. Institutional challenges in putting ecosystem service knowledge in practice. *Ecosyst. Serv.* **2018**, *29*, 579–598. [CrossRef]

34. Buijs, A.; Elands, B.; Havik, G.; Ambrose-Oji, B.; Gerőházi, E.; van der Jagt, A.; Mattijssen, T.; Møller, M.S.; Vierikko, K. *Innovative Governance of Urban Green Spaces: Learning from 18 Innovative Examples across Europe*; EU FP7 GREEN SURGE Project; Department of Geosciences and Natural Resource Management, University of Copenhagen: Copenhagen, Denmark, 2016.

35. Folke, C.; Hahn, T.; Olsson, P.; Norberg, J. Adaptive governance of social-ecological systems. *Annu. Rev. Environ. Resour.* **2005**, *30*, 441–473. [CrossRef]

36. Loft, L.; Mann, C.; Hansjürgens, B. Challenges in ecosystem services governance: Multi-levels, multi-actors, multi-rationalities. *Ecosyst. Serv.* **2015**, *16*, 150–157. [CrossRef]

37. Hammer, M.; Bonow, M.; Petersson, M. The role of horse keeping in transforming peri-urban landscapes: A case study from metropolitan Stockholm, Sweden. *Nor. J. Geogr.* **2017**, *71*, 146–158. [CrossRef]

38. Patenaude, G.; Lautenbach, S.; Paterson, J.S.; Locatelli, T.; Dormann, C.F.; Metzger, M.J.; Walz, A. Breaking the ecosystem services glass ceiling: Realising impact. *Reg. Environ. Chang.* **2019**, *19*, 2261–2274. [CrossRef]

39. Brody, S.D.; Highfield, W.; Carrasco, V. Measuring the collective planning capabilities of local jurisdictions to manage ecological systems in southern Florida. *Landsc. Urban Plan.* **2004**, *69*, 33–50. [CrossRef]

40. Cortinovis, C.; Geneletti, D. Ecosystem services in urban plans: What is there, and what is still needed for better decisions. *Land Use Policy* **2018**, *70*, 298–312. [CrossRef]

41. Rounsevell, M.D.A.; Metzger, M.J.; Walz, A. Operationalising ecosystem services in Europe. *Reg. Environ. Chang.* **2019**, *19*, 2143–2149. [CrossRef]

42. Schoonover, H.A.; Grêt-Regamey, A.; Metzger, M.J.; Ruiz-Frau, A.; Santos-Reis, M.; Scholte, S.S.K.; Walz, A.; Nicholas, K.A. Creating space, aligning motivations, and building trust: A practical framework for stakeholder engagement based on experience in 12 ecosystem services case studies. *Ecol. Soc.* **2019**, *24*, 11. [CrossRef]

43. Braat, L.C.; De Groot, R. The ecosystem services agenda:bridging the worlds of natural science and economics, conservation and development, and public and private policy. *Ecosyst. Serv.* **2012**, *1*, 4–15. [CrossRef]

44. Ministry of Environment. *Swedish Environmental Objectives—Clarification of Environmental Quality Objectives and a First Set of Targets*; Ds 2012:23; Ministry of the Environment: Stockholm, Sweden, 2012.

45. Ministry of Environment. *Making the Value of Ecosystem Services Visible*; SOU 2013:68; Ministry of Environment: Stockholm, Sweden, 2013.

46. SEPA. *Fördjupad Utvärdering av Miljömålen 2019*; Swedish Environment Protection Agency: Stockholm, Sweden, 2019.

47. Hedström, R.T.; Lundström, M.T. Swedish Land-use Planning Legislation. In *Planning and Sustainable Urban Development in Sweden*; Lundström, M.J., Fredriksson, C., Witzell, J., Eds.; Swedish Society for Town & County Planning: Stockholm, Sweden, 2013.

48. Persson, C. Deliberation or doctrine? Land use and spatial planning for sustainable development in Sweden. *Land Use Policy* **2013**, *34*, 301–313. [CrossRef]

49. Lidmo, J.; Bogason, Á.; Turunen, E. *The Legal Framework and National Policies for Urban Greenery and Green Values in Urban Areas—A Study of Legislation and Policy Documents in the Five Nordic Countries and Two European Outlooks*; Nordregio Report 2020:3; Nordregio: Stockholm, Sweden, 2020.

50. Koglin, T.; Pettersson, F. Changes, problems, and challenges in Swedish spatial planning—An analysis of power dynamics. *Sustainability* **2017**, *9*, 1836. [CrossRef]

51. Khoshkar, S.; Hammer, M.; Borgström, S.; Dinnétz, P.; Balfors, B. Moving from vision to action-integrating ecosystem services in the Swedish local planning context. *Land Use Policy* **2020**, *97*, 104791. [CrossRef]

52. SCC. *Regional Utvecklingsplan föˆr Stockholmsregionen. RUFS 2050. Europas Mest Attraktiva Storstadsregion*; Stockholm County Council, SCC: Stockholm, Sweden, 2018.

53. Greene, J.C. *Mixed Methods in Social Inquiry*; Jossey-Bass: San Francisco, CA, USA, 2007.

54. Bryman, A. *Social Research Methods*, 2nd ed.; Oxford University Press: New York, NY, USA, 2012.

55. Peek, L.; Fothergill, A. Using focus groups: Lessons from studying daycare centers, 9/11, and Hurricane Katrina. *Qual. Res.* **2009**, *9*, 31–59. [CrossRef]

56. Palys, T.; Given, L. Purposive Sampling. In *The SAGE Encyclopedia of Qualitative Research Methods*; Sage Publications: Thousand Oaks, CA, USA, 2008; p. 698.

57. Creswell, J.W. *Research Design. Qualitative, Quantitative and Mixed Methods Approaches*; Sage: Thousand Oaks, CA, USA, 2014.

58. Ørngreen, R.; Levinsen, K. Workshops as a research methodology. *Electr. J. e-Learn.* **2017**, *15*, 70–81.

59. Kvale, S.; Brinkmann, S. *Interviews, Learning the Craft of Qualitative Research Interviewing*; Sage: Thousand Oaks, CA, USA, 2009.

60. Wihlborg, M.; Sörensen, J.; Olsson, J.A. Assessment of barriers and drivers for implementation of blue-green solutions in Swedish municipalities. *J. Environ. Manag.* **2019**, *233*, 706–718. [CrossRef]

61. Huddinge Municipality. Environment Barometer. 2019. Available online: http://miljobarometern.huddinge.se/ (accessed on 1 October 2019).

62. Blicharska, M.; Hilding-Rydevik, T. "A thousand flowers are flowering just now"—Towards integration of the ecosystem services concept into decision making. *Ecosyst. Serv.* **2018**, *30*, 181–191. [CrossRef]

63. National Board of Housing, Building and Planning. Ecosystem Services in the Built Environment. 2019. Available online: https://www.boverket.se/sv/PBL-kunskapsbanken/Allmant-om-PBL/teman/ekosystemtjanster/ (accessed on 1 October 2019).

64. Swedish Environmental Protection Agency. *Guide for Valuing Ecosystem Services*; Rapport 6690; Swedish Environmental Protection Agency: Stockholm, Sweden, 2015.

65. Jakobsson, A.; Kofoed Schröder, J.; Balfors, B. *Tools and Working Methods for Managing Ecosystem Services in Detailed Planning*; Report TRITA-ABE-RPT-202; KTH Royal Institute of Technology: Stockholm, Sweden, 2020; ISBN 978-91-7873-538-9. (In Swedish)

66. Borgström, S. Appendix 4: Ecosystem Services perspective in Swedish Environmental policy and practice, potentials, barriers and ways for integration. In *SOU 2013:68 Synliggöra Värdet av Ekosystemtjänster. Åtgärder för Välfärd Genom Biologisk Mångfald Och Ekosystemtjänster*; Regeringen: Stockholm, Sweden, 2013; pp. 245–258.

67. Upplands Väsby. The Point System. 2019. Available online: http://upplandsvasby.se/minisajter/fyrklovern/om-fyrklovern/poangsystemet.html (accessed on 2 October 2019).

68. Kruuse, A. The green space factor and green points system. *Town Country Plan. J.* **2011**, *80*, 287–290.

69. Juhola, S. Planning for a green city: The Green Factor tool. *Urban For. Urban Green.* **2018**, *34*, 254–258. [CrossRef]

70. Schröter, M.; Albert, C.; Marques, A.; Tobon, W.; Lavorel, S.; Maes, J.; Brown, C.; Klotz, S.; Bonn, A. National ecosystem assessments in europe: A review. *BioScience* **2016**, *66*, 813–828. [CrossRef] [PubMed]

71. Bouwma, I.; Schleyer, C.; Primmer, E.; Winkler, K.J.; Berry, P.; Young, J.; Carmen, E.; Špulerová, J.; Bezák, P.; Preda, E.; et al. Adoption of the ecosystem services concept in EU policies. *Ecosyst. Serv.* **2018**, *29*, 213–222. [CrossRef]

72. Helming, K.; Diehl, K.; Geneletti, D.; Wiggering, H. Mainstreaming ecosystem services in European policy impact assessment. *Environ. Impact Assess. Rev.* **2013**, *40*, 82–87. [CrossRef]

73. Slootweg, R. Ecosystem services in SEA: Are we missing the point of a simple concept? *Impact Assess. Proj. Apprais.* **2015**, *34*, 1–8. [CrossRef]

74. Costanza, R.; De Groot, R.; Braat, L.; Kubiszewski, I.; Kubiszewski, I.; Sutton, P.C.; Farber, S.; Grasso, M. Twenty years of ecosystem services: How far have we come and how far do we still need to go? *Ecosyst. Serv.* **2017**, *28*, 1–16. [CrossRef]

75. Keenan, R.J.; Pozza, G.; Fitzsimons, J.A. Ecosystem services in environmental policy: Barriers and opportunities for increased adoption. *Ecosyst. Serv.* **2019**, *38*, 100943. [CrossRef]

76. Gustafsson, S.; Andréen, V. *Local Spatial Planning Processes and Integration of Sustainability Perspective Through a Broad Systems Perspective and Systematic Approach*; Springer: Cham, Switzerland, 2017; pp. 567–580.

 *land*

Article

Urban Planning and Design for Building Neighborhood Resilience to Climate Change

Katarzyna Rędzińska * and Monika Piotrkowska

Department of Spatial Planning and Environmental Sciences, Faculty of Geodesy and Cartography,
Warsaw University of Technology, Plac Politechniki 1, 00-661 Warsaw, Poland; monika.piotrkowska@pw.edu.pl
* Correspondence: katarzyna.redzinska@pw.edu.pl

Received: 17 August 2020; Accepted: 8 October 2020; Published: 12 October 2020

Abstract: The aim of the paper was to present the procedure of building neighborhood resilience to climate threats, embedded in planning (from the strategic to local level) and design process and focused on usage of natural adaptive potential. The presented approach encompasses: (1) the strategic identification of focal areas in terms of climate adaptation needs, (2) comprehensive diagnosis of local ecological vulnerability and natural adaptive potential to build adaptive capacity, and (3) incorporation of natural adaptive potential through an identified set of planning and design tools. For diagnosis and strategic environmental impact assessment, the multicriteria analysis has been elaborated. The described procedure is applied to the City of Warsaw on the strategic level, by elaboration of the ranking of districts in terms of priority to take adaptation actions based on climatic threats, demographic vulnerability, and assessment of Warsaw Green Infrastructure potential. For further analysis at the planning and design stage, the district with the most urgent adaptation needs has been chosen, and within its borders, two neighborhoods (existing and planned one) with diagnosed ecological sensitivity were selected. Both case studies were analyzed in terms of environmental conditions, urban structure, and planning provisions. It enabled identification of existing natural adaptive potential and assessment of its use. As a result, propositions for enhancing neighborhood resilience to climate change were suggested.

Keywords: environmental planning; nature-based solutions; urban adaptive capacity

1. Introduction

Cities and their inhabitants are particularly vulnerable to threats related to climate change (thermal and hydrological in particular), which have a negative impact on human health, quality of life, and urban infrastructure. Creating resilient cities is a major challenge for city builders [1,2]. It applies not only to new development but also to existing urban structures. Creating resilient neighborhoods should be the result of properly implemented urban planning and design [3]; for this to happen, planning documents should contribute significantly, because they shape the natural performance of planned areas. However, this is problematic in terms of the conceptualization of resilience and its implementation in the urban realm.

Meerow et al. [4] and Masnavi et al. [5] stated that the concept of resilience related to the urban realm is inconsistent and contested, but crucial in order to develop the adaptive capacity of urban socioecological systems. For a resilient city to be understood as a socioecological system, it should consist of physical and social sub-systems. The physical sub-systems encompass the natural and built components of an urban structure. The social sub-system is built by human societies. Since system structure determines overall system behavior, systems should not be managed only for productivity but also for resilience [6].

Masnavi et al. [5] indicated three conceptual approaches of resilience thinking present in literature: (1) resilience as recovery, (2) resilience as compatibility or adaptation capacity, and (3) resilience as change. There are also two levels of resilience: general and specific. A strategic evaluation of urban resilience focused on general resilience proprieties before considering specific resilience. The authors pointed out the importance of the relationship between urban form and urban resilience. The role of spatial planning in building urban resilience to climate change was demonstrated by Jabareen [7] as one of four drivers in his resilient city planning framework.

Adaptive instruments aimed at building urban resilience to climate change should be implemented systemically and strategically, through planning tools to design and technical solutions. The strategic level focuses on a city's general resilience and corresponds to its ecological and social vulnerability [8,9]. The planning and design level relates to the specific urban physical system resilience. According to Aguiar et al. [10], who compared local adaptation strategies in Europe, spatial planning was considered as one of the priority sectors for adaptation. However, there is a little knowledge of how it is implemented in practice at the planning level [11]. Masnavi et al. [5] highlighted the need for further research on spatial morphology and urban spatial structures as tools to build urban resilience. While opportunities at the strategic and technical levels have already been recognized, there is a gap at the planning and design level [5,11–13]. The reason may be internal constraints related to the legal context, for example, available tools and scope of mandatory regulations, or the level of awareness of local authorities and the awareness and skills of designers [11,14].

Such a problem can be observed in Poland, where, at the strategic level, adaptation plans are developed both at the national level [15] in relation to EU policy [16,17] and at the municipal level. Local adaptation plans to climate change were produced in three separate projects including: the Urban Adaptation Plans (MPA44) [18] and the Adaptcity [19], which resulted in plans for 44 cities with over 100,000 residents and the capital city of Warsaw, and the CLIMCITIES [20], which provide training on climate change issues and developing an urban climate adaptation plan for local authorities in cities with populations from 50,000 to 99,000 residents. This paper discusses only the Adaptcity project in more detail as an exemplification.

However, even though climate change adaptation is well recognized at the strategic level, Polish spatial planning system does not directly address this issue. Instead, it requires the provision of proper living conditions and the maintenance of biological balance. The key instruments (the study of conditions and directions of spatial development at the municipal scale and local spatial development plans) are accompanied by environmental study and strategic environmental impact assessment (EIA). Environmental study and strategic EIA in general refer to the natural performance and very rarely comprehensively consider the issue of climatic threats. Nevertheless, planning documents must compulsorily set the principles of environmental protection to include: (1) a rational use of the earth's surface, (2) ensuring protection of landscape values of the environment and climatic conditions, and (3) comprehensive solutions for urban development problems with particular emphasis on: (a) water management and (b) arranging and shaping green areas.

From the climate adaptation perspective, a problem occurs in terms of available tools on the planning level. For instance, even though the biologically active area index is the most important indicator, it lacks sufficient legal clarification and authorization. The index does not reflect the impact of vegetation structure and adopted technical solutions on natural performance. Moreover, stormwater management is not mandatory to planning provisions and appears mostly as facultative recommendations.

The implementation of adaptation actions to climate change postulated at the strategic level requires the use of urban planning and design tools, but owing to the flawed spatial planning system in Poland as mentioned above, there is a problem in the integration of activities between strategic and planning levels.

This article is focused on the issue of urban planning and design as tools for building specific urban resilience to climate change with reference to urban form. Due to the problems indicated in the introduction, it aims to fill the gap in the implementation of adaptation measures that exists at the planning and design level. To fill that gap, we propose the procedure of building neighborhood resilience to climate threats embedded in planning (from the strategic to the local level) and designing while focusing on usage of natural adaptive potential. It will be applicable both when planning new investments and when evaluating the natural condition of existing neighborhoods in order to improve their resilience. The practical implementation of the procedure is described on the example of Warsaw, Poland.

To achieve the objective of the paper, a literature review describing resilience implementation with the use of natural adaptation potential and a set of tools for urban planning and design are presented. Next, at the strategic level, a ranking of districts in terms of priority to take adaptation actions has been enumerated, while at the planning and design level a multicriteria analysis to diagnose the natural functioning of the neighborhoods in their existing and planned states has been elaborated. Then, the results and discussion related to the case studies and literature are provided. The paper concludes with proposal of a procedure to integrate the strategic level with the planning and design level.

2. Literature Review

2.1. Resilience Implementation

The first step in dealing with the system is to get a deep understanding of its structure and behavior [6]. Since cities constitute socioecological systems, the integration of ecology with urban planning and design has been recommended to build urban resilience, particularly to climate change. This way of thinking about urban planning already has a long tradition that fits with existing environmental approaches [1,21–29]. According to McHarg and Steiner [30], the design process should start with a comprehensive ecological inventory focused on natural processes in order to integrate them into planning and design. Ecological factors constitute determinants of the environmental capacity to support human activity and suitability for a particular type of land use. The idea is to use nature as a strategic ally through planning and designing around ecosystems services. To achieve natural and social sub-system compatibility, Pickett et al. [1] indicated understanding and using spatial heterogeneity. Ahern [24] proposed five strategies to build urban resilience capacity: (1) biodiversity, (2) multifunctionality, (3) multiscale networks, (4) modularity, and (5) adaptive design. Nature-based solutions (e.g., green infrastructure) are recommended as best practices in adaptation by the European Commission [31]. Nature could be integrated into built components of urban systems by incorporating its forms and features, natural processes, and entire living systems through planning and design [1,32]. In relation to hydrological and thermal hazards resulting from climate change, two natural processes and their determinants are crucial to build adaptive capacity: hydrological cycle and air circulation. Natural adaptation potential for building adaptive capacity of urban physical sub-systems consists of environmental features of the area such as geology, soils, water, and vegetation. These features enable rainwater management based on natural hydrological processes and favorable climatic conditions (in particular, optimal thermal conditions). These properties can be employed to minimize hydrological and climatic hazards. Moreover, entire living systems (ecosystems) like forests or wetlands should be integrated to build natural adaptation potential. A set of the most useful tools for urban planning and design level is presented in the next section.

2.2. Tools for Urban Planning and Design

There is a wide range of nature-based and technical adaptation solutions to climate change suitable for urban planning and design. First, the proper zoning of the area corresponding to its natural predispositions have to be established [21,25,28]. Next, three types of adaptation tools for urban planning and design should be taken into consideration: (1) urban development indicators, (2) urban structure (morphology), and (3) technical solutions (Table 1). These tools are useful for building resilience to thermal and hydrological threats resulting from climate change; their effectiveness has been supported by numerous published researches (Table 1).

Table 1. Tools to build resilience at the planning and design level and its impact on climate threats.

Adaptation Actions	Threat		References	
	Thermal	Hydrological	Thermal	Hydrological
Planning tools				
1. Urban development indicators				
Biologically active area index	+	+	Szulczewska et al. [33], Gill et al. [34]	Szulczewska et al. [33], Gill et al. [34], Ellis [35]
Surface runoff indicator	-	+	-	Gill et al. [34], Meng et al. [36]
Maximum building height	+	-	Stewart and Oke [37], Krautheim et al. [38]	-
2. Urban structure (morphology)				
Building structure and layout	+	-	Stewart and Oke [37]; Krautheim et al. [38], Middel et al. [39]	-
Vegetation structure	+	+	Zölch et al. [40], Hertel and Schlink [41]	Deutscher et al. [42]
Green areas layout and size	+	+	Stewart and Oke [37], Asgarian et al. [43], Doick et al. [44], Morini et al. [45]	Kim and Park [46]
Design tools				
3. Technical solutions				
Cool roofs or facades	+	-	Alexandri and Jones [47], Zhang et al. [48]	-
Cool pavements	+	-	Taleghani et al. [49]	-
Green roofs green-blue roofs	+	+	Zölch et al. [40], Alexandri and Jones [47], Berardi [50]	Gill et al. [34], Song et al. [51], Pęczkowski et al. [52], Wang et al. [53], Zhang et al. [54].
Green facades	+	+/-	Zölch et al. [40], Alexandri and Jones [47]	Lau and Mah [55], Tiwary et al. [56]
Permeable surfaces	+/-	+	Hertel and Schlink [41]	Ahiablame and Shakya [57], Liao et al. [58]
Infiltration basins and trenches, bioretention basins and trenches, swales	-	+	-	Ahiablame and Shakya [57], Liao et al. [58], Hua et al. [59], Haghighatafshar et al. [60]
Detention ponds	-	+	-	Haghighatafshar et al. [60], Pereira Souza et al. [61]
Retention ponds	+	+	Stewart and Oke [37]	Baird et al. [62], Chrétien et al. [63]
Wetlands, constructed wetlands	+	+	Sun et al. [64], Thomas and Zachariah [65]	Rizzo et al. [66]
Rainwater barrels, tanks	-	+	-	Ahiablame and Shakya [57], Liao et al. [58], Hua et al. [59]

Zoning allows the incorporation of the natural ecosystems into building resilient neighborhoods and cities as well as using the natural properties of the areas to create suitable functions of the development. The effectiveness of this ecological approach to urban planning and design has been supported by the Woodlands Neighborhood, designed by McHarg [27].

The urban development indicators support the zoning tool in terms of fitting the development intensity to natural conditions in order to manage natural processes and to provide well-being. The most significant indicators used to shape climatic conditions of the urbanized areas are the biologically active area index (BAAI), the surface runoff indicator, and the maximum building height. According to Szulczewska et al. [33], there is a threshold of 45% of the biologically active area's share to enable proper natural performance in the neighborhoods, especially to provide sustainable stormwater management. The BAAI and surface runoff indicator could be integrated into one indicator as it is implemented in Berlin (Biotope Area Factor), Malmö (Green Space Factor), and Seattle (Seattle Green Factor). Scott et al. [32] stated that one of the methods of adapting cities to future high temperatures is to increase the presence of green spaces. Gill et al. [34] indicated the effectiveness of the increase of BAAI by 10% in his case study of Manchester. The maximum building height in relation to the separation width between buildings shapes the areas' roughness and air circulation conditions in terms of wind flow, velocity, turbulences, and dispersion. The following intervals of building height as obstacles for air circulation can be established: 3–10, 10–15, 15–25, above 25 m [37,67].

Urban morphology is a matter of urban design, which is a crucial tool to incorporate natural adaptive potential into urban composition, and to design natural processes, and adjust them to local environmental performance. The building and vegetation structure and layout, as well as green areas' layout and size, should be considered, in particular, to provide proper climatic conditions. The relationships between both are also important. Stewart and Oke [37] and Krautheim et al. [38] pointed out the width/height ratio (the distance between buildings in relation to their height) as the most useful indicator for climatic conditions in urbanized areas. The best air circulation conditions are in areas where the width/height ratio is above 2.4, between 1.4 and 2.4 air circulation conditions are limited, while below 1.4 they are strongly limited.

Vegetation structure modifies not only the roughness, but also evapotranspiration, which is a key process resulting in cooling surface temperature. This is the reason why vegetation structure plays a significant role as a resilience building tool for both thermal and hydrological threats. The areas covered by trees have higher evapotranspiration than grass surfaces and, consequently, a higher cooling effect [40,41]. However, since the coverage of trees has higher roughness and limits horizontal air circulation, while improving convection as a thermally contrasting patch, it is more appropriate to introduce it among intensely urbanized areas rather than in ventilation corridors.

In contrast, it is difficult to give precise specifications as to how many, how big, and where the green areas should be established, because there are too many variables determining climatic conditions in cities [29]. Therefore, the configuration should always be considered for each case, using existing natural forces, processes, and features. Nevertheless, some data have been provided. According to Asgarian et al. [43], composition, configuration, and structure of green space patches considerably affect the nearest urban land surface temperature of built areas. They pointed out that the patches should be homogeneously dispersed, stating that the buffer zone of lower surface temperature reaches up to 200 m. Stewart and Oke [37] indicated the existence of the thermal transitions zones between thermally contrasting local climate zones like green and built-up areas of 200–500 m, depending on surface roughness, building geometry, and atmospheric stability conditions.

The size of green patches is also important: the greater the size, the higher the reduction of surface temperature [43]. Nonetheless, Kensington Garden (100 ha) has the relatively small buffer zone of a width of 400 m. Thus, networks of small (2–3 ha) green spaces were recommended by Doick et al. [44] for effective cooling of urban environments. Also, Hough [29] stated that a fine net of small green areas, distributed homogenously, is more effective than a few large spaces. Apart from size, the crucial features of each green cover patch are: (1) the perimeter-to-area ratio must be minimal—the optimal patches are compact, circular, and rectangular shapes, as well as (2) the core area index—areas with more irregular shapes, which contain more core area, are better than simple, linear shapes [43].

Green areas are also crucial to provide proper hydrological functioning and can be used in stormwater management to reduce the risk of flooding in urban areas dominated by impervious surfaces. Generally, green areas minimize runoff volume; however, the effectiveness of the process greatly depends on vegetation structure as well as the layout and size of a patch. Deutscher et al. [42] has shown that areas covered with trees can intercept up to five times as much water as lawns and produce half as much runoff. Kim and Park [46] indicated that larger and less-fragmented patterns are more likely to decrease peak runoff. Additionally, the effect is amplified by vegetation abundance, especially trees or shrubs, as they increase the storage capacity of an area during flooding.

Finally, technical solutions are to be considered. For adaptation to thermal threats, the modification of the albedo is the point. It could be achieved by technical (cool roofs, facades, or pavements) or nature-based solutions (green roofs, facades, infiltration and bioretention basins and trenches, swales, detention and retention ponds, constructed wetlands, etc.). Adaptation to hydrological threats concerns mostly the sustainable storm water management. Nature-based solutions have a positive impact for both hydrological and thermal threats due to the evapotranspiration process.

3. Materials and Methods

This study consisted of two stages to comply with the aim of the paper, which was the integration of adaptation activities between strategic and planning and design levels. The first stage presents the strategic level of the planning process, and its research area encompasses the city of Warsaw within its administrative borders. The second stage corresponds to the local level of planning and design, and it was conducted on two neighborhoods (existing and planned one) chosen from a district with the most urgent adaptation needs based on the results from the first stage.

3.1. Strategic Level

The first stage of the study involved exploring adaptation needs of Warsaw's districts in context of their potential to implement nature-based solutions. In order to indicate the priority areas for implementing adaptation actions, a ranking of districts was developed. These areas include districts with climatic and demographic risks as well as limited potential for creating green infrastructure. To assess climatic risk, nine indicators were used according to the Warsaw Adaptation Plan [68]: (1) Flood risk in the Vistula valley (Flood), (2) Risk of local flooding after heavy rainfall (Local Flooding), (3) Urban Heat Island, (4) Number of hot nights with minimal temperature above 18 degrees (Hot Nights), (5) Impervious surface coverage (Impervious Surface), (6) Urban Density, (7) Share of built-up areas (Built-up Areas), (8) Estimated increase in residential units (Projected Development), (9) Green areas and forests share (Green Areas).

Demographic vulnerability was estimated based on age structure of inhabitants in districts (percent of the population considered to be vulnerable including people under 4 and over 65 years old) [69]. The potential of green infrastructure was evaluated by eight indicators determining quantitative and spatial potential [70,71]: (1) Share of green infrastructure area in district area (GI Area), (2) Green infrastructure area per inhabitant in district (GI per Inhabitant), (3) Share of recreational green areas in district area (Recreational GI), (4) Share of recreational green areas per inhabitant in district (Recreational GI per Inhabitant), (5) Share of housing areas with recreational green areas within a 500-m distance in district area (Housing with Recreational GI within 500 m), (6) Share of housing areas with recreational green areas above a 500-m distance in district area (Housing with Recreational GI above 500 m), (7) Length of planned bike lanes per 1000 ha of housing area (Planned Bike Lanes Density), (8) Share of potential areas for creating green infrastructure in district area (Potential areas for GI).

The indicators were assessed on a point scale, where those increasing climatic risk scored negative points, while those decreasing the risk gained positive points. Each criterion was scored separately to create three sub-rankings; here, districts were assigned to five classes to facilitate comparison of the results. Final summarized ratings of all three criteria determined the priority of taking adaptation actions in the districts. For further analysis, within the district with the highest adaptation priority, two neighborhoods (one existing and one planned) were chosen based on development plans, exposure to climatic risk and similar natural conditions.

3.2. Local Planning and Design Level

The second stage of the study included determination of natural adaptation potential and an analysis of the existing urban layout and planning provisions of the chosen neighborhoods. We proposed to identify current and future conditions for ventilation, air regeneration and cooling, infiltration, and surface runoff using multicriteria analysis in which criteria were inspired by planning tools (Table 1) derived from literature review presented in Section 2. The method consists of the following steps:

1. Factors selection and division into classes based on ranking criteria (Table 2).
2. Evaluation of classes in a gradient of values from 0 to 1 considering relationships between classes and their importance for the factor in terms of the objective of the analysis (expert method). Class of the highest importance (class I) receives the highest value.

3. Calculation of factor's values for planning units (units from local spatial development plans) using weighted arithmetic mean (WAM) as in Equation (1):

$$WAM = \frac{C_1 \times A_1 + C_2 \times A_2 + \cdots + C_i \times A_i}{A_1 + A_2 + \ldots + A_i} \tag{1}$$

where C_i for $i = 1, 2, \ldots n$ is the class value assigned separately for each factor in the expert method and A_i for $i = 1, 2, \ldots n$ is the area of class i in the planning unit.

4. Assigning weights to factors (expert method) taking into account relative magnitude of the impact the factors have on natural conditions analyzed. For example, in terms of infiltration conditions, land cover obtains lower weight than hydrogeologic factors, because it is easily modifiable by human activity, and therefore its impact also changes.

5. Combining factors' values within each analysis with factors' weights (assigned in step 4) using weighted geometric mean (WGM) as in Equation (2):

$$WGM = K_1^{w_1} \times K_2^{w_2} \times \cdots \times K_i^{w_i} \tag{2}$$

where K_i f $i = 1, 2, \ldots n$ o r is the factor rating (WAM) in the planning unit and w_i for $i = 1, 2, \ldots n$ - the factor weight. This function allows to model synergistic interaction between factors [72]. The step results in average conditions for: ventilation, air regeneration and cooling, infiltration, and average surface runoff in each planning unit.

Next, obtained values (WGM) were classified into four classes of climatic and hydrological conditions (very good, good, moderate, bad). To assess the impact of planned development on hydrological and climatic conditions, the classes in existing and planned state were compared. As a result, three types of changes were distinguished: improvement, no change, deterioration. Criteria values for the existing state were calculated based on data derived from the Warsaw Environmental Atlas [73], topographic objects database (BDOT10k), aerial imagery (orthophoto), and on-site visits.

Criteria values for the planned state were estimated according to urban indicators from planning provisions and included biologically active area index, maximum building height, and percentage of impermeable built-up area. All calculations referred to the planning units specific to local spatial development plans.

The analysis determined natural performance of case studies as well as their sensitivity to hydrological and thermal hazards. In a further step, the degree of usage of the natural potential in adaptation to climate change was assessed.

Table 2. Methodology for natural performance analysis of climatic and hydrological functioning of Warsaw neighborhoods.

Analysis Type	Factor	Ranking Criteria	Class					Factor Weight
			I	II	III	IV	V	
Ventilation conditions	Land cover	Height (min ▶ max)	less than 4 m	4–8 m	8–13 m	above 13 m		0.45
	Local climate zones	Building structure (lowest ▶ highest roughness)	absence	open midrise, open low-rise	compact low-rise, compact midrise	open high-rise	compact high-rise	0.20
	Vegetation cover	Area (max ▶ min)	over 10,000 sq. m		1000–10,000 sq. m,		less than 1000 sq. m,	0.30
	Ventilation corridor	Access	presence				absence	0.05
Air regeneration and cooling conditions	Vegetation	Vertical structure (high ▶ low)	high	mixed	low	seasonal low	surface water	0.50
	Biologically active area index	Share (%) (max ▶ min)	over 60%	46–60%	25–45%	less than 25%		0.50
Infiltration conditions	Soil	Permeability (highest ▶ lowest)	coarse sands, medium sands	fine sands, loess	silty sands, loamy sands, silts	loams, sandy clay	clay, clay loam	0.4
	Water table	Height (high ▶ low)	0–2 m				under 2 m	0.25
	Land cover	Infiltration capacity (high ▶ low)	low, seasonal low vegetation	mixed vegetation	uncovered soils	high vegetation	surface water, impervious surfaces	0.35
Surface runoff indicator	Land cover	Runoff volume (max ▶ min)	black roofs, asphalt and concrete roads	permeable pavement, gravel, or dirt paths	vegetation	uncovered soils	surface water	1.00

4. Results and Discussion–The City of Warsaw Case Study

4.1. Priority of Adaptation to Climate Change in Warsaw

Warsaw will be negatively affected by climate change, in particular the increase in number and intensity of hot days [74] and frequency of precipitation that causes local flooding [68]. The ranking of the adaptation priority shows differences among districts.

Very-high priority was identified in Wola, Praga Południe, Mokotów, and Żoliborz, while high priority in Ursus, Ochota, Śródmieście, Praga Północ (Figure 1, Table 3). These are densely built-up inner parts of the city populated by a vulnerable group of inhabitants, mostly in old age. Climatic threat in those districts comprises urban heat island, high risk of flooding in the event of levee failure in the Vistula valley, and/or inundation due to an overloaded sewage system during heavy rainfall. Simultaneously, the existing amount of greenery and potential areas for creating new green infrastructure is insufficient to compensate the risks.

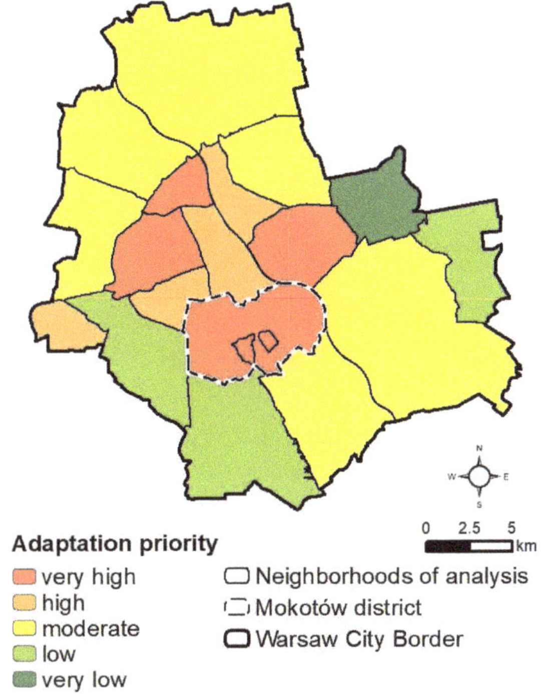

Figure 1. Adaptation priority of Warsaw's districts.

Table 3. Adaptation priority rating of Warsaw's districts.

District	Climatic Threat												Potential of Green Infrastructure (GI)												
	Flood	Local Flooding	Urban Heat Island	Hot Nights	Impervious Surface	Urban Density	Built-up Areas	Projected Development	Green Areas Share	Total Climatic Threat	RATE OF CLIMATIC THREAT (A)	RATE OF DEMOGRAPHIC VULNERABILITY (B)	GI Area	GI Area per Inhabitant	Recreational GI Area	Recreational GI per Inhabitant	Housing with Recreational GI within 500 m	Housing with Recreational GI above 500 m	Density of Planned Bike Lanes fin Housing Areas	Potential Areas for GI	Total Potential	RATE OF GI POTENTIAL (C)	FINAL RATING (A+B+C)	Priority of Taking Adaptation Actions
PRAGA PD.	−4	−4	−1	−4	−4	−5	−5	−3	3	−27	−5	−4	3	1	3	2	5	4	3	1	22	2	−7	very high
WOLA	0	−4	−5	−5	−4	−5	−5	−5	5	−28	−5	−4	2	1	3	2	4	4	4	1	21	2	−7	very high
MOKOTÓW	−2	−3	−1	−4	−4	−5	−4	−5	3	−25	−4	−5	3	2	4	3	4	2	3	3	24	3	−6	very high
ŻOLIBORZ	−1	−4	−5	−4	−4	−5	−5	−2	4	−26	−5	−5	3	2	4	3	5	5	4	1	27	4	−6	very high
OCHOTA	0	−4	−3	−5	−4	−5	−5	−2	4	−24	−4	−4	2	1	5	3	5	5	3	1	25	3	−5	high
PRAGA PN.	−3	−4	−5	−3	−4	−5	−5	−3	3	−29	−5	−3	4	2	2	1	3	5	5	1	23	3	−5	high
ŚRÓDMIEŚCIE	−1	−4	−5	−5	−4	−5	−5	−2	4	−27	−5	−5	4	2	5	3	4	5	5	1	29	5	−5	high
URSUS	0	−4	0	−3	−4	−5	−4	−3	2	−21	−3	−3	1	1	2	1	5	2	3	2	17	1	−5	high
BEMOWO	0	−2	−1	−3	−3	−5	−4	−4	4	−18	−3	−4	4	3	5	4	3	2	3	2	26	4	−3	moderate
TARGÓWEK	0	−3	0	−3	−3	−5	−3	−4	4	−17	−2	−4	3	2	5	4	4	3	2	2	25	3	−3	moderate
BIAŁOŁĘKA	−3	−2	−1	−2	−2	−3	−2	−5	3	−17	−2	−1	3	3	1	2	1	1	2	5	18	1	−2	moderate
BIELANY	−1	−2	−1	−3	−3	−5	−3	−4	5	−17	−2	−5	5	3	4	4	3	5	3	2	29	5	−2	moderate
WAWER	−2	−3	0	−2	−2	−3	−2	−3	5	−12	−1	−3	5	5	1	2	1	3	2	3	22	2	−2	moderate
WILANÓW	−4	−2	0	−3	−2	−3	−2	−4	2	−18	−3	−2	3	4	2	4	1	3	3	5	25	3	−2	moderate
URSYNÓW	0	−5	−1	−3	−3	−3	−3	−4	5	−17	−2	−3	3	3	5	5	3	1	2	4	26	4	−1	low
WESOŁA	0	−2	0	−2	−2	−3	−2	−2	5	−8	−1	−2	5	5	1	1	1	5	1	1	20	2	−1	low
WŁOCHY	0	−2	−1	−4	−3	−5	−4	−2	2	−19	−3	−3	4	4	3	5	3	3	4	3	29	5	−1	low
REMBERTÓW	0	−2	0	−2	−2	−5	−3	−2	5	−11	−1	−2	5	5	1	2	2	5	3	2	25	3	0	low

Moderate priority to take adaptation action concerns the outer districts of Bemowo, Targówek, Bielany, Białołęka, Wawer, and Wilanów, with low or moderate climatic threat characterized by less frequent or no occurrence of urban heat island and only local risk of flooding. This group comprises both older districts with high share of built-up areas and a high amount of well-designed greenery (Bielany, Bemowo, Targówek), and developing districts which lack accessible green infrastructure but have potential areas for creating it (Białołęka, Wilanów). The former are inhabited predominantly by an older population, the latter by younger people.

Low priority was also diagnosed in the outer districts of Włochy, Ursynów, Wesoła and Rembertów, where the climatic threat is very low to moderate due to lower urban density and a larger share of open green areas or forests, and share of vulnerable groups among inhabitants is not significant.

For further analysis in the case studies formula, two neighborhoods were chosen from the district of Mokotów (Figure 1):

1. Sadyba is the existing neighborhood, in need of modernization (Planning provisions LXVIII/1817/2013) [75],
2. Pod Skocznią is the planned neighborhood (Planning provisions NR XLII/1299/2008) [76].

According to existing planning provisions [75,76], the biggest changes will concern the Pod Skocznią, where, as a result of the development of residential and service buildings, the green area will be significantly reduced and it will take the form of a linear park with a water system consisting of retention ponds and water canals. In Sadyba, the plan allows for more dense housing and the development of service functions within existing housing units.

4.2. Neighborhood Resilience in Question

Case studies are located on the upper terrace of the Vistula valley. For both analyzed neighborhoods, we diagnosed hydrological threats, such as flooding caused by an overloaded sewage system during heavy rainfall. Moreover, there is a risk of inundation in Sadyba related to levees breaking in the Vistula valley and groundwater ponding in Pod Skocznią. Thermal threats mainly concern Sadyba due to its location away from the ventilation corridor, land cover albedo, and urban structures that impede air exchange. Whereas in Pod Skocznią, there are no thermal hazards because of the location in the ventilation corridor and the large share of biologically active areas. According to the Warsaw Environmental Atlas [73] the function of the ventilation corridor is to be maintained.

The analysis of natural performance showed that both neighborhoods have high natural adaptive potential, which can be used to minimize thermal and hydrological hazards. In Pod Skocznią, the adaptive potential is aided by mostly flat relief, water table less than 2 m below ground level, predominantly good soil permeability, good quality soil with water retention capacity (peat), and high proportion of biologically active area. Moreover, the existence of a hydrographic system and location within the ventilation corridor can be considered beneficial. Similarly, the natural adaptive potential of Sadyba comprises flat relief, high proportion of green areas with tree dominance, water table less than 2 m below ground level, and very good soil permeability.

As the natural performance analysis indicated, climatic and hydrological disorders occur in Sadyba but not in the undeveloped area of Pod Skocznią (Figure 2). However, in the case of implementation of planning provisions, the deterioration of the natural performance in terms of climatic and hydrological functioning had been predicted for both areas. Furthermore, the usage of natural adaptive potential of the study areas and technical adaptation solutions in the planning provisions [75,76] had been relatively low and not compulsory.

The possibilities of implementing adaptation tools varied between the newly designed housing estate and the modernized one. While in Sadyba, the possibility to engage its natural adaptive potential had been limited; in Pod Skocznią it had been neglected. In Sadyba, a large proportion of impervious surfaces implied constrained infiltration capacity. In Pod Skocznią, the existing planning provisions allow for development in areas with organic soils (peats) that are crucial for water retention in the context of adaptation to climate change.

Figure 2. Changes in hydrological and climatic performance in neighborhoods: (**A**) Change of ventilation conditions, (**B**) Change of air regeneration and heat reduction conditions, (**C**) Change of infiltration conditions, (**D**) Change of surface runoff volume.

Some adaptation tools identified in Table 1 were introduced in the neighborhoods. These tools included biologically active area index, maximum building height, urban structure, and selected technical solutions (green roofs, permeable pavements) for which optional recommendations were made. Moreover, sustainable rainwater management has been introduced as a rule in Pod Skocznią, while it remains only a recommendation for new investments in Sadyba.

4.3. Building Resilient Neighborhoods

The implementation of the procedure on the strategic level requires identifying focal areas for adaptive interventions and their spatial distribution. In order to properly locate adaptation actions, the ranking of priority among districts was developed. The method presented in the paper differs from the one in The Warsaw Adaptation Plan [68] (strategy of adaptation to climate change), which identifies priority areas based on data related only to threats resulting from climate change. Our approach, however, also considers demographic vulnerability; its importance was pointed out by Meerow and Newell [8], Shokry et al. [9], and Błażejczyk et al. [69] and the potential to implement green infrastructure [70,71]. Consequently, there was a need to integrate strategic information from different municipal documents [68,71]. The results obtained from both rankings slightly differ, having taken into account the additional criteria allowed for more holistic assessment of districts in terms of their needs and their potential to adapt. Considering evaluation of green infrastructure makes it possible to assign higher priority to districts which lack areas for creating new greenery to sufficiently compensate the climatic risk. With a lesser effect, the social vulnerability index also influenced the position of some districts in the ranking, exposing those inhabited by an older population.

The Warsaw Adaptation Plan [68] recommends that adaptation to climate change should be considered during urban planning, particularly in local spatial development plans. Although plans are potentially powerful instruments for building adaptive capacity at the local level, available planning tools in the Polish legal context are deficient. Moreover, it is also a matter of the designer's skills [14] as well as investor and local authorities' awareness [11,27].

In the literature, there are some guidelines and recommendations on how to design with respect to climatic and hydrological processes, but they are dispersed [13], as evidenced by extensive literature (Table 1). The key for selecting the literature for this paper was the effectiveness of the planning and design tools in shaping climatic and hydrological conditions proved by empirical studies. Defined tools like land cover height, vegetation vertical structure and size, biologically active area index, surface runoff volume, and building structure (expressed by local climate zones [37]) impacting terrain roughness were used to design the criteria for the multicriteria analysis. Due to the fact that there are many interdependent tools to build resilience of neighborhoods and that it is desirable to use them simultaneously to obtain the expected effect, the multicriteria analysis method was the best to take into account these interdependencies [72].

The multicriteria analysis described in this paper aims to fill the gap of implementing solutions for building urban resilience which exists at the planning and design level. Moreover, it helps to visualize the possible consequences of planning decisions. The results of the analysis were aggregated on the planning unit level which allowed the assessment of the potential impact of specific planning provisions on climatic and hydrological processes in the unit. Conducting such an evaluation enhances the planning process by introducing the issue of adaptation to climate change.

Nevertheless, in the analysis, some limitations occur connected with the availability of data. The choice of criteria depended on the possibility of estimating the value of indicators in the selected analysis scale (planning level). Furthermore, the scope and accuracy of spatial development plans which comprise zoning, a set of urban indicators, and building lines for planning units do not allow the extraction of detailed information about future urban composition and planned vertical structure of vegetation. This required making assumptions about those properties of the site that will shape the climatic and hydrological conditions in the future. Consequently, the method has the potential to be further extended. Introducing the floor area ratio to differentiate the impact of various development scenarios (for example, a scenario with maximal building coverage ratio or maximal building height) may allow the selection of an optimal combination of built-up and biologically active areas on the site in compliance with the planning provisions.

Our analysis of spatial development plans shows that constructing adaptive capacity of the neighborhoods was not the priority. The implementation of planning provisions negatively affects the conditions of hydrological and climatic functioning. In case of the Pod Skocznią neighborhood, where permeable land cover prevailed, the possibility for infiltration will decrease because of an increase in soil sealing. In addition, both housing estates may experience an increase in surface runoff as a result of fragmentation of green areas and change in the vertical structure of vegetation, which shape the amount of runoff, as demonstrated by Kim and Park [46] and Deutscher et al. [42]. Moreover, failure to maintain the existing tall trees in the housing estates will result in worse conditions for air regeneration, because, as shown by Zölch et al. [40] and Hertel and Schlink [41], areas covered by trees have a higher cooling effect than grass surfaces. Therefore, an update of existing planning provisions in the focal areas is suggested.

For Sadyba and Pod Skocznią, we recommend correction of existing spatial development plans oriented towards the utilization of natural adaptation potential (Figure 3B). In reference to environmental approach recommendations [1,21–26,28], it comprises modifying zoning, increasing the index of the biologically active area, and adjusting technical solutions to the natural conditions. The modification of zoning refers to the units where identified natural potential was ignored and to the units located within the climatic corridor. To sustain its function in the city's natural system, the protection of a 200-m wide strip of open land in the Warsaw Escarpment areas has been proposed. Zoning as green areas has been suggested for all these units [27]. The increase of the index of the biologically active area was motivated by the need to maintain good climatic and hydrological performance in Sadyba and Pod Skocznią [33]. Technical solutions aimed at stormwater management have been suggested in order to reduce runoff. Research-supported guidelines include the application of permeable pavements and combinations of infiltration and bioretention devices [57,58,60].

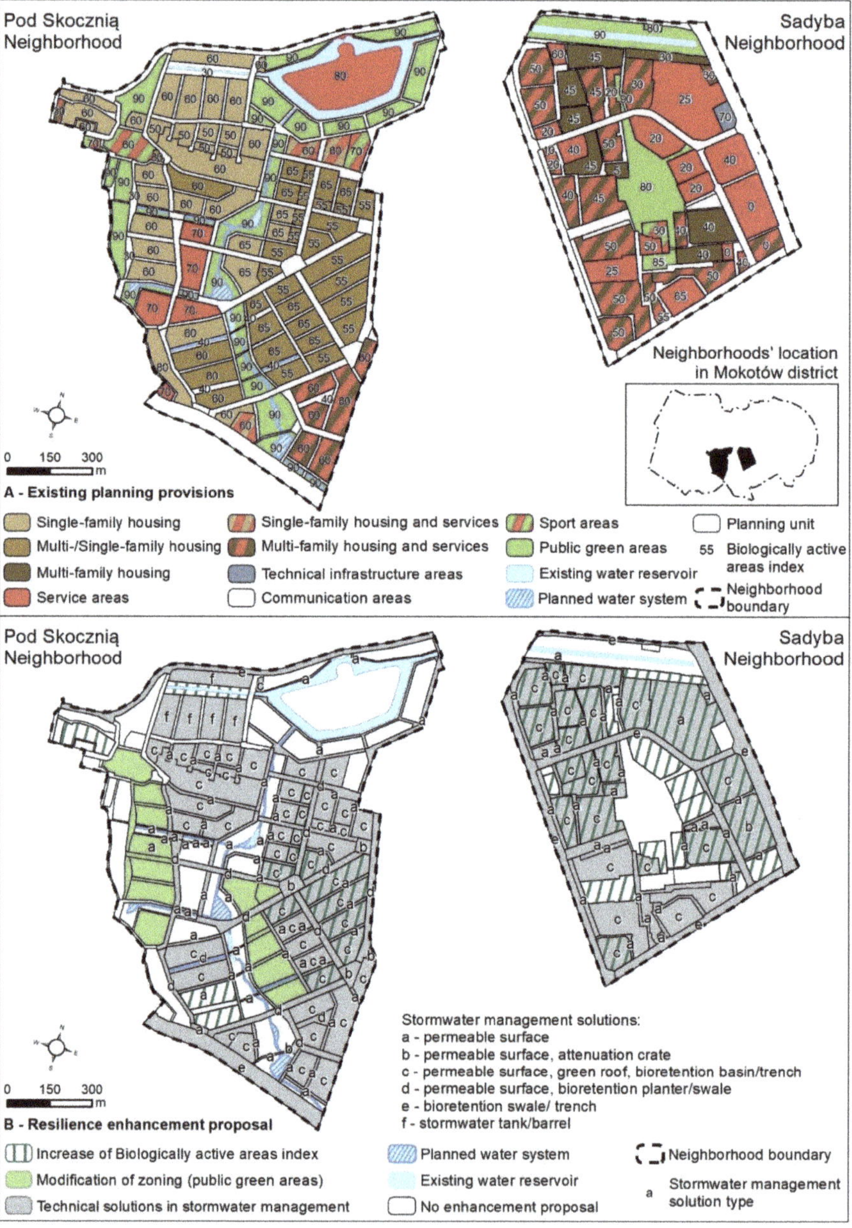

Figure 3. (**A**) Existing planning provisions in case studies, (**B**) Resilience enhancement proposal.

5. Summary and Conclusions

Redefining cities to build resilience to climate threats in urban neighborhoods should be carefully conducted from the strategic to the local level. While the strategic and technical tools are well developed, the planning and design phase needs to be more considered. The relationship between the strategic and planning levels ought to be strongly established.

Polish law related to spatial planning addresses climate adaptation indirectly, providing some deficient tools to implement adaptation actions. The planning provisions aimed to build resilience to climate threats are rather facultative than obligatory. Analyzed case studies have shown that, despite available instruments, local spatial development plans in Sadyba and Pod Skocznią were not formulated to build neighborhood resilience to climate change.

The adaptation of urban areas to climate change by spatial planning and urban design should prioritize the engagement of natural adaptive potential and the adjustment of adopted solutions to natural conditions. If this is not possible, or if the potential has been limited because of existing or planned development, compensation measures should be implemented (i.e., nature-based solutions, see Table 1—technical solutions). However, we argue that, among a set of adaptation tools provided in this paper, the most important one for building neighborhood resilience is properly conducted urban design.

The procedure proposed in this paper could be a useful, simple method within the planning process to build neighborhoods' resilience to climate threats. It is applicable both when planning new investments and when evaluating the natural performance of existing neighborhoods to enhance their resilience. The method utilizes the indicators from literature like land cover height, vegetation vertical structure and size, biologically active area index, building structure, and surface runoff volume. Still, it has the potential to be further developed considering the local context and tools available. The procedure consists of the following steps:

1. Strategic level

    (a) Identification of needs in response to climatic threats, social vulnerability, and possibilities of implementing adaptation solutions based on ecosystem services. At this stage, rankings of assessment could be helpful, as shown in the paper.

    (b) Selection of the most vulnerable areas, which have limited potential to benefit from ecosystem services engagement and are threatened by rapid urbanization.

    (c) Planning and design level

2. Diagnosis of the existing state of the neighborhood including local ecological sensitivity (in terms of existing climatic threats and disturbances in natural performance), natural adaptive potential, and natural functioning (climatic and hydrological). It should be performed in environmental studies for spatial development plans.

    (a) Evaluation of the impact of planning provisions on natural performance (climatic and hydrological functioning) using planning tools embedded in the multicriteria analysis presented in this paper.

    (b) Implementing adaptation solutions with reference to diagnosed needs and possibilities. Available planning and design tools comprise zoning, urban development indicators, urban morphology, and technical solutions.

The application of this procedure in the Sadyba and Pod Skocznią neighborhoods in Warsaw has shown various possibilities of using natural adaptation potential and tools. In Sadyba, the natural adaptation potential has already been limited by development, so the possibilities of its incorporation were smaller compared to Pod Skocznią. This implies the need for compensation measures, including nature-based solutions. However, implementation of these usually requires undertaking renewal actions while adjusting to environmental and technical conditions of existing buildings. In Pod Skocznią, the range of possibilities was much broader, as the mostly undeveloped area allowed incorporation of natural ecosystems, processes, and features to build a resilient neighborhood. Unfortunately for both neighborhoods, local spatial development plans ignored or neglected natural adaptation potential and only a few solutions to increase the neighborhoods' resilience to climate threats were used. This poor planning will trigger the deterioration of the neighborhoods' natural performance; therefore, enhancement of existing planning provisions is highly recommended.

Author Contributions: Conceptualization, K.R.; methodology, K.R. and M.P.; formal analysis, M.P.; investigation, K.R. and M.P.; writing—original draft preparation, K.R. and M.P.; writing—review and editing, K.R. and M.P.; visualization, M.P. All authors have read and agreed to the published version of the manuscript.

Funding: The research was carried out as part of the cooperation of the City of Warsaw and the science sector. "The APC was funded by WARSAW UNIVERSITY OF TECHNOLOGY".

Acknowledgments: We would like to thank Warsaw City Hall for providing access to data related to Warsaw Green Infrastructure.

Conflicts of Interest: The authors declare no conflict of interest. The funders had no role in the design of the study; in the collection, analyses, or interpretation of data; in the writing of the manuscript, or in the decision to publish the results.

References

1. Pickett, S.T.; Cadenasso, M.L.; McGrath, B. *Resilience in Ecology and Urban Design: Linking Theory and Practice for Sustainable Cities*; Springer: Berlin/Heidelberg, Germany, 2013; Volume 3.
2. Bai, X.; Dawson, R.J.; Ürge-Vorsatz, D.; Delgado, G.C.; Barau, A.S.; Dhakal, S.; Dodman, D.; Leonardsen, L.; Masson-Delmotte, V.; Roberts, D.C. Six research priorities for cities and climate change. *Nature* **2018**, *555*, 23–25. [CrossRef] [PubMed]
3. Biesbroek, G.R.; Swart, R.J.; Van der Knaap, W.G. The mitigation–adaptation dichotomy and the role of spatial planning. *Habitat Int.* **2009**, *33*, 230–237. [CrossRef]
4. Meerow, S.; Newell, J.P.; Stults, M. Defining urban resilience: A review. *Landsc. Urban Plan.* **2016**, *147*, 38–49. [CrossRef]
5. Masnavi, M.R.; Gharai, F.; Hajibandeh, M. Exploring urban resilience thinking for its application in urban planning: A review of literature. *Int. J. Environ. Sci. Technol.* **2019**, *16*, 567–582. [CrossRef]
6. Meadows, D.H.; Wright, D. *Thinking in Systems: A Primer*; Chelsea Green Pub: White River Junction, VT, USA, 2008; 218p.
7. Jabareen, Y. Planning the resilient city: Concepts and strategies for coping with climate change and environmental risk. *Cities* **2013**, *31*, 220–229. [CrossRef]
8. Meerow, S.; Newell, J.P. Spatial planning for multifunctional green infrastructure: Growing resilience in Detroit. *Landsc. Urban Plan.* **2017**, *159*, 62–75. [CrossRef]
9. Shokry, G.; Connolly, J.J.T.; Anguelovski, I. Understanding climate gentrification and shifting landscapes of protection and vulnerability in green resilient Philadelphia. *Urban Clim.* **2020**, *31*, 100539. [CrossRef]
10. Aguiar, F.C.; Bentz, J.; Silva, J.M.N.; Fonseca, A.L.; Swart, R.; Santos, F.D.; Penha-Lopes, G. Adaptation to climate change at local level in Europe: An overview. *Environ. Sci. Policy* **2018**, *86*, 38–63. [CrossRef]
11. Kim, H.; Tran, T. An evaluation of local comprehensive plans toward sustainable green infrastructure in US. *Sustainability* **2018**, *10*, 4143. [CrossRef]
12. Kabisch, N.; Korn, H.; Stadler, J.; Bonn, A. (Eds.) *Nature-Based Solutions to Climate Change Adaptation in Urban Areas*; Springer International Publishing: Cham, Switzerland, 2017; p. 337.
13. Bartesaghi Koc, C.; Osmond, P.; Peters, A. Evaluating the cooling effects of green infrastructure: A systematic review of methods, indicators and data sources. *Sol. Energy* **2018**, *166*, 486–508. [CrossRef]
14. Eliasson, I. The use of climate knowledge in urban planning. *Landsc. Urban Plan.* **2000**, *48*, 31–44. [CrossRef]
15. Polish National Strategy for Adaptation to Climate Change (NAS 2020). 2013. Available online: https://klimada.mos.gov.pl/wp-content/uploads/2014/12/ENG_SPA2020_final.pdf (accessed on 5 July 2020).
16. Commission of the European Communities. *White Paper Adapting to Climate Change: Towards a European Framework for Action*; COM (2009) 147 final; European Commission: Brussels, Belgium, 2009.

17. COMMUNICATION FROM THE COMMISSION TO THE EUROPEAN PARLIAMENT, THE COUNCIL, THE EUROPEAN ECONOMIC AND SOCIAL COMMITTEE AND THE COMMITTEE OF THE REGIONS *An EU Strategy on Adaptation to Climate Change*; (COM(2013) 216); European Commission: Brussels, Belgium, 2013.

18. MPA 44. Available online: http://www.44mpa.pl/?lang=en (accessed on 5 July 2020).

19. ADAPTCITY. Available online: http://www.adaptcity.pl/english/about/ (accessed on 5 July 2020).

20. Climcities. Available online: http://www.climcities.ios.gov.pl/about-the-project (accessed on 5 July 2020).

21. McHarg, I.L. *Design with Nature*; American Museum of Natural History: New York, NY, USA, 1969.

22. Spirn, A.W. *The Granite Garden: Urban Nature and Human Design*; Basic Books: New York, NY, USA, 1984; 334p.

23. Steiner, F. Commentary: Healing the earth: The relevance of Ian McHarg's work for the future. *Philos. Geogr.* **2004**, *7*, 141–149. [CrossRef]

24. Ahern, J. Urban landscape sustainability and resilience: The promise and challenges of integrating ecology with urban planning and design. *Landsc. Ecol.* **2013**, *28*, 1203–1212. [CrossRef]

25. McHarg, I.L. An ecological method for landscape architecture. In *The Ecological Design and Planning Reader*; Ndubisi, F.O., Ed.; Island Press: Washington, DC, USA, 2014; pp. 341–347.

26. Spirn, A.W. Ecological urbanism: A framework for the design of resilient cities (2014). In *The Ecological Design and Planning Reader*; Ndubisi, F.O., Ed.; Island Press: Washington, DC, USA, 2014; pp. 557–571.

27. Yang, B.; Li, S. Design with Nature: Ian McHarg's ecological wisdom as actionable and practical knowledge. *Landsc. Urban Plan.* **2016**, *155*, 21–32. [CrossRef]

28. Steiner, F.R. *The Living Landscape: An Ecological Approach to Landscape Planning*; Island Press: Washington, DC, USA, 2012.

29. Hough, M. *Cities and Natural Process*; Routledge, Taylor & Francis e-Library: Oxford, UK, 2004.

30. McHarg, I.L.; Steiner, F.R. *To Heal the Earth: Selected Writings of Ian L. McHarg*; Island Press: Washington, DC, USA, 1998.

31. European Commission. *Building a Green Infrastructure for Europe*; Publication Office of the European Union: Luxembourg, 2013; p. 23.

32. Scott, M.; Lennon, M.; Haase, D.; Kazmierczak, A.; Clabby, G.; Beatley, T. Nature-based solutions for the contemporary city/Re-naturing the city/Reflections on urban landscapes, ecosystems services and nature-based solutions in cities/Multifunctional green infrastructure and climate change adaptation: Brownfield greening as an adaptation strategy for vulnerable communities?/Delivering green infrastructure through planning: Insights from practice in Fingal, Ireland/Planning for biophilic cities: From theory to practice. *Plan. Theory Pract.* **2016**, *17*, 267–300.

33. Szulczewska, B.; Giedych, R.; Borowski, J.; Kuchcik, M.; Sikorski, P.; Mazurkiewicz, A.; Stańczyk, T. How much green is needed for a vital neighbourhood? In search for empirical evidence. *Land Use Policy* **2014**, *38*, 330–345. [CrossRef]

34. Gill, S.E.; Handley, J.F.; Ennos, A.R.; Pauleit, S. Adapting cities for climate change: The role of the green infrastructure. *Built Environ.* **2007**, *33*, 115–133. [CrossRef]

35. Ellis, J.B. Sustainable surface water management and green infrastructure in UK urban catchment planning. *J. Environ. Plan. Manag.* **2013**, *56*, 24–41. [CrossRef]

36. Meng, M.; Dąbrowski, M.; Chan, F.K.S.; Stead, D. Chapter 19—Spatial planning for climate adaptation and flood risk: Development of the sponge city program in Guangzhou. In *Smart, Resilient and Transition Cities*; Galderisi, A., Colucci, A., Eds.; Elsevier: Amsterdam, The Netherlands, 2018; pp. 153–162.

37. Stewart, I.D.; Oke, T.R. Local climate zones for urban temperature studies. *Bull. Am. Meteorol. Soc.* **2012**, *93*, 1879–1900. [CrossRef]

38. Krautheim, M.; Pasel, R.; Pfeiffer, S.; Schultz-Grandberg, J. *City and Wind: Climate as An Architectural Instrument*; DOM Publishers: Berlin, Germany, 2014.

39. Middel, A.; Häb, K.; Brazel, A.J.; Martin, C.A.; Guhathakurta, S. Impact of urban form and design on mid-afternoon microclimate in Phoenix Local Climate Zones. *Landsc. Urban Plan.* **2014**, *122*, 16–28. [CrossRef]

40. Zölch, T.; Maderspacher, J.; Wamsler, C.; Pauleit, S. Using green infrastructure for urban climate-proofing: An evaluation of heat mitigation measures at the micro-scale. *Urban For. Urban Green.* **2016**, *20*, 305–316. [CrossRef]

41. Hertel, D.; Schlink, U. Decomposition of urban temperatures for targeted climate change adaptation. *Environ. Model. Softw.* **2019**, *113*, 20–28. [CrossRef]

42. Deutscher, J.; Kupec, P.; Kučera, A.; Urban, J.; Ledesma, J.L.J.; Futter, M. Ecohydrological consequences of tree removal in an urban park evaluated using open data, free software and a minimalist measuring campaign. *Sci. Total Environ.* **2019**, *655*, 1495–1504. [CrossRef] [PubMed]
43. Asgarian, A.; Amiri, B.J.; Sakieh, Y. Assessing the effect of green cover spatial patterns on urban land surface temperature using landscape metrics approach. *Urban Ecosyst.* **2015**, *18*, 209–222. [CrossRef]
44. Doick, K.J.; Peace, A.; Hutchings, T.R. The role of one large greenspace in mitigating London's nocturnal urban heat island. *Sci. Total Environ.* **2014**, *493*, 662–671. [CrossRef] [PubMed]
45. Morini, E.; Touchaei, A.G.; Rossi, F.; Cotana, F.; Akbari, H. Evaluation of albedo enhancement to mitigate impacts of urban heat island in Rome (Italy) using WRF meteorological model. *Urban Clim.* **2018**, *24*, 551–566. [CrossRef]
46. Kim, H.W.; Park, Y. Urban green infrastructure and local flooding: The impact of landscape patterns on peak runoff in four Texas MSAs. *Appl. Geogr.* **2016**, *77*, 72–81. [CrossRef]
47. Alexandri, E.; Jones, P. Temperature decreases in an urban canyon due to green walls and green roofs in diverse climates. *Build. Environ.* **2008**, *43*, 480–493. [CrossRef]
48. Zhang, J.; Li, Y.; Tao, W.; Liu, J.; Levinson, R.; Mohegh, A.; Ban-Weiss, G. Investigating the Urban Air Quality Effects of Cool Walls and Cool Roofs in Southern California. *Environ. Sci. Technol.* **2019**, *53*, 7532–7542. [CrossRef]
49. Taleghani, M.; Crank, P.J.; Mohegh, A.; Sailor, D.J.; Ban-Weiss, G.A. The impact of heat mitigation strategies on the energy balance of a neighborhood in Los Angeles. *Solar Energy* **2019**, *177*, 604–611. [CrossRef]
50. Berardi, U. The outdoor microclimate benefits and energy saving resulting from green roofs retrofits. *Energy Build.* **2016**, *121*, 217–229. [CrossRef]
51. Song, U.; Kim, E.; Bang, J.H.; Son, D.J.; Waldman, B.; Lee, E.J. Wetlands are an effective green roof system. *Build. Environ.* **2013**, *66*, 141–147. [CrossRef]
52. Pęczkowski, G.; Kowalczyk, T.; Szawernoga, K.; Orzepowski, W.; Żmuda, R.; Pokładek, R. Hydrological performance and runoff water quality of experimental green roofs. *Water* **2018**, *10*, 1185. [CrossRef]
53. Wang, X.; Tian, Y.; Zhao, X. The influence of dual-substrate-layer extensive green roofs on rainwater runoff quantity and quality. *Sci. Total Environ.* **2017**, *592*, 465–476. [CrossRef] [PubMed]
54. Zhang, Q.; Miao, L.; Wang, X.; Liu, D.; Zhu, L.; Zhou, B.; Sun, J.; Liu, J. The capacity of greening roof to reduce stormwater runoff and pollution. *Landsc. Urban Plan.* **2015**, *144*, 142–150. [CrossRef]
55. Lau, J.T.; Mah, D.Y.S. Green Wall for Retention of Stormwater. *Pertanika J. Sci. Technol.* **2018**, *26*, 283–298.
56. Tiwary, A.; Godsmark, K.; Smethurst, J. Field evaluation of precipitation interception potential of green façades. *Ecol. Eng.* **2018**, *122*, 69–75. [CrossRef]
57. Ahiablame, L.; Shakya, R. Modeling flood reduction effects of low impact development at a watershed scale. *J. Environ. Manag.* **2016**, *171*, 81–91. [CrossRef]
58. Liao, Z.L.; He, Y.; Huang, F.; Wang, S.; Li, H.Z. Analysis on LID for highly urbanized areas' waterlogging control: Demonstrated on the example of Caohejing in Shanghai. *Water Sci. Technol.* **2013**, *68*, 2559–2567. [CrossRef]
59. Hua, P.; Yang, W.; Qi, X.; Jiang, S.; Xie, J.; Gu, X.; Li, H.; Zhang, J.; Krebs, P. Evaluating the effect of urban flooding reduction strategies in response to design rainfall and low impact development. *J. Clean. Prod.* **2020**, *242*, 118515. [CrossRef]
60. Haghighatafshar, S.; Nordlöf, B.; Roldin, M.; Gustafsson, L.-G.; la Cour Jansen, J.; Jönsson, K. Efficiency of blue-green stormwater retrofits for flood mitigation—Conclusions drawn from a case study in Malmö, Sweden. *J. Environ. Manag.* **2018**, *207*, 60–69. [CrossRef] [PubMed]
61. Pereira Souza, F.; Leite Costa, M.E.; Koide, S. Hydrological modelling and evaluation of detention ponds to improve urban drainage system and water quality. *Water* **2019**, *11*, 1547. [CrossRef]
62. Baird, J.; Hunt, I.; Winston, R. Evaluating the Hydrologic and Water Quality Performance of Infiltrating Wet Retention Ponds. In Proceedings of the World Environmental and Water Resources Congress, Portland, OH, USA, 1–5 June 2014; pp. 145–154.
63. Chrétien, F.; Gagnon, P.; Thériault, G.; Guillou, M. Performance analysis of a wet-retention pond in a small agricultural catchment. *J. Environ. Eng.* **2016**, *142*, 04016005. [CrossRef]
64. Sun, R.; Chen, A.; Chen, L.; Lü, Y. Cooling effects of wetlands in an urban region: The case of Beijing. *Ecol. Indic.* **2012**, *20*, 57–64. [CrossRef]

65. Thomas, G.; Zachariah, E. Urban heat island in a tropical city interlaced by wetlands. *J. Environ. Sci. Eng.* **2011**, *5*, 234–240.

66. Rizzo, A.; Bresciani, R.; Masi, F.; Boano, F.; Revelli, R.; Ridolfi, L. Flood reduction as an ecosystem service of constructed wetlands for combined sewer overflow. *J. Hydrol.* **2018**, *560*, 150–159. [CrossRef]

67. Britter, R.; Hanna, S. Flow and dispersion in urban areas. *Annu. Rev. Fluid Mech.* **2003**, *35*, 469–496. [CrossRef]

68. *Uchwała Rady Miasta Stołecznego Warszawy NR XV/339/2019 z dnia 4 lipca 2019 r. w sprawie przyjęcia Strategii adaptacji do zmian klimatu dla m.st. Warszawy do roku 2030 z perspektywą do roku 2050. Miejski Plan Adaptacji", stanowiącej politykę m.st. Warszawy w zakresie podejmowania działań zapobiegających i łagodzących negatywne skutki zmian klimatu*; Rada m.st. Warszawy: Warszawa, Poland, 2019.

69. Błażejczyk, K.; Kuchcik, M.; Milewski, P.; Dudek, W.; Kręcisz, B.; Błażejczyk, A.; Szmyd, J.; Degórska, B.; Pałczyński, C. *Miejska Wyspa Ciepła w Warszawie: Uwarunkowania Klimatyczne i Urbanistyczne*; Wydawnictwo Akademickie Sedno: Warszawa, Poland, 2014; p. 171.

70. Szulczewska, B.; Giedych, R.; Maksymiuk, G. Can we face the challenge: How to implement a theoretical concept of green infrastructure into planning practice? Warsaw case study. *Landsc. Res.* **2017**, *42*, 176–194. [CrossRef]

71. Szulczewska, B.; Adamczyk-Jabłońska, J.; Cieszewska, A.; Giedych, R.; Janus, A.; Maksymiuk, G.; Pirowski, A.; Szumański, M.; Szumilas, H.; Wałdykowski, P.; et al. *Potencjał do Kształtowania Zielonej Infrastruktury w Warszawie—Raport*; Miasto Stołeczne Warszawa, Biuro Architektury i Planowania Przestrzennego: Warszawa, Poland, 2016; p. 144.

72. Langhans, S.D.; Reichert, P.; Schuwirth, N. The method matters: A guide for indicator aggregation in ecological assessments. *Ecol. Indic.* **2014**, *45*, 494–507. [CrossRef]

73. Fogel, P.; Szulczewska, B.; Kiczyńska, A.; Fic, M.; Lewiński, S. *Atlas ekofizjograficzny Miasta st. Warszawy. Miasto Stołeczne Warszawa*; Biuro Architektury i Planowania Przestrzennego: Warszawa, Poland, 2018.

74. Kuchcik, M. The attempt to validate the applicability of two climate models for the evaluation of heat wave related mortality in Warsaw in the 21st century. *Geogr. Pol.* **2013**, *86*, 295–311. [CrossRef]

75. *Uchwała Rady Miasta Stołecznego Warszawy NR LXVIII/1817/2013 z dnia 17 października 2013 r. w sprawie miejscowego planu zagospodarowania przestrzennego obszaru Sadyby Północnej—część pierwsza. Dz. Urz. Woj. Mazowieckiego z dnia 26.11.2013 poz. 12259*; Rada m.st. Warszawy: Warszawa, Poland, 2013.

76. *Uchwała Rady Miasta Stołecznego Warszawy NR XLII/1299/2008 z dnia 23 października 2008 r. w sprawie uchwalenia miejscowego planu zagospodarowania przestrzennego rejonu pod Skocznią—część I. Dz. Urz. Woj. Mazowieckiego nr 210 z dnia 5.12.2008 poz. 8300*; Rada m.st. Warszawy: Warszawa, Poland, 2008.

 *land*

Article

A Guide to Public Green Space Planning for Urban Ecosystem Services

Evan Elderbrock [1], Chris Enright [1], Kathryn A. Lynch [2] and Alexandra R. Rempel [2,*]

[1]   Department of Landscape Architecture, University of Oregon, Eugene, OR 97403, USA;
      eelderbr@uoregon.edu (E.E.); cenright@uoregon.edu (C.E.)
[2]   Environmental Studies Program, University of Oregon, Eugene, OR 97403, USA; klynch@uoregon.edu
*   Correspondence: arempel@uoregon.edu; Tel.: +1-541-510-7713

Received: 25 September 2020; Accepted: 12 October 2020; Published: 14 October 2020

Abstract: Street trees, native plantings, bioswales, and other forms of green infrastructure alleviate urban air and water pollution, diminish flooding vulnerability, support pollinators, and provide other benefits critical to human well-being. Urban planners increasingly value such urban ecosystem services (ES), and effective methods for deciding among alternative planting regimes using urban ES criteria are under active development. In this effort, integrating stakeholder values and concerns with quantitative urban ES assessments is a central challenge; although it is widely recommended, specific approaches have yet to be explored. Here, we develop, apply, and evaluate such a method in the Friendly Area Neighborhood of Eugene, Oregon by investigating the potential for increased urban ES through the conversion of public lawn to alternative planting regimes that align with expressed stakeholder priorities. We first estimated current urban ES from green space mapping and published supply rates, finding lawn cover and associated ES to be dominant. Resident and expert priorities were then revealed through surveys and Delphi analyses; top priorities included air quality, stormwater quality, native plantings, and pollinator habitat, while concerns focused on cost and safety. Unexpectedly, most residents expressed a willingness to support urban ES improvements financially. This evidence then informed the development of planting regime alternatives among which we compared achievable future urban ES delivery, revealing clear differences among those that maximized stakeholder priorities, those that maximized quantitative urban ES delivery, and their integration. The resulting contribution is a straightforward method for identifying planting regimes with a high likelihood of success in delivering desired urban ES in specific local contexts.

Keywords: green infrastructure; urban planning; LiDAR/NDVI; stakeholders; Delphi analysis

---

1. Introduction

Dense networks of streets, buildings, industry, and transportation interfere with numerous ecosystem processes, affecting the local hydrology, quantity and biodiversity of native flora and fauna, biogeochemical cycling, and microclimate stability [1]. Urban ecosystem services (ES), the benefits humans derive from ecological processes in urban and peri-urban areas [2], are therefore often compromised in population centers, resulting in diminished air, water, and soil quality as well as intensified vulnerability to flooding and heatwaves [1,3,4]. As urban populations grow, the importance of urban ES is increasing: over four billion people now live in cities, a 20-fold increase since 1900 [5,6], and by 2050, urban residents are predicted to number six billion [6].

To strengthen urban ES, green infrastructure, or planned networks of urban vegetated land cover ("urban green space"), including parks, right-of-way planting strips, private yards, green roofs, wetlands, and other natural areas, may be deployed [7]. Urban forests, for example, reduce concentrations of air pollutants, including ozone, carbon monoxide, sulfur dioxide, nitrogen oxides, and particulate matter [8,9]; store atmospheric carbon [9,10]; intercept rainfall, thereby reducing stormwater runoff [11];

provide shade and air temperature regulation [12]; increase recreation value [13]; supply diverse nesting and foraging opportunities for birds; diminish soil erosion; and contribute to stormwater purification [14,15]. Recent biophysical, empirical, and GIS-based modeling methods now allow certain urban ES delivery rates to be quantified [16,17], and economic models allow their monetary value to be evaluated (e.g., [18]), facilitating estimation and comparison of urban green space contributions to air quality [9,19], stormwater runoff retention [20], air temperature regulation [21,22], and carbon sequestration [9,23,24]. For example, urban forests removed an estimated 27,000 metric tons of $PM_{2.5}$, 523,000 metric tons of ozone, 68,000 metric tons of nitrogen dioxide, and 33,000 metric tons of sulfur dioxide from the U.S. urban air in 2010, providing an estimated \$4.7 billion in annual health benefits [8]. Such urban ES quantification and valuation are then directly useful in deciding among urban land-use alternatives [14,16].

Currently, the lawn is the dominant green land cover type throughout urban and suburban areas of Europe, Canada, and the USA [25]; in 2005, lawn accounted for nearly half of all urban land cover in the USA [26], an area comparable to half of the total irrigated cropland in the USA [27,28]. Although lawns are relatively easy and inexpensive to maintain, enjoy widespread acceptance, and provide some urban ES, under typical management they consume extensive irrigation water [26] and are treated with fertilizers, pesticides, and herbicides that are harmful to fish, birds, and insects [29]. Additionally, lawns store limited carbon [30], and their mowing leads to both biogenic and fuel-related greenhouse gas emissions [31]. They also contribute less to stormwater retention, air purification, microclimate regulation, and recreation than other vegetative land-cover types [14,25,32,33].

In light of this evidence, urban land use planners face crucial decisions regarding the continuation of public lawn maintenance, complicated by pressures of cost, restrictive land-use codes, and uncertain public support, as well as limited land area with which to provide urban ES [34]. In these decisions, the perspectives of stakeholders such as policymakers, environmental managers, and affected residents are critical [2,16,35–37], revealing ES priorities, design preferences, and barriers to green infrastructure development [34,38–43]. The value of stakeholder input to ES planning was first emphasized by the Millennium Ecosystem Assessment in 2005, and the integration of urban ES quantification with stakeholder-expressed urban ES priorities emerged as a central urban environmental planning prescription [2,37].

The essential nature of stakeholder input in ensuring long-term green infrastructure success, combined with the characteristic urban ES provided by specific land cover types (e.g., woodland, trees, shrubs, native grasses, stormwater filtration facilities, etc.), require effective decision-making processes to integrate several lines of evidence. Specifically, quantitative urban ES delivery potential must be evaluated in the context of a possibly conflicting set of stakeholder perspectives [2,4,16,34,37,44], involving an approach that is widely advocated but has not, to our knowledge, been further investigated. To address this need, here we develop and evaluate such a method. We begin by establishing the urban ES currently provided in the study area and surveying diverse stakeholders to reveal their ES priorities. These data next inform the selection of alternative planting regimes that address individual stakeholder priorities and quantitative urban ES delivery, respectively. Comparison of priority ES delivery among these alternatives then guides their integration, yielding a composite regime that improves upon each initial alternative's likelihood of local acceptance while increasing delivery of the desired ES. Notably, this integrated regime could not have been clearly identified by either stakeholder priorities or quantitative urban ES assessments alone.

2. Methods

2.1. Study Area

The City of Eugene (Figure 1; population 156,000; median income \$44,859; area 113 km^2) sits within the southern Willamette Valley in western Oregon [45], an area with a Mediterranean climate (Köppen *Csb*) of long cool rainy winters and warm dry summers. The valley surroundings promote winter temperature inversions and summer wildfire smoke collection, causing Eugene to rank among the twenty worst cities in the USA for short-term small particulate ($PM_{2.5}$) air pollution [46]. The City of Eugene is also currently required by its National Pollution Discharge Elimination System (NPDES) permit to reduce the waterborne discharge of pollutants from the municipal system to the maximum extent possible [47]. The City of Eugene Park system possesses nearly 2000 ha of natural areas and open space, but their aggregation on the outskirts of town [48] limits their contributions to ES in urban neighborhoods.

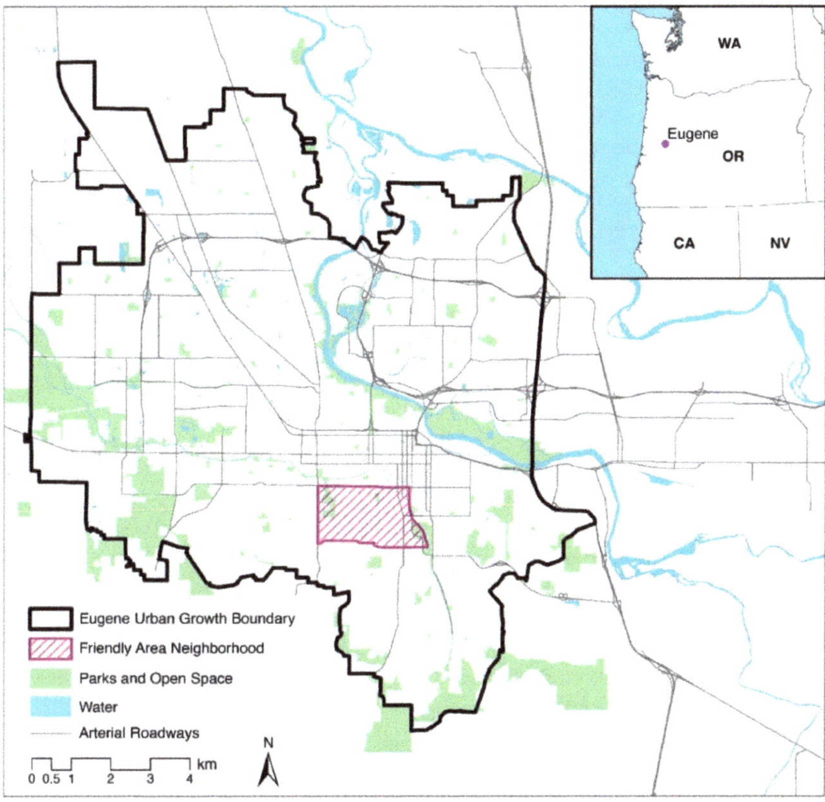

Figure 1. Location of the Friendly Area Neighborhood in Eugene, Oregon.

Within the city, the Friendly Area Neighborhood (FAN; Figure 1; population 7000; area 3.7 km^2) is zoned primarily (~75%) for low-density residential development (8–10 dwelling units/ha) and consists largely of single-detached units, with a median tax lot parcel area slightly below the USA median (0.073 ha vs. 0.083 ha) [45]. Nearly all streets in the neighborhood contain vegetated planting strips within city right-of-way easements, while sidewalks are intermittent. The FAN median annual household income ($46,300) is $7000 below state and $11,300 below national medians [49,50], but its access to public green space is above average, with >10% of its land devoted to public parks and schoolyards and ~95% of residents living within a five-min walking distance along roads (i.e., <400 m) of a public park or schoolyard (Figure S1); the neighborhood is therefore comparable to top cities in the United States for such access [51].

2.2. Public Green Space Mapping and Urban ES Quantification

Although privately owned land is important in providing urban ES [52,53], this study focuses on public green space in which urban ES delivery is managed by the City. To characterize this space, we used multiple mapping strategies to inventory five distinct vegetated land cover types in the neighborhood (Table S1). Each street was traversed on foot in 2017 to identify lawn within the public right-of-way, and lawn without tree canopy cover was geospatially located and measured using a Garmin GPSMAP 62S handheld Global Positioning System (Garmin Ltd., Olathe, KS, USA). In tax lot parcels without adjacent sidewalks, right-of-way boundaries were assumed to extend 3 m on either side of the roadway. The boundaries of woodlands, classified as clustered trees clearly distinguishable from the U.S. Department of Agriculture's 2016 National Agriculture Imagery Program (NAIP) imagery, were assessed visually in ESRI ArcMap 10.7 (ESRI, Redlands, CA, USA) [54] and confirmed in the field. All other vegetation classifications (i.e., trees, tall shrubs, and short shrubs, as well as lawns located in parks) were made using normalized difference vegetation indices (NDVIs) and height; the NDVI was calculated on a continuum from -1 to $+1$ using the NAIP four-band imagery with 1 m resolution. The NDVI range for each vegetation class was determined by comparing NDVI and color composite images [55]. The minimum NDVI threshold for all vegetation classes was set at 0.25, with the exception of lawn, which was identified using a minimum NDVI threshold of 0.0 (Table S1). Vegetation height was derived from 2015 light detection and ranging (LiDAR) point-cloud data [56] that were used to generate digital elevation and digital surface models. Digital elevation model values were subtracted from the digital surface model to create a digital height model at 1 m resolution, and vegetation was classified by combining NDVI thresholds with height ranges determined by Derkzen et al. (Table S1) [14].

The accuracy of each NDVI/LiDAR-derived land cover classification was evaluated through a process in which four hundred points, or 100 for each of the four vegetation types classified using NDVIs and LiDAR, were randomly selected and validated visually with NAIP imagery. Air photo interpretation was used to determine land cover type for all points clearly and obviously identifiable from the air photo. Land cover types for all remaining unidentified points were confirmed in the field (Table S2). NDVI/LiDAR-derived public green space land cover quantities were adjusted using validation proportions from Table S2 (see Table S3 footnotes), and five urban ES were quantified from these adjusted spatial data using indicators and supply rates compiled by Derkzen et al. for each of the five green cover types—vegetative ground cover (i.e., lawn), short shrub, tall shrub, tree, and woodland (Table S1) [14].

2.3. Urban ES Supply Rates

For the existing land cover, supply rates of five urban ES (air purification, carbon storage, runoff retention, cooling fraction, and outdoor recreation) provided by the five green cover types described above were estimated according to Derkzen et al. [14], in which urban ES supply rates from numerous studies were integrated for the analogous Mediterranean (*Csb*) climate of Rotterdam, NL (Table S1). Although numerous modeling techniques exist for urban ES assessment [16], we chose this straightforward approach, consistent with recent recommendations and used by other case studies [57], as one that would be accessible to a wide range of urban planning practices.

In exploring potential future alternative planting regimes (Section 4.2.), we included stormwater filtration facilities (e.g., stormwater planters and rain gardens) that are not currently present in the neighborhood, estimating their stormwater reduction potential using the Simplified Approach described in Eugene's Stormwater Management Manual [58]. Impervious surface area, a necessary input, was calculated for the neighborhood using image segmentation and supervised learning in ESRI ArcGIS Pro 2.6 (ESRI, Redlands, CA, USA) [59] based on infrared, red, and blue bands from 2016 NAIP four-band imagery. To assess the accuracy of impervious and pervious surface classification, 100 random points were selected for each land cover type, and every point was validated visually with the NAIP imagery. The overall accuracy of the supervised segmentation classification was 94.5% (Table S9).

Urban green space also has the potential to provide ecosystem disservices, including pollen production that exacerbates allergies; a volatile organic compound release that contributes to ground-level ozone formation in the presence of automobile exhaust; and growth of tree limbs that may interfere with electricity lines or fall during storms, causing property damage [60,61]. These may also be estimated quantitatively in some cases (e.g., [62]), but we have not included these considerations here.

2.4. Resident Surveys

Non-stratified random sample surveys were administered to residents of the FAN to determine their urban ES priorities for public green space and the potential for increased funding for green infrastructure development. A random sample of 500 residential tax lot parcels was selected using county tax lot parcel data for the FAN as a sampling frame. Each selected lot was visited once on a weekday between 5 and 7 PM, and 19.4% of these visits yielded a completed survey (*n* = 97). The majority of the recorded non-responses resulted from resident absences, suggesting that repeated visits could have increased the response rate, and homes with posted "Do Not Disturb" or "No Soliciting" signs were also recorded as non-responses. Among residents who answered their doors, over half agreed to participate. Surveys were conducted orally in a format approved by the University of Oregon's Institutional Review Board. To minimize the survey's perceived invasiveness, sociodemographic information was not collected, although it could have been informative.

Residents were asked to rate 17 randomly ordered urban ES according to their importance for public green space in their neighborhood using a five-point Likert scale from 1 ("very unimportant") to 5 ("very important") (detailed in Supplementary Materials Section S2). They were then asked whether they supported the management of public green space to increase urban ES delivery and whether they would be willing to support such efforts financially, through personal donations or taxes, and through direct contribution of volunteer time.

Resident priorities for public green space urban ES were evaluated using Pearson's chi-square tests for both pairwise and aggregate comparisons, and chi-square tests were further used to compare priorities among urban ES classification types (i.e., provisioning, regulating, cultural, and supporting). Results for each urban ES classification type were tested for internal consistency using Cronbach's alpha (α), and values above 0.7 were regarded as acceptable [63]. "Priority" urban ES were defined as those with Likert responses of 4 ("moderately important") or 5 ("very important"), and Likert responses were reclassified as either priority (values 4 and 5) or non-priority (values 1–3) for data analysis. Descriptive statistics were used to compare residents' willingness to support green infrastructure development. All statistical analyses were conducted in R [64].

2.5. Delphi Method

We used an iterative survey process, known as a Delphi analysis, to consult with a group of individuals with specific knowledge of the planning and management of public green space in Eugene [65,66]. Of the 34 people invited to participate on the basis of their expertise in public policy and green space management, 15 agreed, including nine members of the Eugene Public Works Department (including Parks and Open Space, Stormwater Management, and Urban Forestry), two City Planning and Development members, two local environmental non-profit representatives, one City Council member, and one University of Oregon Landscape Architecture faculty member.

In the first survey, participants ranked the 17 urban ES used in the resident survey in order of importance for public green space management in Eugene; those urban ES with mean and median rankings below the top 10 were eliminated from the second survey. In addition, seven open-ended questions asked participants to describe and explain their perspectives on urban ES opportunities and barriers further. In the second round, participants were asked to review the collective results and representative responses from the first round before again selecting the urban ES they considered to be priorities and expressing their levels of agreement with responses to the open-ended questions of the first round. To reflect differences in management, safety, and ecological benefit potential, these questions distinguished between parks and right-of-way planting strips (Supplementary Materials Section S3).

No particular proportion of agreement defines "consensus" in the Delphi method, and documented thresholds have varied from a simple majority to 95% [67–70]. Here, we chose a consensus threshold of two-thirds (67%), consistent with practices in many city governments [71,72].

3. Results

3.1. Public Green Space Inventory

To understand current urban ES delivery rates, we first inventoried public green space in the FAN through a combination of ground survey and NDVI/ LiDAR green cover assessment methods. These revealed that lawn was the dominant green cover type (Figures 2 and 3; Tables S3 and S4), typical of low-density residential development [29]. Of the 57.4 ha of public green space, approximately 55% was covered by lawn without tree canopy, 30% by trees with unidentified understory, 9% by woodland, 4% by tall shrubs, and 3% by short shrubs (Figure 3; Table S4). All woodland and most lawn (>85%) were located in municipal parks and public schoolyards, while ~80% of all non-woodland tree canopy, short shrubs, and tall shrubs were located in right-of-way zones (Table S3).

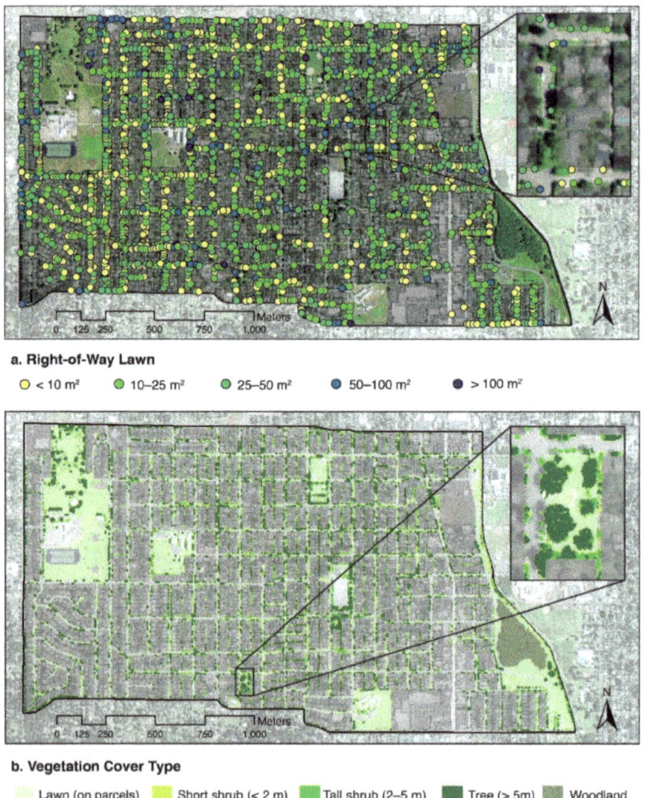

Figure 2. Land cover type and distribution. (**a**) Lawn size and distribution in right-of-way zones in the Friendly Area Neighborhood, evaluated by ground measurement, and (**b**) vegetation classes and distributions on all public lands in the neighborhood identified using NDVI and LiDAR data.

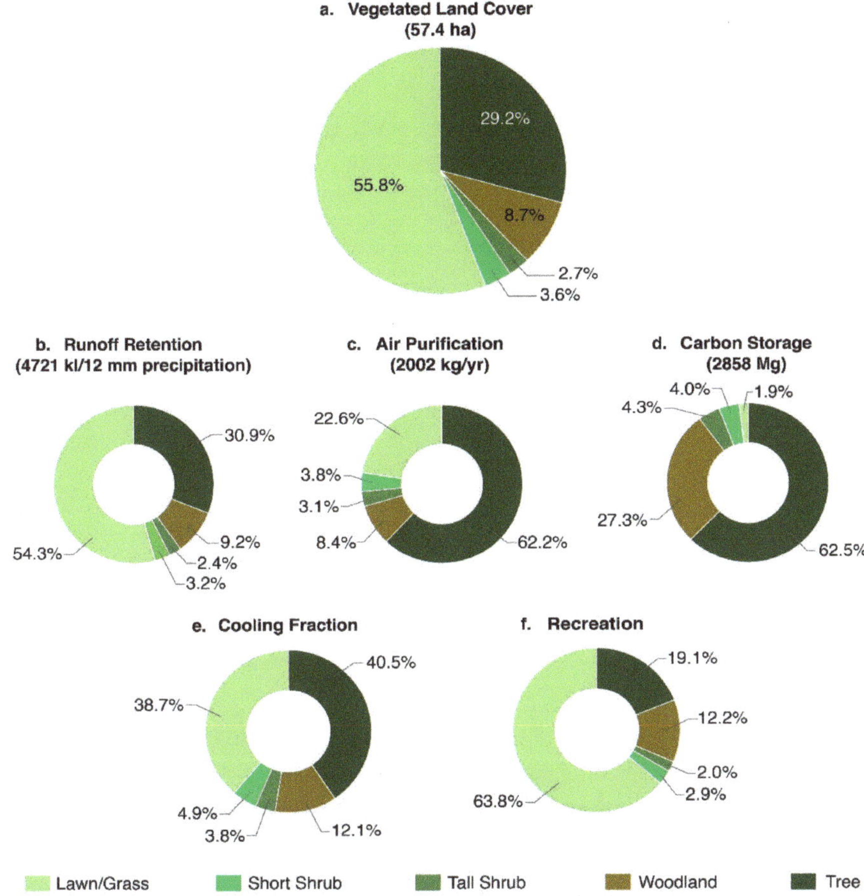

Figure 3. Urban ecosystem services (ES) provided by existing vegetated land cover. Existing vegetated land cover distribution (**a**), detailed in Figure 2, and the corresponding provision of urban ES by vegetated land cover type: (**b**) runoff retention; (**c**) air purification; (**d**) carbon storage; (**e**) cooling fraction; and (**f**) recreation, as evaluated by supply rates compiled by Derkzen et al. [14], summarized in Table S4.

The approach combining LiDAR and NDVI was most accurate in identifying tree cover and lawn (98% and 86% accuracy, respectively), while 66% of the area classified as "tall shrub" was found to be tree canopy cover, and 33% of the area classified as "short shrub" was found to be tall shrubs, lawn, or tree canopy cover (Table S2).

Using published supply rates [14] (Table S1), we next estimated that this public green space provides nearly 2900 metric tons of carbon storage, removes over 2000 kg per year of atmospheric particulate matter (PM_{10}), and retains over 4.7 million liters of stormwater during each 12 mm storm event (Figure 3; Table S4). Lawn covers over 50% of the total public green space and provides more than half of the runoff retention and recreation value but less than one-quarter of the air purification services and 2% of the carbon storage (Figure 3; Table S4). By comparison, trees cover less than 30% of the total public green space but supply over half of all air purification and carbon storage, as well as over 40% of cooling services; trees provide runoff retention roughly proportional to their coverage area but only one-fifth of all recreation services. Woodland covers less than one-tenth of the total public green space yet provides over one-tenth of the recreation value and over one-quarter of the carbon storage, or more than 14× that provided by lawn. Tall and short shrubs, in comparison, cover the least area of the total public green space but provide urban ES approximately proportional to their coverage area.

3.2. Resident Surveys—Urban ES Priorities

To understand residents' urban ES priorities for public green space, we asked a random sample ($n = 97$) to rate 17 individual urban ES on a scale from 1 ("very unimportant") to 5 ("very important"). Responses showed that outdoor recreation, stormwater quality, air quality, pollinator habitat, and native species were the top priorities (Figure 4; Table S5), showing a clear preference for supporting services; except for outdoor recreation, cultural and provisioning services were rated as relatively unimportant (Table S5).

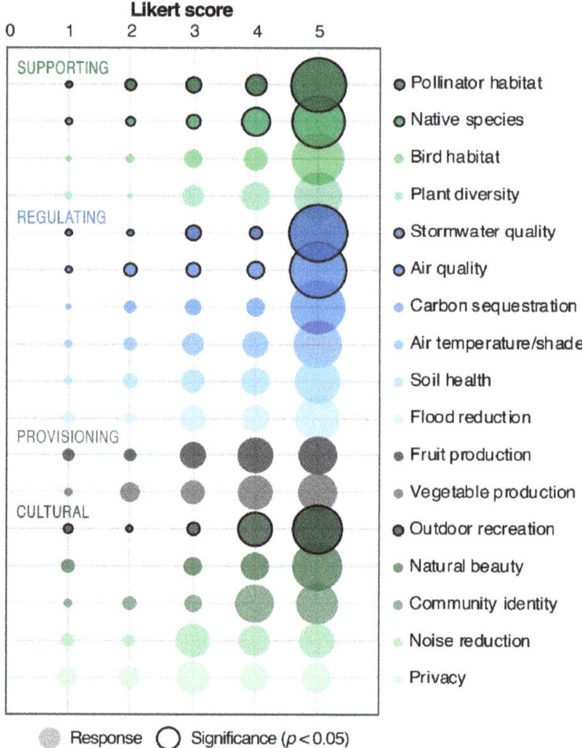

Figure 4. Ratings by Friendly Area Neighborhood residents ($n = 97$), from 1 (very unimportant) to 5 (very important), of 17 urban ecosystem services (ES). Colors designate urban ES categories (green: supporting; blue: regulating; brown: provisioning; olive: cultural); bubble size designates frequency of the indicated response; outer black line indicates significance ($p < 0.05$) according to chi-square tests in which responses of 1–3 and 4–5 were binned to compare each individual urban ES to overall urban ES. Data, including Cronbach's alpha values for each urban ES domain, are tabulated in Table S5.

This survey also investigated residents' willingness to support public green infrastructure development for urban ES improvement through contributions of time and/or money. Unexpectedly, most respondents (>85%) expressed willingness to contribute financially to urban ES projects in parks, with over one-quarter supporting direct, "out-of-pocket" payments and over 80% supporting tax measures to fund public works projects (Figure 5; Table S6). Support for such projects on right-of-way strips was lower but still substantial, with over 70% stating willingness to contribute financially; again, over one-quarter supported direct payments, but in this case, only 65% supported corresponding tax measures. Additionally, a large majority (>80%) expressed the willingness to volunteer for green infrastructure projects in the neighborhood, and over half stated interest in contributing five or more hours per year (Figure 5; Table S7).

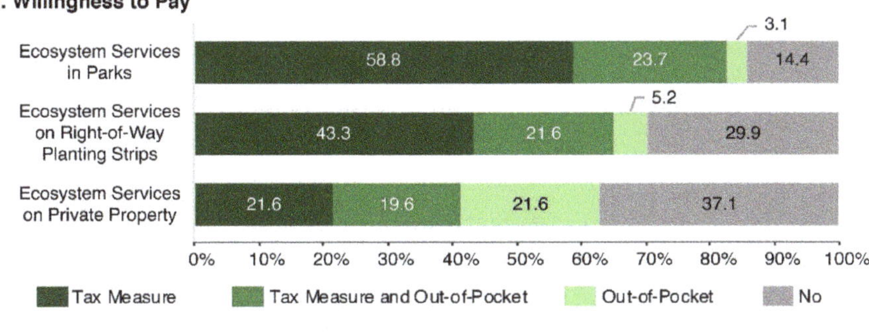

Figure 5. (**a**) Residents' stated willingness by in-person survey (*n* = 97) to financially support green infrastructure development in parks, in the public right-of-way, and on private property that increases urban ecosystem services (ES) through tax measures alone; tax measures combined with personal contributions; personal contributions alone; or none of the above. (**b**) Residents' stated willingness by in-person survey (*n* = 97) to volunteer time toward the development of public urban ES projects from 0 to 12+ h per year. Data are provided in Tables S6 and S7.

3.3. Delphi Analysis

To understand the perspectives of stakeholders involved in the planning, implementation, and management of public green space, with the potential to differ substantially from those of residents, we used a Delphi analysis to seek consensus (greater than two-thirds agreement) regarding urban ES priorities, perceived benefits of and concerns regarding lawn cover, benefits of and barriers to green infrastructure development, and strategies for overcoming these barriers. In the first-round survey, six urban ES—noise reduction, community identity, vegetable production, fruit production, improved soil health, and privacy—received sufficiently low rankings that they were excluded from the second round (Supplementary Materials Section S3). In the second-round survey, participants viewed the reduction of stormwater pollution as the top priority for both parks and for right-of-way planting strips, with over 80% agreement (Table 1; Figure 6; Figure S2). Improving air quality, supporting native species, increasing carbon sequestration, providing natural beauty, and reducing flooding were also consensus priorities for both parks and right-of-way planting strips. Providing shade for cooling was a strong priority for right-of-way strips but did not reach the consensus threshold in parks; instead, parks were most valued for providing habitat and educational opportunities. Outdoor recreation, plant diversity, erosion control, and physical and mental health benefits did not reach the two-thirds consensus threshold and were classified as non-priorities.

Table 1. Urban ecosystem services that generated consensus [a] among Delphi participants (*n* = 15).

Priority Ecosystem Services	First Survey		Second Survey [c]	
	Ranking Mean [b]	Ranking Median [b]	Priority for Right-of-Way (%)	Priority for Parks (%)
Stormwater Purification	3.2	2.0	86.7	80.0
Carbon Sequestration	6.1	4.0	66.7	80.0
Air Purification	3.9	3.0	73.3	66.7
Native Species	6.6	7.5	73.3	73.3
Aesthetic/Natural Beauty	7.7	8.0	73.3	66.7
Flood Reduction	7.9	6.0	66.7	66.7
Air Temperature Regulation	5.4	4.5	73.3	_ [d]
Habitat for Birds/Pollinators	8.0	7.0	_ [d]	73.3
Educational Opportunities	N/A	N/A	_ [d]	66.7

[a] Consensus was defined as a ≥66.7% agreement. [b] 1 = highest; 17 = lowest. [c] Data are shown graphically in Figure S2. [d] Consensus was not reached.

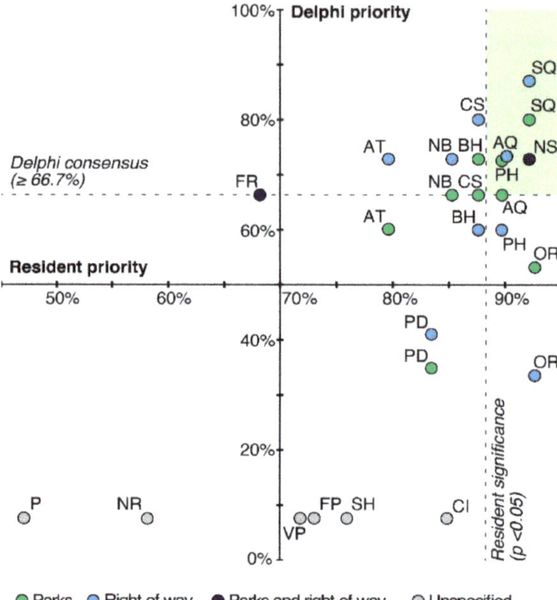

Figure 6. Comparison of Delphi stakeholder responses in favor of each urban ecosystem service (ES; vertical axis), detailed in Figure S2, with resident priorities (horizontal axis), detailed in Figure 4 and Table S5. Delphi stakeholder priority was defined as two-thirds or greater consensus approval; residential priority was established by significance of Fisher's exact test at the $p < 0.05$ level (*n* = 97); green shaded region represents urban ES prioritized by both stakeholder groups. AQ = air quality; AT = air temperature; BH = bird habitat; CI = community identity; CS = carbon sequestration; FP = fruit production; FR = flood reduction; NB = natural beauty; NS = native species; OR = outdoor recreation; P = privacy; PD = plant diversity; PH = pollinator habitat; SH = soil health; SQ = stormwater quality; VP = vegetable production.

Participants viewed the primary benefits of public lawn in parks as providing recreational and gathering space (93% and 73% agreement, respectively) and ease of maintenance (67% agreement) (Table S8); on right-of-way strips, safety and sightlines were the only benefits that reached a consensus, with over 85% agreement. The principal concerns, in turn, both in parks and on right-of-way strips, were lawn's limited ability to provide regulating services (i.e., air and water filtration, carbon sequestration, and flood reduction), irrigation requirements, and lack of biodiversity (Table S8). Additionally, two-thirds agreed that fertilizer, pesticide, and herbicide impacts were a concern for right-of-way planting strips (Table S8).

Accordingly, participants agreed that replacing lawn with alternative planting regimes could increase biodiversity and improve the habitat in parks while reducing stormwater runoff and improving aesthetics along right-of-way planting strips (Table S8). The possibility of impaired sightlines remained a safety concern, however, and emerged as the only consensus barrier to green infrastructure development on right-of-way planting strips (Table S8). While over half agreed that converting lawn to alternative planting regimes would increase maintenance time, complexity, and cost during the transition period, they did not reach a consensus regarding the importance of these barriers. Still, to address them, the consensus recommendation was to install attractive, easily maintained plantings and to implement educational and outreach efforts to promote support. Overall, a substantial majority (>85%) of participants supported the conversion of at least some lawn to alternative planting regimes both on right-of-way planting strips and in parks.

4. Integration of Stakeholder Priorities with Quantitative Urban ES Estimates

Although the dual values of stakeholder priorities and quantitative understanding of urban ES potential in municipal decision-making have been widely discussed [2,4,16,34,37], methods to accomplish their integration have not previously been explored. To undertake this integration, we considered a series of questions planners might ask in making urban ES-motivated vegetated land cover decisions; developed a set of alternative planting regimes that responds to these questions in the context of the FAN; and evaluated them according to the local evidence collected, yielding a single integrated result.

4.1. Planning Considerations

4.1.1. What Urban ES are Available from the Landscape?

Comprehensive ES assessments and contemporary literature addressing the location of interest are expected to reveal relevant urban ES for most locations; here, such resources (e.g., [3,37–41,73]) were used to identify the 17 urban ES considered in our survey (Figure 4). Since urban ES vary with climate and biome, however, analogous resources might emphasize very different services for other locations, potentially including insect or disease control, provision of raw materials, production of fresh drinking water, etc. [4,37].

4.1.2. What Land Cover Types Thrive in This Location?

Climate, soil, and existing land uses are expected to limit the land cover types eligible for consideration. Here, Eugene's climate and the neighborhood's existing land use and cover types (Figure 2) focused our exploration on combinations of woodlands, dispersed trees, tall shrubs, short shrubs, and grasses, including lawn.

4.1.3. What are the Urban ES Priorities of Multiple Stakeholder Groups?

Stakeholder perspectives can be revealed through interviews; in-person, mail, or online surveys; focus group discussions; and/or Delphi analyses, each with their own benefits and limitations (e.g., [74–76]). Here, we chose in-person surveys to reveal resident perspectives and to ensure a sufficiently large, random distribution of responses, despite the time-intensive nature of this approach, and we chose Delphi analyses to bring coherence to the input of diverse green space managers.

4.1.4. Which Urban ES can be Quantified According to Land Cover Type?

Quantitative evidence documenting the ES provided by different land cover types is growing rapidly (e.g., [14,16,77–83]), and where it exists, it can be used to inform decisions among alternatives.

Additionally, urban ES delivery without published land cover supply rates may be evaluated qualitatively with guidance from locally or regionally available information (e.g., [84,85]), while others (e.g., natural beauty, pollinator habitat, and native plant species) may still be factored into design decisions, particularly through species choice. Here, the priorities of stormwater quality and air quality were among those with supply rates published by land cover type (e.g., [14,58]), allowing their urban ES to be quantified. Pollinator and native species habitat urban ES had not been similarly quantified, but local guidance existed in the form of a City resolution [86] and in regional lists of recommended native tree, shrub, vine, grass, and forb species (e.g., [87,88]). Using resources such as these, new plantings designed to meet quantifiable urban ES priorities may generally be chosen to meet non-quantifiable priorities as well.

4.1.5. What Barriers or Constraints Exist?

Finally, various barriers are expected to limit the resulting green infrastructure development options, particularly including lack of funds for establishment, expansion, or maintenance of green infrastructure; insufficient social support resulting from conflicting stakeholder desires; and safety or accessibility concerns (e.g., [89]). Here, Delphi participants expressed concerns consistent with those found elsewhere, focusing on cost and safety (Table 1).

4.2. Alternative Planting Regimes

The considerations above guided the following investigation of alternative planting regimes with which to provide urban ES through the conversion of public lawn, illustrating the way in which integration of quantitative urban ES supply rates with stakeholder priorities leads to a different result than that obtained by reliance on any one line of evidence alone. The status quo, to which the others were compared, represents the result of current decision-making processes that have yielded lawn-dominated public spaces, with substantial outdoor playing field area as well as several hectares of dispersed trees and one prominent woodland park. The *"Forest and Stream"* alternative planting regime maximizes the provision of quantifiable, locally relevant urban ES in the study area, named to reflect the resulting emphasis on woodlands and stormwater filtration facilities. The *"Birdland"* regime, in contrast, represents Delphi priority urban ES, showing the value placed on bird habitat and air quality; *"Flower Sports"* represents resident priority urban ES, distinguished by an emphasis on pollinator habitat and outdoor recreation; and *"Integration"* capitalizes upon the multiple urban ES provided by individual land cover types to address both Delphi and resident priority urban ES with minimal compromise to either one. Urban ES supply rates expected of each alternative planting regime were estimated as described in Methods, with the inclusion of an additional "recreational lawn" metric reflecting the local importance of soccer and other playing fields [90].

The first alternative planting regime, *Forest and Stream*, maximizes the quantifiable, locally relevant urban ES of air quality, carbon storage, cooling, and runoff retention and purification, independent of stakeholder priorities. All park and schoolyard lawns are therefore converted to woodlands except for the 0.5 ha devoted to rain gardens, and nearly 4.6 ha of stormwater planters, as well as an additional 0.3 ha of trees, are added to right-of-way planting strips, sufficient to intercept stormwater runoff pollution from all public and private impervious surfaces in the neighborhood (Figure 7, Tables S10 and S11). Estimated from published supply rates [14,58], this regime would increase air purification by nearly 40%, carbon sequestration by over 150%, and runoff retention by 3.5%, as well as reduce runoff pollutant loading by 80% (Table 2). At the same time, Delphi responses suggest that the conversion of such a large area would encounter cost barriers as well as safety concerns associated with dense vegetation.

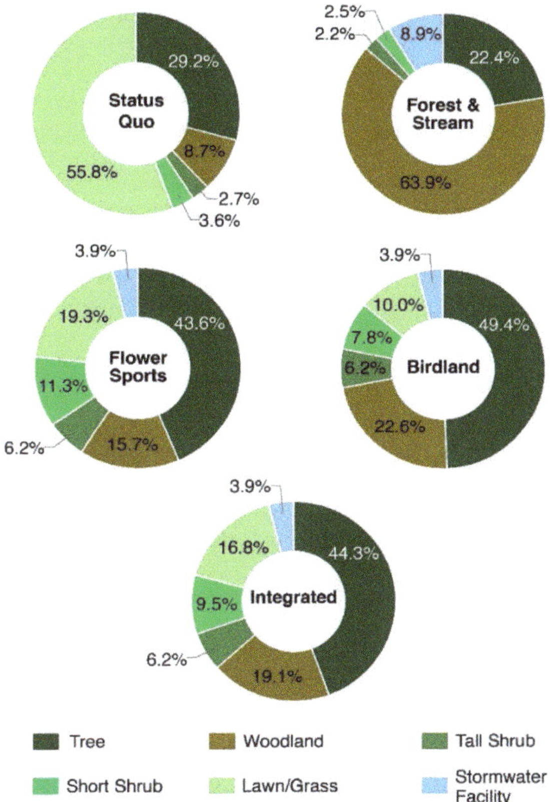

Figure 7. Land cover distributions for alternative planting regimes. Proportions of public green space (57.4 ha total) devoted to dispersed trees, woodland, tall shrubs, short shrubs, lawn or grass, and stormwater facilities, respectively. *Status Quo* describes the existing condition in the neighborhood (Section 3.1.); *Forest and Stream* maximizes locally-relevant, quantifiable urban ecosystem services (ES); *Birdland* maximizes delivery of Delphi respondents' priority urban ES; *Flower Sports* maximizes delivery of residents' priority urban ES; and *Integration* maximizes delivery of the urban ES prioritized by both Delphi respondents and residents.

Table 2. Urban ecosystem service delivery associated with alternative planting regimes.

Alternative Planting Regime	Area Converted ha	Air Purification Tonnes yr^{-1} (% change)	Carbon Storage Tonnes (% change)	Runoff Retention [b] kL/12 mm storm event (% change)	Stormwater Pollutant Filtration [c] (%)
Status Quo [a]	0	2.0 (0%)	2860 (0%)	4720 (0%)	(0%)
Forest and Stream	32.0	2.8 (+38.5%)	7570 (+164.9%)	4960 (+3.5%)	(80%)
Birdland	26.3	2.9 (+44.2%)	5710 (+99.9%)	4810 (+1.9%)	(33.2%)
Flower Sports	21.0	2.7 (+33.4%)	4860 (+70.0%)	4750 (+0.5%)	(33.2%)
Integration	22.4	2.7 (+36.7%)	5160 (+80.5%)	4770 (+1.0%)	(33.2%)

[a] Supply rates were calculated according to Derkzen et al. [14] unless otherwise specified. [b] Retention by woodlands, trees, tall shrubs, short shrubs, and lawn only. [c] Filtration by stormwater facilities, calculated using the Simplified Approach as described in the City of Eugene Stormwater Manual [58]; accounts for stormwater pollutants from impervious surfaces removed by stormwater planters and rain gardens on both publicly—and privately—owned land (see Table S10).

The second planting regime, *Birdland*, maximizes the response to the Delphi priorities of carbon storage, bird habitat in parks, air temperature regulation (i.e., cooling), and natural beauty, as well as the priorities held in common with residents (i.e., air quality, stormwater quality, and native species throughout the neighborhood, as well as pollinator habitat in parks). Clear sightlines for safety and moderate cost were prominent Delphi concerns, expressed in part as a desire to retain some existing lawn, and *Birdland*, therefore, converts only about one-quarter as much existing park lawn to woodland as *Forest and Stream*, or ~8 ha, envisioned as patches of native oak woodland and restoring native willow and ash woodland for bird habitat in the area designated as Westmoreland wetlands [91]. To address air quality and cooling priorities while maintaining ground-level openness, *Birdland* adds ~8 ha of dispersed trees to parks and schoolyards, capitalizing on the superior air pollutant removal rates of trees near roadways [14]. Like *Forest and Stream*, this regime adds 0.5 ha of rain gardens and ~5 ha of short and tall shrubs to parks, again removing all recreational lawn (i.e., softball fields) but leaving ~6 ha of other lawn intact, responding to Delphi safety concerns. On right-of-way planting strips, *Birdland* reduces the ~5 ha of stormwater planters proposed by *Forest and Stream* to ~2 ha, sufficient to manage the publicly-owned impervious area in the neighborhood (Table S10) and responding to Delphi participants' cost concerns. The remaining ~3 ha of right-of-way lawn are then replaced with dispersed trees for air quality (Table 2). This conversion, involving ~5 fewer ha than *Forest and Stream* (Table S11), is estimated to increase existing air purification by over 40%, carbon storage by ~100%, and runoff retention by ~2%, as well as to provide pollutant filtration for about one-third of the neighborhood's total stormwater runoff (Table 2).

The substantial conversion of playing-field lawn found in *Forest and Stream* and *Birdland* is reversed in the third planting regime, *Flower Sports*, which maximizes responses to resident priorities of outdoor recreation and pollinator habitat throughout the neighborhood, while accommodating the priorities of air and water quality held in common with Delphi respondents. A recent survey of Eugene residents showed that outdoor playing fields (i.e., recreational lawn areas) were in especially short supply compared to resident desires, providing specific, local evidence that superseded the outdoor recreation supply rates compiled by Derkzen et al. [14]. In parks and schoolyards, *Flower Sports*, therefore, converts only half as much lawn to native oak and ash woodland around the Westmoreland wetlands (4 ha) and ~15% less lawn (~7 ha) to dispersed trees as *Birdland*, while preserving the full 4 ha of existing sports fields (Table S11). Like *Birdland*, this regime adds 0.5 ha of rain gardens and ~5 ha of tall and short shrubs to parks, as well as ~2 ha of stormwater planters to the right-of-way, for stormwater purification; in contrast, however, it adds 2 ha of flowering shrubs to right-of-way plantings for additional pollinator habitat in place of dispersed trees. This regime converts ~5 fewer ha of lawn than *Birdland* but still increases air purification over the existing condition by about one-third and carbon storage by 70% while adding the ability to remove about one-third of the neighborhood's stormwater runoff pollution.

The fourth planting regime, *Integration*, prioritizes the urban ES held in common by both resident and Delphi stakeholders (i.e., stormwater quality, air quality, park pollinator habitat, and native species), using quantitative supply rates to indicate the most effective land cover types for each priority and allowing other priorities to be addressed through species selection. *Integration*, therefore, converts an area of existing park lawn to woodland between those of *Birdland* and *Flower Sports* (6 ha), representing a significant compromise that diminishes the outdoor playing field area by one-quarter in the interest of greater air quality, cooling, bird habitat, carbon storage, and native species urban ES. *Integration* also includes less dispersed tree area in parks (~6 ha) than either stakeholder-driven scheme, accommodating both the outdoor playing field area prioritized by residents and woodland urban ES prioritized by Delphi participants. Like *Birdland* and *Flower Sports*, this scheme converts 0.5 ha of park lawn to rain gardens and ~5 ha to flowering shrubs. To compensate for tree loss in parks, *Integration* increases tree cover and diminishes flowering shrubs relative to *Flower Sports* on right-of-way strips; stormwater planters are maintained at the level of both stakeholder-driven schemes. Compared to *Forest and Stream*, which maximizes quantifiable urban ES, *Integration* converts ~30% less land area but provides 95% of its air quality improvement and over one-third of its stormwater pollutant filtration, while retaining over 3 ha of outdoor playing field area (Table 2). *Integration* also provides clear but unquantified increases in pollinator habitat and native species diversity through the inclusion of flowering shrubs and woodland, and it addresses concerns of cost and safety raised in the Delphi analysis by converting less total lawn and maintaining greater openness at ground level than *Forest and Stream* or even *Birdland* (Table S11).

5. Discussion

The investigation above addresses an emerging issue in urban ES development and planning: the integration of stakeholder perspectives with quantitative estimates of urban ES provision to inform decisions regarding future land cover [4,16,35,44]. The new method that results is then applied to an urban neighborhood through the evaluation of existing public green space and current urban ES supply rates, the collection of resident and city-wide stakeholder priorities, and the translation of these lines of evidence into a set of alternative planting regimes.

5.1. Public Green Space Inventory

The population density and land cover distribution in the study area is broadly representative of urban residential neighborhoods in the USA [92], with lawn as the dominant vegetated land cover type (Tables S3 and S4), consistent with other urban areas studied in the USA [29] and in Europe [93]. Approximately one-eighth of this area is covered by trees, again consistent with tree coverage of European research sites [93]. At the same time, resident access to public green space is substantially above the USA average [51] (Figure S1), an unusual situation in a neighborhood with below-median income [49,50] that may partly explain the high value residents placed on numerous urban ES (Figure 4). In the evaluation of existing vegetated land cover types, the combined LiDAR and NDVI-based analysis identified tree cover with high accuracy but was less accurate in identifying shrubs (Table S2); it also obscured the co-occurrence of tree cover over herbaceous ground cover and shrubs. We, therefore, recommend the incorporation of further image analysis rules and waveform LiDAR processing in future green cover analyses for the accurate distinction of these land cover types (e.g., [94,95]).

Currently, lawn provides the majority of runoff retention and recreation value in the neighborhood, while trees and woodlands provide the majority of air purification, carbon storage, and cooling, despite their much smaller area. As a result, the question faced by those responsible for increasing urban ES in Eugene (Section 5.2.) is whether existing public lawn in the neighborhood should be replaced, and if so, with what.

5.2. Stakeholder Priorities

Alignment with local values is known to be critical to the successful planning, development, and management of public urban green space [89,96], and because resident urban ES priorities are locally idiosyncratic (e.g., [40–42,97]), local input is necessary. Accordingly, Eugene's Parks and Open Space Division surveyed thousands of residents and hundreds of city government and operations employees from 2015–2018, culminating in a vision and implementation plan for future green space development [48]. Although the plan focused on the recreational importance of public parks, responding to expressed desires for public gathering space and for additional sports fields [48,90], another prominent goal was to "further the parks and recreation system's capacity to serve as critical infrastructure for clean air, clean water, flood control, carbon sequestration, and climate resilience" [48] (p. 42). Strategies to provide these urban ES have not yet been determined, but an economic assessment of current park value has been completed [98], and a $50 million tax bond was passed in 2018 to support park operations and development in preparation for future green infrastructure development to expand urban ES delivery [99], showing the timeliness and relevance of the decision-making process we contemplate here.

5.2.1. Resident Priorities

FAN residents valued urban ES highly overall, with most respondents rating 16 of the 17 listed urban ES as "moderately" to "very" important (Table S5). These primarily represented supporting or regulating services, showing the importance of ecological resilience as represented by native plant species, bird and pollinator habitat, stormwater purification, and carbon sequestration [3]. Similarly, the importance of human health and well-being was shown by the priority of air purification. These findings contrasted strongly, however, with results from Paris and Angers, France, and from Porto and Lisbon, Portugal, in which residents valued cultural and provisioning urban ES most highly [41]; Table S5. They also differed from the results of six cities in the USA, in which residents viewed native species and pollution mitigation as low priorities [100], and from results of Barcelona, Spain, in which pollination and biodiversity were low priorities as well [97]. Air purification, however, seems to be one of the few urban ES that shows consistently high priority among city dwellers globally, and the high priority of air quality found here is consistent with recent results from China, Portugal, Spain, France, and the Netherlands [39–41,101].

Outdoor recreation was the single cultural urban ES given high priority by FAN residents, consistent with the previous local survey [90] as well as with recent results from Finland and China [42,101]. In contrast, noise reduction and community identity were low priorities among FAN residents, consistent with findings in European cities [41] but unlike those of Guangzhou, China, in which residents ranked noise abatement as a high priority [101]. FAN residents also assigned a low priority to vegetable and fruit production, in contrast with results from cities such as Barcelona [97], revealing an unexpected lack of support given Eugene's promotion of urban farming [102,103]. Still, this result may show that the numerous community and private gardens in the neighborhood have already met this need.

5.2.2. Delphi Analysis

The perspectives of stakeholders who would either be involved in making and influencing green infrastructure development decisions, or responsible for implementing and maintaining any changes, differed from each other substantially in the first round but partly converged during the second round, reflecting the exchange of ideas allowed by the Delphi procedure (Supplementary Materials Section S2; Table S8) and showing the promise of this method in reaching consensus among other diverse stakeholder groups. Among the consensus priorities, Delphi participants valued air and stormwater purification most highly, reflecting current urgent needs for these services at the city level (Section 2.1.). Although residential neighborhoods typically contribute fewer total suspended solids and other stormwater pollutants than do roads with heavy traffic [104], low-density development typically contributes greater runoff volume and affects more of its watershed than high-density development [105]. Delphi respondents also assigned high value to carbon sequestration, natural beauty, and native plant species (Table 1), but the prospect of dense vegetation raised safety concerns (Table S8), illustrating an internal conflict in the responses that would require eventual resolution.

In contemplating land cover changes, Delphi participants described the recreational and ease-of-maintenance benefits of lawn in neighborhood and city parks; one participant noted that city operations management is currently converting long-established landscape beds into lawn to realize these benefits, adding that lawn is viewed as "neat-looking". These views, combined with the acknowledgment of lawn's minimal delivery of regulating and supporting services, are consistent with those found among municipal land managers in three Swedish cities [106]; however, Delphi participants did not express concern about the costs and greenhouse gas emissions of mowing, as Swedish stakeholders did [106]. Instead, most Delphi participants viewed the "cost and complexity" associated with the installation and maintenance of alternative planting regimes as a greater set of barriers, although they did not elaborate (Table S8). Still, most Delphi participants (>85%) supported the conversion of some public lawn to alternative plantings to increase regulating and supporting services, agreeing that the primary limitation would be the preservation of sightlines along right-of-way planting strips.

5.2.3. Resident and Delphi Comparison

The differences between resident and Delphi urban ES priorities appear to have reflected their frames of reference: while Delphi participants contemplated the city scale, residents considered only their own neighborhood, as intended. Illustrating distinctly city-level concerns, one Delphi participant explained:

> "What drives public agencies such as our Public Works Department is regulatory requirements—meeting the Federal Clean Water Act and our NPDES Permit requirements (e.g., reducing pollution into local waterways, reducing flooding, and improving air quality). Other items are secondary to the basic welfare and safety of the general public." (Supplementary Materials Section S3, p. 17).

Consistent with the diversity of stakeholder opinion found in related studies, the partial divergence among Delphi and resident priorities supports the inclusion of neighborhood-level input even in city-wide urban ES planning.

A notable discrepancy was found in the stakeholders' view of cost barriers: while Delphi participants agreed that the cost of lawn conversion was an important barrier (Table S7), over 85% of residents expressed willingness to pay for green infrastructure development that would increase urban ES delivery through tax measures or private donations (Figure 5; Table S6). While stated willingness is not a guarantee of future payment, these findings are consistent with survey results regarding willingness to pay for urban ES in Palm Beach, FL, USA, involving biodiversity, outdoor recreation, and flood protection [107], as well as in Wuhan, Changsha, and Nanchang, China, involving climate regulation, cultural services, air quality, erosion prevention, and habitat services [108]. In each of these cases, the willingness to pay exceeded the expected cost per capita despite heterogeneity in the responses. In contrast, residents of Veneto, Italy expressed a willingness to pay for recreational services but felt that biodiversity conservation and landscape quality should be provided without taxes [109], illustrating the importance of personal history and the governmental context in such attitudes [110] and suggesting that urban planners survey their target neighborhoods before assuming particular cost barriers.

5.3. Integration of Stakeholder Priorities with Potential for Local Benefit

Recent studies have emphasized the need to integrate the ES priorities of multiple stakeholder groups with the quantitative and qualitative potential for the desired ES to be realized [2,16,35,37,44], but none have explored this further. Here, we found that such integration could be readily accomplished by evaluating questions typical of a green space planning process (Section 4.1.) in light of the corresponding stakeholder survey and urban ES supply rate evidence, yielding a set of alternative planting regimes representing each line of evidence. The ability of each vegetated land cover type to provide multiple urban ES then facilitated the creation of an integrated scheme with limited compromise to stakeholder priorities (Section 4.2.). Of central importance, the contrasts among the individual lines of evidence (i.e., quantitative considerations, resident priorities, and Delphi priorities) illustrated the value of consulting all three. In particular, safety and sightline concerns of Delphi participants moderated the extent of woodland required to maximize air and water quality services, while cost concerns limited stormwater planter area; likewise, residents' outdoor recreation priorities limited the conversion of lawn-based playing fields and emphasized widespread inclusion of native flowering shrubs, carrying particular importance because of residents' financial responsibility for any changes as well as their willingness to bear additional costs. As a result, the Integration planting regime is one that could not have been found by any one approach alone, adding strong support to previous recommendations that diverse stakeholder views and quantitative evidence should be considered together in developing urban ES delivery plans [37].

6. Conclusions

This study illustrates a straightforward approach for deciding among green infrastructure alternatives based on their quantitative and qualitative potential to provide urban ES of high priority to diverse stakeholders in a particular location. Here, investigating a neighborhood in Eugene, Oregon, we show how a combination of surveys, Delphi analyses, land-cover analysis, and urban ES quantification can be integrated to reveal a clear direction for urban green space development.

The importance of consulting multiple stakeholders was emphasized by the prominent areas of agreement as well as disagreement between residents and decision-makers, consistent with other studies that have queried multiple stakeholder types [111,112]. Since stakeholder support is essential to successful urban ES provision [36], and since these priorities are locally and regionally idiosyncratic (e.g., [38,40,41,101,113]), results here suggest that revealing these consensus priorities is a necessary first step and highlight the value of the Delphi technique as a method for finding consensus among diverse stakeholders. Subsequent evaluation of existing land cover and urban ES delivery showed that, despite the generally high value residents assigned to supporting and regulating urban ES, lawn currently dominates the neighborhood public green space, reflecting the priority that decision-makers have given to ease of maintenance, lawn-related outdoor recreation, and safety perceptions (Supplementary Materials Section S3; Table S8). While lawn prevalence is consistent with that found in other urban settings [29,93], the value residents placed on habitat and regulating urban ES differed from that of European residents, who generally prioritized cultural over environmental urban ES [41]. Additionally, the willingness residents expressed to support urban ES-related changes financially was unexpected, although others have documented similar results (e.g., [107]), showing that resident views should be explored before options are limited for financial reasons alone.

Evidence such as this, whether qualitative or quantitative, allows green space development to focus on the priority urban ES that can be meaningfully delivered by the land area of interest; here, air quality and stormwater quality were the clearest stakeholder priorities and had the highest potential for local delivery. Such urban ES priorities can then inform the development and evaluation of alternative planting regimes, as illustrated above (Section 4.2.), to reveal the relative benefits of each. Once large-scale land cover decisions are made, additional urban ES priorities can be addressed through species selection during green space design, as in the accommodation of pollinator habitat and native species coverage priorities above. Together, these steps provide a straightforward, flexible method suitable for widespread application in local planting decisions with the goal of increasing urban ES delivery on public land.

Supplementary Materials: The following are available online at http://www.mdpi.com/2073-445X/9/10/391/s1. Supplementary Materials, including Figures S1 and S2, Tables S1–S11, and related discussion are included in a separate file. Survey data, Delphi analysis responses, and green space quantification GIS data are available at doi:10.5281/zenodo.4044203. All other materials are available upon request.

Author Contributions: Conceptualization, E.E., A.R.R., C.E., and K.A.L.; methodology, E.E., A.R.R., C.E., and K.A.L.; formal analysis, E.E.; investigation, E.E., A.R.R., and C.E.; resources, A.R.R. and C.E.; data curation, E.E.; writing—original draft preparation, E.E. and A.R.R.; writing—review and editing, E.E., A.R.R., C.E., and K.A.L.; visualization, E.E., A.R.R., and C.E.; supervision, A.R.R., C.E., and K.A.L.; project administration, E.E., A.R.R., C.E., and K.A.L. All authors have read and agreed to the published version of the manuscript.

Funding: This research received no external funding.

Acknowledgments: We thank Peg Boulay for lending us the Garmin GPSMAP 62S for field data collection. We thank Scott Altenhoff and Justin Overdevest for early project conceptualization and for their help connecting us with Delphi participants. We also are appreciative of the Friendly Area Neighborhood residents who participated in the survey, as well as the individuals who put their time and effort into responding to the two rounds of Delphi surveys.

Conflicts of Interest: The authors declare no conflict of interest.

References

1. Grimm, N.B.; Faeth, S.H.; Golubiewski, N.E.; Redman, C.L.; Wu, J.; Bai, X.; Briggs, J.M. Global change and the ecology of cities. *Science* **2008**, *319*, 756–760. [CrossRef] [PubMed]

2. Luederitz, C.; Brink, E.; Gralla, F.; Hermelingmeier, V.; Meyer, M.; Niven, L.; Panzer, L.; Partelow, S.; Rau, A.-L.; Sasaki, R.; et al. A review of urban ecosystem services: Six key challenges for future research. *Ecosyst. Serv.* **2015**, *14*, 98–112. [CrossRef]

3. Gómez-Baggethun, E.; Gren, Å.; Barton, D.N.; Langemeyer, J.; McPhearson, T.; O'Farrell, P.; Andersson, E.; Hamstead, Z.; Kremer, P. Urban Ecosystem Services. In *Urbanization, Biodiversity and Ecosystem Services: Challenges and Opportunities*; Springer: Dordrecht, The Netherlands, 2013; pp. 175–251. ISBN 978-94-007-7087-4.

4. Millennium Ecosystem Assessment. *Ecosystems and Human Well-Being*; Island Press: Washington, DC, USA, 2005.

5. United Nations. *World Urbanization Prospects: The 2005 Revision*; United Nations: New York, NY, USA, 2006.

6. United Nations. *World Urbanization Prospects: The 2014 Revision, Highlights*; Department of Economic and Social Affairs: New York, NY, USA, 2014.

7. Demuzere, M.; Orru, K.; Heidrich, O.; Olazabal, E.; Geneletti, D.; Orru, H.; Bhave, A.G.; Mittal, N.; Feliu, E.; Faehnle, M. Mitigating and adapting to climate change: Multi-functional and multi-scale assessment of green urban infrastructure. *J. Environ. Manag.* **2014**, *146*, 107–115. [CrossRef] [PubMed]

8. Nowak, D.J.; Hirabayashi, S.; Bodine, A.; Greenfield, E. Tree and forest effects on air quality and human health in the United States. *Environ. Pollut.* **2014**, *193*, 119–129. [CrossRef] [PubMed]

9. Nowak, D.J.; Dwyer, J.F. Understanding the Benefits and Costs of Urban Forest Ecosystems. In *Urban and Community Forestry in the Northeast*; Springer: Amsterdam, The Netherlands, 2007; pp. 25–46. ISBN 978-1-4020-4288-1.

10. Nowak, D.J.; Greenfield, E.J.; Hoehn, R.E.; Lapoint, E. Carbon storage and sequestration by trees in urban and community areas of the United States. *Environ. Pollut.* **2013**, *178*, 229–236. [CrossRef]

11. Xiao, Q.; McPherson, E.G. Rainfall interception by Santa Monica's municipal urban forest. *Urban Ecosyst.* **2002**, *6*, 291–302. [CrossRef]

12. Akbari, H.; Pomerantz, M.; Taha, H. Cool surfaces and shade trees to reduce energy use and improve air quality in urban areas. *Sol. Energy* **2001**, *70*, 295–310. [CrossRef]

13. Jim, C.Y.; Chen, W.Y. Recreation–amenity use and contingent valuation of urban greenspaces in Guangzhou, China. *Landsc. Urban Plan.* **2006**, *75*, 81–96. [CrossRef]

14. Derkzen, M.L.; van Teeffelen, A.J.A.; Verburg, P.H. Quantifying urban ecosystem services based on high-resolution data of urban green space: An assessment for Rotterdam, the Netherlands. *J. Appl. Ecol.* **2015**, *52*, 1020–1032. [CrossRef]

15. Mexia, T.; Vieira, J.; Principe, A.; Anjos, A.; Silva, P.; Lopes, N.; Freitas, C.; Santos-Reis, M.; Correia, O.; Branquinho, C.; et al. Ecosystem services: Urban parks under a magnifying glass. *Environ. Res.* **2018**, *160*, 469–478. [CrossRef]

16. Haase, D.; Larondelle, N.; Andersson, E.; Artmann, M.; Borgstrom, S.; Breuste, J.; Gomez-Baggethun, E.; Gren, A.; Hamstead, Z.; Hansen, R.; et al. A quantitative review of urban ecosystem service assessments: Concepts, models, and implementation. *Ambio* **2014**, *43*, 413–433. [CrossRef] [PubMed]

17. Paulin, M.J.; Remme, R.P.; van der Hoek, D.C.J.; de Knegt, B.; Koopman, K.R.; Breure, A.M.; Rutgers, M.; de Nijs, T. Towards nationally harmonized mapping and quantification of ecosystem services. *Sci. Total Environ.* **2020**, *703*, 134973. [CrossRef] [PubMed]

18. Elmqvist, T.; Setälä, H.; Handel, S.; van der Ploeg, S.; Aronson, J.; Blignaut, J.; Gómez-Baggethun, E.; Nowak, D.; Kronenberg, J.; de Groot, R. Benefits of restoring ecosystem services in urban areas. *Curr. Opin. Environ. Sustain.* **2015**, *14*, 101–108. [CrossRef]

19. Jim, C.Y.; Chen, W.Y. Assessing the ecosystem service of air pollutant removal by urban trees in Guangzhou (China). *J. Environ. Manag.* **2008**, *88*, 665–676. [CrossRef] [PubMed]

20. Tratalos, J.; Fuller, R.A.; Warren, P.H.; Davies, R.G.; Gaston, K.J. Urban form, biodiversity potential and ecosystem services. *Landsc. Urban Plan.* **2007**, *83*, 308–317. [CrossRef]

21. Goldenberg, R.; Kalantari, Z.; Cvetkovic, V.; Mortberg, U.; Deal, B.; Destouni, G. Distinction, quantification and mapping of potential and realized supply-demand of flow-dependent ecosystem services. *Sci. Total Environ.* **2017**, *593–594*, 599–609. [CrossRef] [PubMed]

22. Wang, Y.; Akbari, H. The effects of street tree planting on Urban Heat Island mitigation in Montreal. *Sustain. Cities Soc.* **2016**, *27*, 122–128. [CrossRef]

23. Nowak, D.J.; Stevens, J.C.; Sisinni, S.M.; Luley, C.J. Effects of Urban Tree Management and Species Selection on Atmospheric Carbon Dioxide. *J. Arboric.* **2002**, *28*, 113–122.

24. Townsend-Small, A.; Czimczik, C.I. Carbon sequestration and greenhouse gas emissions in urban turf. *Geophys. Res. Lett.* **2010**, *37*. [CrossRef]

25. Ignatieva, M.; Ahrné, K.; Wissman, J.; Eriksson, T.; Tidåker, P.; Hedblom, M.; Kätterer, T.; Marstorp, H.; Berg, P.; Eriksson, T.; et al. Lawn as a cultural and ecological phenomenon: A conceptual framework for transdisciplinary research. *Urban For. Urban Green.* **2015**, *14*, 383–387. [CrossRef]

26. Milesi, C.; Running, S.W.; Elvidge, C.D.; Dietz, J.B.; Tuttle, B.T.; Nemani, R.R. Mapping and modeling the biogeochemical cycling of turf grasses in the United States. *Environ. Manag.* **2005**, *36*, 426–438. [CrossRef] [PubMed]

27. United States Department of Agriculture. *2018 Irrigation and Water Management Survey*; 2017 Census of Agriculture; US Department of Agriculture: Washington, DC, USA, 2019.

28. United States Department of Agriculture. Land Use and Land Cover Estimates for the United States. Available online: https://www.ers.usda.gov/about-ers/partnerships/strengthening-statistics-through-the-icars/land-use-and-land-cover-estimates-for-the-united-states/#d (accessed on 27 May 2020).

29. Robbins, P.; Birkenholtz, T. Turfgrass revolution: Measuring the expansion of the American lawn. *Land Use Policy* **2003**, *20*, 181–194. [CrossRef]

30. Qian, Y.; Follett, R.F.; Kimble, J.M. Soil Organic Carbon Input from Urban Turfgrasses. *Soil Sci. Soc. Am. J.* **2010**, *74*. [CrossRef]

31. Lerman, S.B.; Contosta, A.R. Lawn mowing frequency and its effects on biogenic and anthropogenic carbon dioxide emissions. *Landsc. Urban Plan.* **2019**, *182*, 114–123. [CrossRef]

32. Monteiro, J.A. Ecosystem services from turfgrass landscapes. *Urban For. Urban Green.* **2017**, *26*, 151–157. [CrossRef]

33. Nowak, D.J.; Heisler, G.M. *Air Quality Effects of Urban Trees and Parks*; National Recreation and Park Association: Ashburn, VA, USA, 2010.

34. Matthews, T.; Lo, A.Y.; Byrne, J.A. Reconceptualizing green infrastructure for climate change adaptation: Barriers to adoption and drivers for uptake by spatial planners. *Landsc. Urban Plan.* **2015**, *138*, 155–163. [CrossRef]

35. De Groot, R.S.; Fisher, B.; Christie, M.; Aronson, J.; Braat, L.; Gowdy, J.; Haines-Young, R.; Maltby, E.; Neuville, A.; Polasky, S.; et al. Integrating the Ecological and Economic Dimensions in Biodiversity and Ecosystem Service Valuation. In *The Economics of Ecosystems and Biodiversity: Ecological and Economic Foundations*; Kumar, P., Ed.; Routledge: Abingdon-on-Thames, UK, 2010; pp. 9–40.

36. Menzel, S.; Teng, J. Ecosystem services as a stakeholder-driven concept for conservation science. *Conserv. Biol.* **2010**, *24*, 907–909. [CrossRef]

37. TEEB. *TEEB Manual for Cities—Ecosystem Services in Urban Management*; TEEB: Geneva, Switzerland, 2011.

38. Castro, A.J.; Martín-López, B.; García-Llorente, M.; Aguilera, P.A.; López, E.; Cabello, J. Social preferences regarding the delivery of ecosystem services in a semiarid Mediterranean region. *J. Arid Environ.* **2011**, *75*, 1201–1208. [CrossRef]

39. Castro, A.J.; Verburg, P.H.; Martín-López, B.; Garcia-Llorente, M.; Cabello, J.; Vaughn, C.C.; López, E. Ecosystem service trade-offs from supply to social demand: A landscape-scale spatial analysis. *Landsc. Urban Plan.* **2014**, *132*, 102–110. [CrossRef]

40. Derkzen, M.L.; van Teeffelen, A.J.A.; Verburg, P.H. Green infrastructure for urban climate adaptation: How do residents' views on climate impacts and green infrastructure shape adaptation preferences? *Landsc. Urban Plan.* **2017**, *157*, 106–130. [CrossRef]

41. Madureira, H.; Nunes, F.; Oliveira, J.V.; Cormier, L.; Madureira, T. Urban residents' beliefs concerning green space benefits in four cities in France and Portugal. *Urban For. Urban Green.* **2015**, *14*, 56–64. [CrossRef]

42. Tyrväinen, L.; Mäkinen, K.; Schipperijn, J. Tools for mapping social values of urban woodlands and other green areas. *Landsc. Urban Plan.* **2007**, *79*, 5–19. [CrossRef]

43. Young, R.F. Managing municipal green space for ecosystem services. *Urban For. Urban Green.* **2010**, *9*, 313–321. [CrossRef]

44. Müller, F.; De Groot, R.; Willemen, L. Ecosystem Services at the Landscape Scale: The Need for Integrative Approaches. *LO* **2010**, *23*, 1–11. [CrossRef]

45. US Census Bureau. *Oregon: 2010 Population and Housing Unit Counts*; U.S. Department of Commerce: Washington, DC, USA, 2012.

46. American Lung Association. *State of the Air 2019*; American Lung Association: Chicago, IL, USA, 2019.

47. Benninghoff, B. *NPDES Municipal Separate Storm Sewer System Permit: Evaluation and Fact Sheet*; Oregon Department of Environmental Quality: Portland, OR, USA, 2010.

48. City of Eugene. *Picture. Plan. Play. A Vision and Implementation Plan for Eugene's Parks and Recreation System*; City of Eugene: Eugene, OR, USA, 2018.

49. Diebel, J.; Norda, J.; Kretchmer, O. Overview of Friendly Area, Eugene, Oregon. Available online: https://statisticalatlas.com/neighborhood/Oregon/Eugene/Friendly-Area/Overview (accessed on 8 January 2019).

50. U.S. Census Bureau. U.S. Census Bureau QuickFacts. Available online: https://www.census.gov/quickfacts/fact/table/US/PST045219 (accessed on 14 August 2020).

51. The Trust for Public Land ParkScore index: The most comprehensive evaluation of park access and quality in the 100 largest U.S. cities. Available online: https://www.tpl.org/parkscore (accessed on 13 October 2020).

52. Davies, Z.G.; Edmondson, J.L.; Heinemeyer, A.; Leake, J.R.; Gaston, K.J. Mapping an urban ecosystem service: Quantifying above-ground carbon storage at a city-wide scale. *J. Appl. Ecol.* **2011**, *48*, 1125–1134. [CrossRef]

53. Zulian, G.; Liekens, I.; Broekx, S.; Kabisch, N.; Kopperoinen, L.; Geneletti, D. Mapping Urban Ecosystem Services. In *Mapping Ecosystem Services*; Burkhard, B., Maes, J., Eds.; Pensoft Publishers: Sofia, Bulgaria, 2017; pp. 312–318.

54. ESRI. *ArcMap 10.7*; Environmental Systems Research Institute: Redlands, CA, USA, 2019.

55. Senanayake, I.P.; Welivitiya, W.D.D.P.; Nadeeka, P.M. Urban green spaces analysis for development planning in Colombo, Sri Lanka, utilizing THEOS satellite imagery—A remote sensing and GIS approach. *Urban For. Urban Green.* **2013**, *12*, 307–314. [CrossRef]

56. Oregon Department of Geology and Mineral Industries. *OLC Middle Fork Willamette River 2015 Lidar Project Lidar QC Report*; Oregon Department of Geology and Mineral Industries: Portland, OR, USA, 2017.

57. Geneletti, D.; Cortinovis, C.; Zardo, L.; Esmail, B.A. *Planning for Ecosystem Services in Cities*; SpringerBriefs in Environmental Science; Springer International Publishing: Cham, Switzerland, 2020; ISBN 978-3-030-20023-7.

58. City of Eugene. *Stormwater Management Manual*; City of Eugene: Eugene, OR, USA, 2014.

59. ESRI. *ArcGIS Pro 2.6*; Environmental Systems Research Institute: Redlands, CA, USA, 2020.

60. Cariñanos, P.; Calaza-Martínez, P.; O'Brien, L.; Calfapietra, C. The Cost of Greening: Disservices of Urban Trees. In *The Urban Forest*; Pearlmutter, D., Calfapietra, C., Samson, R., O'Brien, L., Krajter Ostoić, S., Sanesi, G., Alonso del Amo, R., Eds.; Future City; Springer International Publishing: Cham, Switzerland, 2017; Volume 7, pp. 79–87. ISBN 978-3-319-50279-3.

61. Escobedo, F.J.; Kroeger, T.; Wagner, J.E. Urban forests and pollution mitigation: Analyzing ecosystem services and disservices. *Environ. Pollut.* **2011**, *159*, 2078–2087. [CrossRef]

62. Juanita, A.-D; Ignacio, P.; Jorgelina, G.-A.; Cecilia, A.-S.; Carlos, M.; Francisco, N. Assessing the effects of past and future land cover changes in ecosystem services, disservices and biodiversity: A case study in Barranquilla Metropolitan Area (BMA), Colombia. *Ecosyst. Serv.* **2019**, *37*, 100915. [CrossRef]

63. Loewenthal, K. *An Introduction to Psychological Tests and Scales*; Taylor & Francis Inc.: New York, NY, USA, 2001; Volume 42.

64. R Core Team. *R: A Language and Environment for Statistical Computing*; R Foundation for Statistical Computing: Vienna, Austria, 2019.

65. Dalkey, N.C. *The Delphi Method: An Experimental Study of Group Opinion*; The RAND Corporation: Santa Monica, CA, USA, 1969.

66. Thrall, G.I.; McCartney, J.W. Keeping the Gardbage Out: Using Delphi Method for GIS Criteria. *Geogr. Inf. Syst.* **1991**, *1*, 46–52.

67. Loughlin, K.G.; Moore, L.F. Using Delphi to achieve congruent objectives and activities in a pediatrics department. *Acad. Med.* **1979**, *54*, 101–106. [CrossRef]

68. Putnam, J.W.; Spiegel, A.N.; Bruininks, R.H. Future Directions in Education and Inclusion of Students with Disabilities: A Delphi Investigation. *Except. Child.* **2016**, *61*, 553–576. [CrossRef]

69. Stewart, J.; O'Halloran, C.; Harrigan, P.; Spencer, J.A.; Barton, J.R.; Singleton, S.J. Identifying appropriate tasks for the preregistration year: Modified Delphi technique. *BMJ* **1999**, *319*, 224–229. [CrossRef] [PubMed]

70. Von der Gracht, H.A. Consensus measurement in Delphi studies. *Technol. Forecast. Soc. Chang.* **2012**, *79*, 1525–1536. [CrossRef]

71. City of Eugene. *Eugene Charter*; City of Eugene: Eugene, OR, USA, 2002.

72. City of Ashland. *City Charter*; City of Ashland: Ashland, OR, USA, 2007.

73. De Groot, R.S.; Wilson, M.A.; Boumans, R.M.J. A typology for the classification, description and valuation of ecosystem functions, goods and services. *Ecol. Econ.* **2002**, *41*, 393–408. [CrossRef]

74. Bidegain, I.; Cerda, C.; Catalán, E.; Tironi, A.; López-Santiago, C. Social preferences for ecosystem services in a biodiversity hotspot in South America. *PLoS ONE* **2019**, *14*, e0215715. [CrossRef]

75. Schmidt, K.; Walz, A.; Martín-López, B.; Sachse, R. Testing socio-cultural valuation methods of ecosystem services to explain land use preferences. *Ecosyst. Serv.* **2017**, *26*, 270–288. [CrossRef]

76. Stålhammar, S.; Pedersen, E. Recreational cultural ecosystem services: How do people describe the value? *Ecosyst. Serv.* **2017**, *26*, 1–9. [CrossRef]

77. Costanza, R.; d'Arge, R.; de Groot, R.; Farber, S.; Grasso, M.; Hannon, B.; Limburg, K.; Naeem, S.; O'Neill, R.V.; Paruelo, J.; et al. The value of the world's ecosystem services and natural capital. *Nature* **1997**, *387*, 253–260. [CrossRef]

78. Hardelin, J.; Lankoski, J. *Land use and ecosystem services*; OECD Food, Agriculture and Fisheries Papers; OECD Publishing: Paris, France, 2018; Volume 114.

79. Hasan, S.; Shi, W.; Zhu, X. Impact of land use land cover changes on ecosystem service value—A case study of Guangdong, Hong Kong, and Macao in South China. *PLoS ONE* **2020**, *15*, e0231259. [CrossRef]

80. Kreuter, U.P.; Harris, H.G.; Matlock, M.D.; Lacey, R.E. Change in ecosystem service values in the San Antonio area, Texas. *Ecol. Econ.* **2001**, *39*, 333–346. [CrossRef]

81. Lopes, L.F.G.; dos Santos Bento, J.M.R.; Arede Correia Cristovão, A.F.; Baptista, F.O. Exploring the effect of land use on ecosystem services: The distributive issues. *Land Use Policy* **2015**, *45*, 141–149. [CrossRef]

82. Martínez-Harms, M.J.; Balvanera, P. Methods for mapping ecosystem service supply: A review. *Int. J. Biodivers. Sci. Ecosyst. Serv. Manag.* **2012**, *8*, 17–25. [CrossRef]

83. Zardo, L.; Geneletti, D.; Pérez-Soba, M.; Van Eupen, M. Estimating the cooling capacity of green infrastructures to support urban planning. *Ecosyst. Serv.* **2017**, *26*, 225–235. [CrossRef]

84. Lee-Mäder, E.; Fowler, J.; Vento, J.; Hopwood, J. *100 Plants to Feed the Bees: Provide a Healthy Habitat to Help Pollinators Thrive*; Storey Publishing: North Adams, MA, USA, 2016; ISBN 978-1-61212-701-9.

85. *Attracting Native Pollinators: Protecting North America's Bees and Butterflies: The Xerces Society Guide*; Lee-Mäder, E.; Xerces Society (Eds.) Storey Pub: North Adams, MA, USA, 2011; ISBN 978-1-60342-695-4.

86. City of Eugene. *Council Resolution No. 5240: A Resolution Designating Eugene as a Bee City USA Affiliate*; City of Eugene: Eugene, OR, USA, 2018.

87. Adamson, N.L.; Borders, B.; Cruz, J.K.; Jordan, S.F.; Gill, K.; Hopwood, J.; Lee-Mäder, E.; Minnerath, A.; Vaughan, M. *Pollinator Plants: Maritime Northwest Region*; The Xerces Society: Portland, OR, USA, 2017.

88. Metro. Native plants for Willamette Valley yards booklet. Available online: https://www.oregonmetro.gov/native-plants-willamette-valley-yards-booklet (accessed on 15 May 2020).

89. Biernacka, M.; Kronenberg, J. Classification of institutional barriers affecting the availability, accessibility and attractiveness of urban green spaces. *Urban For. Urban Green.* **2018**, *36*, 22–33. [CrossRef]

90. City of Eugene. *Parks and Recreation System Needs Assessment Report*; City of Eugene: Eugene, OR, USA, 2016.

91. City of Eugene. Water Resource Conservation Maps. Available online: https://www.eugene-or.gov/808/Water-Resource-Conservation-Maps (accessed on 4 August 2020).

92. Romem, I. *Can U.S. Cities Compensate for Curbing Sprawl by Growing Denser?* Buildzoom: San Francisco, CA, USA, 2016.

93. Hedblom, M.; Lindberg, F.; Vogel, E.; Wissman, J.; Ahrné, K. Estimating urban lawn cover in space and time: Case studies in three Swedish cities. *Urban Ecosyst.* **2017**, *20*, 1109–1119. [CrossRef]

94. Casalegno, S.; Anderson, K.; Hancock, S.; Gaston, K.J. Improving models of urban greenspace: From vegetation surface cover to volumetric survey, using waveform laser scanning. *Methods Ecol. Evol.* **2017**, *8*, 1443–1452. [CrossRef]

95. Parent, J.R.; Volin, J.C.; Civco, D.L. A fully-automated approach to land cover mapping with airborne LiDAR and high resolution multispectral imagery in a forested suburban landscape. *ISPRS J. Photogramm. Remote Sens.* **2015**, *104*, 18–29. [CrossRef]

96. Faehnle, M.; Bäcklund, P.; Tyrväinen, L. Looking for the role of nature experiences in planning and decision making: A perspective from the Helsinki Metropolitan Area. *Sustain. Sci. Pract. Policy* **2011**, *7*, 45–55. [CrossRef]

97. Camps-Calvet, M.; Langemeyer, J.; Calvet-Mir, L.; Gómez-Baggethun, E. Ecosystem services provided by urban gardens in Barcelona, Spain: Insights for policy and planning. *Environ. Sci. Policy* **2016**, *62*, 14–23. [CrossRef]

98. Earth Economics. *Nature's Value; An Economic View of Eugene's Parks, Natural Areas and Urban Forest*; City of Eugene: Eugene, OR, USA, 2014.

99. City of Eugene. *Eugene Parks & Recreation Ballot Measures 2018*; City of Eugene: Eugene, OR, USA, 2018.

100. Larson, K.L.; Nelson, K.C.; Samples, S.R.; Hall, S.J.; Bettez, N.; Cavender-Bares, J.; Groffman, P.M.; Grove, M.; Heffernan, J.B.; Hobbie, S.E.; et al. Ecosystem services in managing residential landscapes: Priorities, value dimensions, and cross-regional patterns. *Urban Ecosyst.* **2015**, *19*, 95–113. [CrossRef]

101. Jim, C.Y.; Chen, W.Y. Perception and Attitude of Residents Toward Urban Green Spaces in Guangzhou (China). *Environ. Manag.* **2006**, *38*, 338–349. [CrossRef] [PubMed]

102. City of Eugene. Community Gardens: Gardener's Manual. Available online: https://www.eugene-or.gov/DocumentCenter/View/31611/Community-Gardens-Spring-2017-Manual?bidId= (accessed on 14 August 2020).

103. City of Eugene Friendly Area Neighbors. Available online: https://www.eugene-or.gov/1374/Friendly-Area-Neighbors (accessed on 21 May 2020).

104. Wang, S.; He, Q.; Ai, H.; Wang, Z.; Zhang, Q. Pollutant concentrations and pollution loads in stormwater runoff from different land uses in Chongqing. *J. Environ. Sci.* **2013**, *25*, 502–510. [CrossRef]

105. Richards, L. *Protecting Water Resources with Higher-Density Development*; Environmental Protection Agency: Washington, DC, USA, 2006.

106. Ignatieva, M.; Eriksson, F.; Eriksson, T.; Berg, P.; Hedblom, M. The lawn as a social and cultural phenomenon in Sweden. *Urban For. Urban Green.* **2017**, *21*, 213–223. [CrossRef]

107. Pienaar, E.F.; Soto, J.R.; Lai, J.H.; Adams, D.C. Would County Residents Vote for an Increase in Their Taxes to Conserve Native Habitat and Ecosystem Services? Funding Conservation in Palm Beach County, Florida. *Ecol. Econ.* **2019**, *159*, 24–34. [CrossRef]

108. Tian, Y.; Wu, H.; Zhang, G.; Wang, L.; Zheng, D.; Li, S. Perceptions of ecosystem services, disservices and willingness-to-pay for urban green space conservation. *J. Environ. Manag.* **2020**, *260*, 110140. [CrossRef]

109. Gatto, P.; Vidale, E.; Secco, L.; Pettenella, D. Exploring the willingness to pay for forest ecosystem services by residents of the Veneto Region. *Bio-Based Appl. Econ.* **2013**, 21–43. [CrossRef]

110. Sato, M.; Ushimaru, A.; Minamoto, T. Effect of different personal histories on valuation for forest ecosystem services in urban areas: A case study of Mt. Rokko, Kobe, Japan. *Urban For. Urban Green.* **2017**, *28*, 110–117. [CrossRef]

111. McNally, C.G.; Gold, A.J.; Pollnac, R.B.; Kiwango, H.R. Stakeholder perceptions of ecosystem services of the Wami River and Estuary. *ES* **2016**, *21*, art34. [CrossRef]

112. Zoderer, B.M.; Tasser, E.; Carver, S.; Tappeiner, U. Stakeholder perspectives on ecosystem service supply and ecosystem service demand bundles. *Ecosyst. Serv.* **2019**, *37*, 100938. [CrossRef]

113. Larson, L.R.; Jennings, V.; Cloutier, S.A. Public Parks and Wellbeing in Urban Areas of the United States. *PLoS ONE* **2016**, *11*, e0153211. [CrossRef]

Publisher's Note: MDPI stays neutral with regard to jurisdictional claims in published maps and institutional affiliations.

 *land*

Article

Infiltration Capacity of Rain Gardens Using Full-Scale Test Method: Effect of Infiltration System on Groundwater Levels in Bergen, Norway

Guri Venvik [1] and Floris C. Boogaard [2,3,]*

[1] Department Resources & Environment, Geological Survey of Norway, P.O. Box 6315 Torgarden, 7491 Trondheim, Norway; guri.venvik@ngu.no

[2] Department Research Centre for Built Environment NoorderRuimte, Hanze University of Applied Sciences Groningen, Zernikeplein 7, P.O. Box 30030 Groningen, The Netherlands

[3] Deltares, Daltonlaan 600, 3584 BK Utrecht Postbus, 85467 3508 AL Utrecht, The Netherlands

* Correspondence: floris@noorderruimte.nl; Tel.: +31-641-852-172

Received: 14 November 2020; Accepted: 12 December 2020; Published: 15 December 2020

Abstract: The rain gardens at Bryggen in Bergen, Western Norway, is designed to collect, retain, and infiltrate surface rainfall runoff water, recharge the groundwater, and replenish soil moisture. The hydraulic infiltration capacity of the Sustainable Drainage System (SuDS), here rain gardens, has been tested with small-scale and full-scale infiltration tests. Results show that infiltration capacity meets the requirement and is more than sufficient for infiltration in a cold climate. The results from small-scale test, 245–404 mm/h, shows lower infiltration rates than the full-scale infiltration test, with 510–1600 mm/h. As predicted, an immediate response of the full-scale infiltration test is shown on the groundwater monitoring in the wells located closest to the infiltration point (<30 m), with a ca. 2 days delayed response in the wells further away (75–100 m). Results show that there is sufficient capacity for a larger drainage area to be connected to the infiltration systems. This study contributes to the understanding of the dynamics of infiltration systems such as how a rain garden interacts with local, urban water cycle, both in the hydrological and hydrogeological aspects. The results from this study show that infiltration systems help to protect and preserve the organic rich cultural layers below, as well as help with testing and evaluating of the efficiency, i.e., SuDS may have multiple functions, not only storm water retention. The functionality is tested with water volumes of 40 m^3 (600 L/min for 2 h and 10 min), comparable to a flash flood, which give an evaluation of the infiltration capacity of the system.

Keywords: full-scale infiltration test; MPD infiltration test; boreholes; SuDS; NBS; flood resilience

1. Introduction

Urbanization and climate change effect the water balance in our cities, resulting in challenges such as flooding, droughts, and heat stress. The implementation of Sustainable Drainage Systems (SuDS) or small-scale Nature-Based Solutions (NBS) can help to restore the water balance by capturing, retaining, treating, and infiltrating stormwater that runs off roofs and impermeable surfaces and potentially into the subsurface [1–5]. This will contribute to minimizing flooding, restoring groundwater levels, increasing soil moisture to alleviate drought impacts, and lowering temperatures by evapotranspiration to mitigate heat stress [2,6–11].

As Wakode et al. [12] point out, the urban water cycle is different from that in non-urban areas, where urbanization can influence natural groundwater recharge due to the restriction of infiltration by impermeable surfaces. Even though leakage from water-wastewater infrastructure is known to recharge the groundwater in cities [13,14] this was not substantial enough to recharge and stabilize the groundwater levels under the UNESCO World Heritage site Bryggen Wharf in Bergen, Western Norway (Figure 1) [15–18]. Therefore, the connected infiltration system at Bryggen was intentionally built for that purpose (Figure 1).

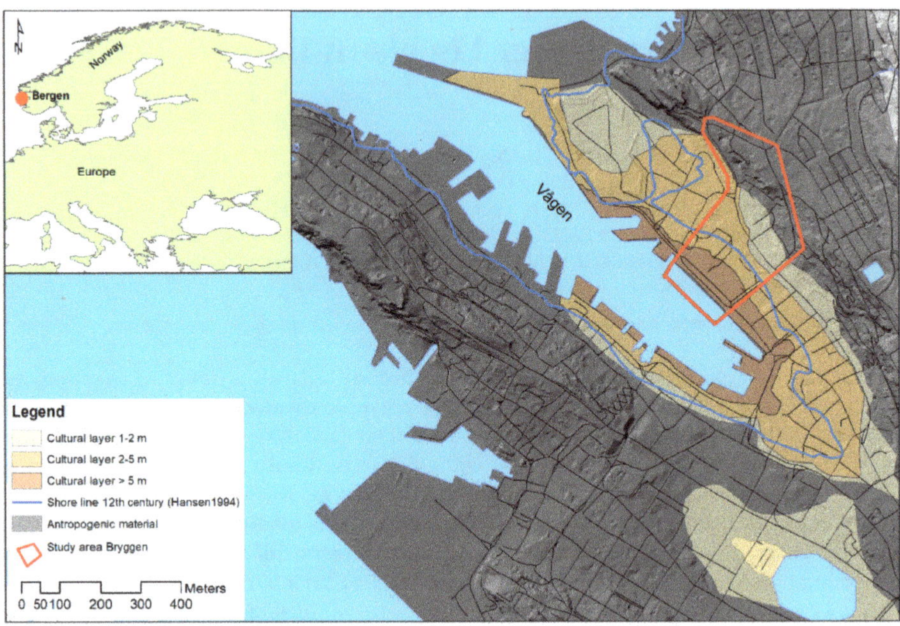

Figure 1. The Medieval city in Bergen is located along the shore of Vågen bay.

The sustainable infiltration and drainage system that has been implemented within the premises of the Bryggen is the largest in Norway (Figure 3). It was built with the purpose of raising and stabilizing the groundwater level and increasing the soil moisture in the cultural heritage layers in the ground below Bryggen [19–22]. The infiltration system has proven its effectiveness for raising the groundwater level to desired levels for preservation [16,22]. However, the infiltration system has not been full-scale tested for its infiltration capacity and interaction with the groundwater below. Such testing of SuDS is commonly executed with small-scale infiltrometer tests [23–25] and further upscaled in modeling tools. However, small-scale testing, such as Modified Phillipe–Dunne (MPD) infiltration tests, does not give a picture of the overfall infiltration capacity of the SuDS. Therefore, a full-scale methodology was first implemented for testing impermeable pavements [26,27] and further used for other infiltration systems such as rain gardens, swales etc. [28–30].

A full-scale infiltration capacity test at Bryggen will contribute to the understanding of the urban water cycle, by quantifying the hydraulic conductivity and infiltration capacity of SuDS, the connectivity to groundwater levels, and thus the overall effectiveness of this system in a larger hydrological and hydrogeological context. SuDS in cold climates require higher infiltration capacity than warm climates to maintain functionality below 0 °C [31,32]. Bryggen is a unique site due to its 35 boreholes with continuous measurements of groundwater level, soil moisture, oxygen content, and other parameters that are essential for the in situ preservation of the cultural heritages below surface. The subsurface data were continuously collected from 2007, gradually expanding with additional boreholes until 2015 and will continue monitoring onwards [19,21].

At present, only a limited part of the catchment area is connected to the infiltration systems (Figure 2). The municipality inquires if the capacity of the rain gardens is acceptable to connect the entire catchment area with stormwater from both roofs and streets to the system. Therefore, the rain gardens were assessed to determine if they work as designed, and if the infiltration capacity and the effectiveness is satisfactory to preserve the cultural layers and thereby enlarge the connected runoff area. The implementation of SuDS at locations where the infiltration of water is a challenge, such as on cultural layers, is a challenge for urban planners, water authorities, and other stakeholders in municipalities. This paper will describe the full-scale infiltration method [26,28] used for testing the hydraulic infiltration capacity of the rain gardens at Bryggen and the response on the groundwater level measured in several monitoring wells.

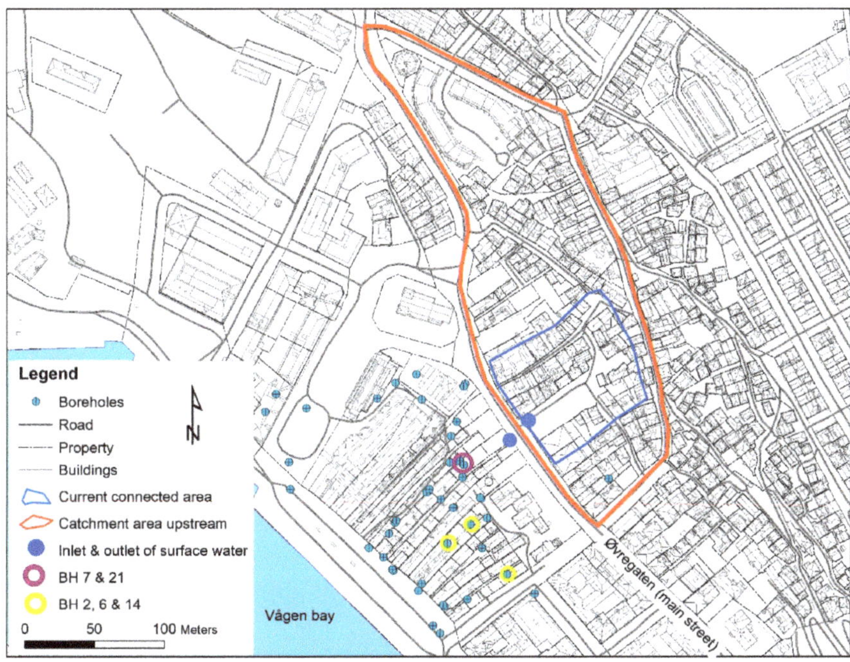

Figure 2. The catchment area and locations of boreholes.

1.1. Study Area

The infiltration system at Bryggen Wharf is located in the Medieval City center of Bergen, the largest city on the west coast and second largest city in Norway (Figure 1). The average temperatures are 23.8 °C in summer and −4.7 °C in winter, giving an annual average temperature of 8.6 °C. During 61 years of data collected, only 17 winters had temperatures below 0 °C [33]. Bergen is one of the wettest places in Europe, with an annual precipitation of 2250 mm/year [33]. The topography is steep hillsides covered with forest vegetation on scares soil, down to flat laying formerly shorelines with thicker natural sediments and anthropogenic layers (Figure 1). The relief goes from 320 m above sea level to 1 m a.s.l. over a distance of 1 km. These natural conditions make surface runoff water abundant.

Bryggen is a Hanseatic Wharf where several of the buildings originate from 1702 [34]. The Medieval city, located along the shore of Vågen bay, is to a large degree built on anthropogenic waste including remains from city fires and industrial and household waste. These have accumulated into abundant anthropogenic cultural heritage layers rich on organic content that locally are more than 10 m thick [19,21]. The reduction of soil moisture or lowering of the groundwater level will introduce oxygen into the organic matter. This will accelerate the oxidation and disintegration of the organic material causing collapse and compaction of the organic layers in the subsurface [17,20]. Due to the slow decay causing damage to the Wharf, the Bryggen project was initiated in 2010 [17,20,21].

The abovementioned processes will further cause subsidence of the ground and damage on buildings and infrastructure [18,35].

The ground beneath Bryggen is characterized by a steeply sloping mountain side, with depth to bedrock from 2 to 12 m. The layers consist of up to 10 m of organic, anthropogenic deposit as described above, on top of beach sand and moraine of ca. 2 m thickness. The recharge of the groundwater is primarily by runoff from the uphill catchment area [16,36]. A 3D hydrogeological model of Bryggen and its subsurface has been made to understand the groundwater movement, hydrogeological characteristics of the subsurface layers and processes linked to water, or the lack of water [16,17,20,36].

A monitoring system was established in 2001, which was expanded in 2010 with a total of 35 monitoring wells [37]. Initially, this network of monitoring wells was placed to understand the complex flow system in the area and to identify the causes of the local groundwater levels and observed, increased subsidence rates [16,18,29,35]. An automated groundwater-monitoring system was installed in the wells, for high frequency of measurements. Some boreholes are dedicated to measuring parameters for archaeological purposes [17] while other boreholes are continuously monitored for groundwater levels [21], using equipment such as Schlumberger Micro diver DI 501 [38]. During the Bryggen project a strong link between the level and stability of the groundwater and the decay of the cultural layers was established [16,17,20,22,39]. Therefore, an infiltration system was installed in the ground in and around the Wharf to infiltrate as much surface water as possible into the ground (Figure 3) [18,29,40]. All measures at Bryggen, including monitoring wells and SuDS, were directed towards raising and stabilizing groundwater levels. The long-term goal for the area was to elevate groundwater levels to about 1 m below the surface [16].

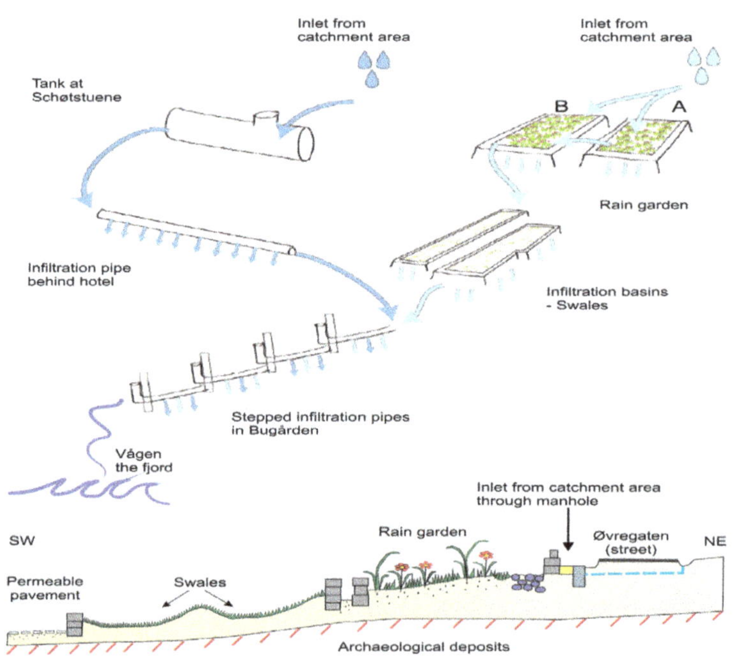

Figure 3. Top: Sketch of the infiltration system at Bryggen (Drawing: Multiconsult AS). Bottom: Cross section of rain garden, swale, and permeable pavement (modified from de Beer & Boogaard [40]).

1.2. The Rain Garden at Bryggen

The rain gardens in Bryggen are a bioretention system that allows runoff to temporarily pond in a shallow planted depression before filtering through vegetation, roots, and underlying soils for infiltration [2,3,41]. The rain gardens have the following primary functions: infiltration, storage, and purification. The catchment area, indicted in Figure 2, is upstream of the main street "Øvregaten" (Figure 2), which is salted during winter to reduce icing and traffic incidence. Plants in the rain gardens are not salt tolerant [29]. To avoid excess salt from the winter salting, water from the watershed is collected in a manhole on the other side of the street and brought in a pipeline underneath the main street to a manhole connected to the rain gardens, as indicated in Figure 3 and with blue points in Figure 2. The infiltration system has two inflow points from the catchment area: into the rain gardens and into the tank, as indicated in Figure 3. Figure 2 shows the current connected area for surface water (blue line) and the total upstream catchment area (red line) for the rain gardens and infiltration system.

2. Methods

The full-scale infiltration test was executed on 6 September 2017, with heavy rain that started the day before with a total of 30.5 mm rain fall [33]. Therefore, the soil was moist and the rain garden saturated with water (Table 1). The days 5th to 14th of September were wet with 28.2 mm precipitation on the 9th [33]. The additional contribution of water through precipitation is reflected in the results from the groundwater level monitoring.

Table 1. Analysis of soil moisture of samples collected before and after the full-scale test. The results are given in precent (%) of water in the soil.

Sample	Water Content in Percent (%) Before Full-Scale Test	Water Content in Percent (%) After Full-Scale Test
1.	30.50	42.40
2.	34.30	56.50
3.	28.60	40.70
4.	28.10	58.50

The compartment B of the rain gardens has an area of 180 m^2, a depth of 78 cm consisting of three layers: a filter medium consisting of sandy soil (38 cm, 60% soil and 40% sand), a drainage layer of sand and gravel (30 cm) and a bottom layer of silty sand (10 cm), with a nonwoven geotextile on top of intact cultural heritage layers (Figure 4). The average porosity is approximately 35%. The water storage capacity of the rain garden is designed for 30 cm above soil surface at deepest point, giving a storage volume of 54 m^3 [29] (Figure 5). When this level is exceeded the water flows into an outlet and down to the swales below (Figure 3). The compartment A has the same construction and layering as B, but an area of 52 m^2. The rain gardens are designed to have an infiltration capacity of 0.5 m/day for a rainfall event with intensity 35–50 mm/day (24-h storm) [28], assuming dry antecedent conditions.

In this study, the infiltration capacity of the rain gardens is compared with international guidelines, such as the FAWB in Australia [42], the MPCA in the USA [43,44], and the CIRIA in the UK [41]. The CIRIA SuDS manual [41] is regarded as the most relevant for Norwegian standards and conditions [45]. To evaluate whether the rain gardens at Bryggen qualifies according to the international guidelines, we compared the measured saturated hydraulic conductivity to both the design values and the measured infiltration capacities. Two infiltration tests were used: Modified Phillipe–Dunne (MPD) [23,24] and full-scale infiltration capacity method [26–28]. The full-scale infiltration capacity test was further correlated with continuous monitoring of the groundwater level in several boreholes (Figure 2).

Before and after the full-scale infiltration test soil samples were collected in the rain gardens, four in compartment A and four in B (Figure 3). The analysis was executed by Vannlaboratoriet Bergen Vann KF [46], where the samples were dried at 100 °C for four days. The samples were analyzed for soil moisture to document the start-up conditions for the full-scale test, and the effect infiltration has on soil moisture.

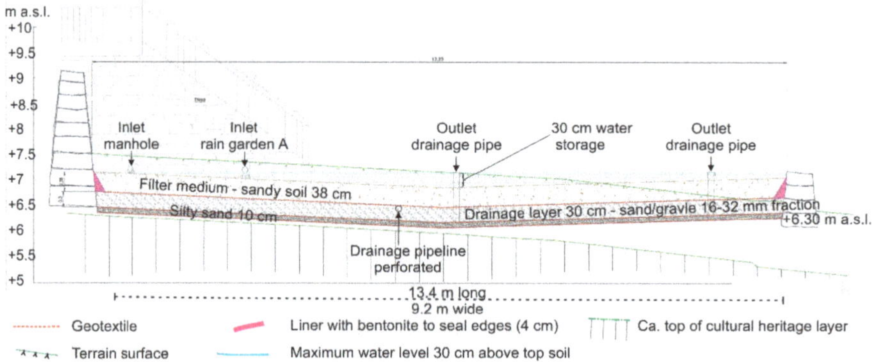

Figure 4. The design of rain garden B. Drawing: Multiconsult AS [21].

Figure 5. Modified Philip–Dunne (MPD) column for infiltration measurements (Photo: Tone M. Muthanna).

2.1. Modified Phillip-Dunne Infiltration (MPD) Method

The Modified Philip–Dunne (MPD) infiltrometer test, which determines the infiltration capacity for saturated hydraulic conductivity [23,24], was executed at four different locations in compartment A and B in the rain garden. The principle for all small-scale infiltrometer tests is that rings or columns are sealed to the surface and filled with water to provide a positive water head. The column has a diameter of 10 cm and length of 50 cm. A measuring tape is attached to the outside of the column in order to measure the height of the water column, as shown in Figure 5. The time recorded for the water to infiltrate through the permeable surface area is used to estimate an average infiltration rate (usually in mm/h) for the test location. Both the constant head and the falling head methods can be utilized in these testing procedures [23,24,30]. These in-situ field methods are easy to facilitate and are time and cost efficient. An illustration of the MPD is given in Figure 5.

The permeability is given by

$$p = K \times \mu/\rho g \tag{1}$$

where p = permeability (cm^2), K = hydraulic conductivity (cm/hr), μ = dynamic viscosity (kg/m*hr), ρ = density of water (kg/m^3), and g = gravitational acceleration (m/s^2) [23,24].

2.2. Full-Scale Infiltration Capacity Test at Bryggen

For the full-scale infiltration test at Bryggen, the total volume of the rain garden is flooded, and the emptying time is measured [26]. The water source was a fire hydrant, which held a constant water flow of ca. 600 L/minute for 2 h and 10 min (Figures 2 and 6). The water influx was continuously measured by an in-line flowmeter, as shown in Figure 7. This translates to a total water volume of ca. 40 m^3, which gives ca. 20,000 L/hour. The water was led through a drainage pipe under the street, Øvregaten, and into a manhole on top of the rain garden (Figure 7) where the water inflow was split into two, into compartment A on the left and compartment B on the right (Figures 3 and 8). All outlets from rain gardens A and B were blocked during the infiltration test to prevent the bypass of water out of the gardens and thus force infiltration. The water influx was kept constant until the infiltration system was completely saturated and water became visible, forming a pond at the surface in both rain gardens and at the swales below (Figures 3, 7 and 8). The rain garden compartment A and B have a confined space which can be filled up to the water level of outflow without any additional restriction to prevent water leaving the rain gardens during the full-scale infiltration test. The maximum water depth at the deepest part of the rain gardens is ca. 30 cm, varying depths due to irregularities of the soil surface (Figure 4). The time from maximum ponded water height to complete infiltration was recorded for both rain garden A and B (Figure 6). To calculate the infiltration rate,

$$K = h/t \tag{2}$$

where K is the infiltration rate (cm/min), h = height of the water column (cm), and t = time (duration) of infiltration (min) [26,28]).

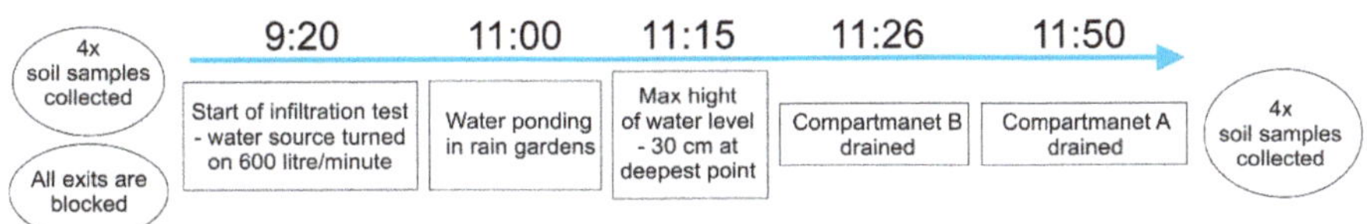

Figure 6. Description of the full-scale infiltration test at Bryggen on 6th of September 2017.

2.3. Continuous Monitoring of Groundwater Level in Boreholes

In the boreholes, the automated data collection is set at a frequency of 1 h, and the instrument has an accuracy of measuring the water height within 0.05% [34]. With this detailed measuring frequency, immediate and short-term effects are detectable [16], and therefore able to show the response of the full-scale infiltration capacity test. All loggers are calibrated for measuring depth relative to surface level, showing the correct groundwater level from surface level. The location of boreholes is shown in Figure 2, where the boreholes used in this study are marked.

Figure 7. A fire hydrant served as a water source, providing ca. 600 L water per minute.

Figure 8. Full-scale infiltration test of the SuDS at Bryggen.

3. Results and Discussions

3.1. Soil Moisture Results

The soil samples had a water content of 28–34% before the infiltration test was executed (Table 1). This saturated condition, ca. 30% water, is explained by the heavy rainfall the previous day. After the full-scale test, four soil samples were collected at the same locations show that the water saturation of the soil had increased to 40–56 percent (%) (Table 1). This is an increase in soil moisture of 12–22 percent (%). Preservation of the organic layers is dependent on the soil moist and prevention of oxygen access, as shown by Matthiesen et al. [17]. An increase in soil moisture by infiltration of water may affect the cultural layers below, by preventing exposure to oxygen. A study of repetitive full-scale testing by Boogaard and Lucke [30], on permeable pavements and swales, show that after refilling the storage volume a second time (simulating a stormwater event after a stormwater event) the infiltration capacity is reduced by 39% from unsaturated to saturated soil conditions. This shows that the infiltration of surface water with the aid of SuDS like rain gardens and swales will increase the soil moist, recharge the groundwater, and further contribute to preservation of the cultural layers below.

3.2. Small-Scale Results

Four MPD infiltration tests were executed 8th of September 2017 in the rain gardens compartment A (tests 3 and 4) and compartment B (tests 1 and 2, Figure 9). The results of the small-scale infiltrometer tests MPD, summarized in Table 2, shows that the infiltration capacity is (1) 245 mm/h, (2) 241 mm/h, (3) 382 mm/h and (4) 404 mm/h (Figure 10). The small-scale infiltration tests verify that the rain gardens qualify according to the international guidelines for SuDS, demanding an infiltration capacity in the range of 100–300 m/h [41–45]. Field tests verify the infiltration capacity, which is a recommendation by the CIRIA guideline [41], due to local variation, if the SuDS is built according to design and possible clogging [41]. The MPD has a small surface, commonly 10 cm diameter (Figures 6 and 9), where the variation of the heterogeneous soil layer has large influence on the measurements. Ahmed et al. [47] show in their study that for the MPD to be representative, a minimum of 20 tests on different locations should be executed to get a representative average. Unfortunately, only four MPDs are collected in this study and are not statistically representative for infiltration rates in the rain gardens. Since the MPD can be inaccurate because of heterogeneity of the filter soil [26,27], the small-scale tests were compared with a full-scale test (Table 2).

Figure 9. MPD infiltration tests where MPD 3 and 4 were measured in the smaller compartment A (picture on the left) and MPD 1 and 2 were measured in the larger compartment B (picture on the right). Photo: Torstein Dalen, Bergen Municipality.

Table 2. The results of the small-scale and full-scale infiltration test show different infiltration capacity for the two rain gardens, compartment A and B, tested in this study. Results compared to international guidelines [41–45].

	Small-Scale Infiltration Tests	Full-Scale Infiltration Test	Requirement 100–300 mm/h and Empty Time of Max 48 h
The large rain garden B	MPD 1: 245 mm/h MPD 2: 241 mm/h	Ca. 1600 mm/h	5 times the requirement of infiltration 11 min empty time
The small rain garden A	MPD 3: 382 mm/h MPD 4: 404 mm/h	Ca. 510 mm/h	1.7 times the requirement of infiltration 35 min empty time

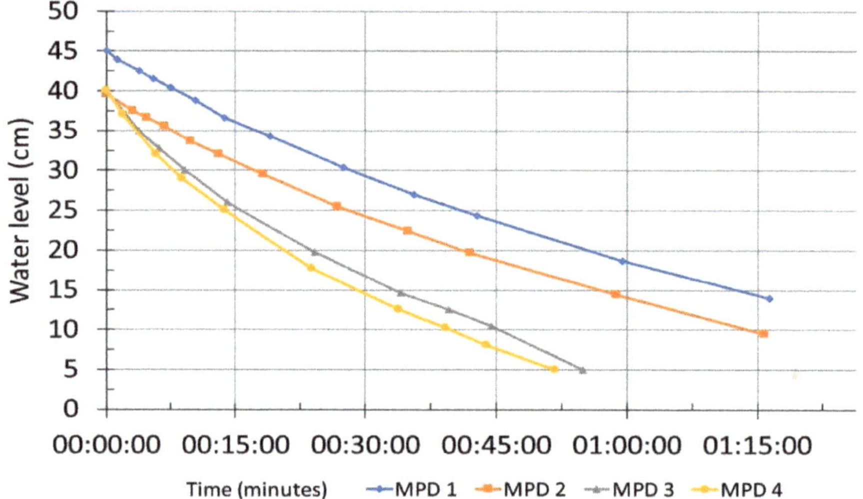

Figure 10. Results of the MPD infiltration test in rain garden A and B. MPD 3 and 4 were measured in compartment A and MPD 1 and 2 in compartment B.

3.3. Full-Scale Results

The results from the full-scale infiltration test are given in Table 2. Based on ca. 30 cm ponding depth and drainage time, which constitutes infiltrating all visible water, the emptying time of the large compartment B was 11 min and 35 min for the smaller compartment A. The infiltration capacity is ca. 1600 mm/h for the large compartment (B) and 510 mm/h for the smaller compartment (A, Table 2). Therefore, both rain gardens A and B meet the minimum requirement of the international SuDS standards [41–45]. Both rain gardens have considerable infiltration capacity, and the capacity is larger than the amount of water coming from the presently connected watershed upstream. The infiltration time was considerably longer in the smaller rain garden A (Table 2). The total water volume for the full-scale test is of ca. 40 m^3, which gives ca. 20,000 L/hour and is larger than any known return period of an extreme event [48].

3.4. Comparison between Small-Scale and Full-Scale Results

The hydraulic efficiency of SuDS such as rain gardens and swales rely on two main standards, which are infiltration and retention capacity [41,49]. The infiltration capacity is usually estimated by measuring the rate at which water infiltrates from small test pits, boreholes [50,51] or as ring infiltrometer tests [23–25,52]. International guidelines recommend a design that enables bioretention, such as rain gardens that can infiltrate stormwater at a rate of 100–300 mm/h. These guidelines are based on several factors such as the limited availability of space in urban areas, low native permeability of the soil, shallow groundwater tables, limited public health concerns, and often safety factors, such as mosquitoes and risk of drowning. The guidelines also take into consideration that the infiltration capacity of rain gardens may reduce over time by clogging [53–55]. Vegetation that is resistant to long inundation can prevent clogging of the topsoil, due to root canalling [42]. Further, the implemented SuDS should be tested for its infiltration capacity in the field [1–3,10,41].

The small-scale infiltration test methods are based on the infiltration rate through a very small area that is used to represent the total area of infiltration. Using such small areas for testing could potentially lead to erroneous results as studies have demonstrated a high degree of spatial variability between different infiltration measurements undertaken in the same area [26–28]. A study by Ahmed et al. [47], based on 722 small-scale infiltrations test using the MPD test in five large swales, shows that the hydraulic conductivity has a high spatial variability within the same swale. The full-scale infiltration test shows that both rain gardens have a much higher infiltration capacity than the results from the MDPs indicate (Table 2). The infiltration rate for the full-scale test is increased by a factor of 6.5 for the large compartment B and 0.8 for the smaller compartment A, compared to results from the small-scale tests. The difference between the large and small compartments may be explained by the coarser material on the surface of the larger and different plants in the smaller compartment (Figures 8 and 9). This reflect the results by Boogaard et al. [26] and Lucke et al. [27] where small-scale (single ring) infiltration test only gives the local condition for the SuDS, independent whether a rain garden, swale or permeable pavement is being tested. Ahmed et al. [47] show that the infiltration capacity or permeability in swales can vary by a factor of 100, giving a large uncertainty if only one or a few small-scale tests are used for testing the infiltration capacity. The full-scale method will demonstrate the infiltration capacity compared to small-scale tests, which is especially important in cold climates [31,32].

3.5. Continuous Monitoring of Groundwater

Monitoring wells MB 07 and MB 21 are located <30 m downstream of the swales (purple circles in Figure 4). These show an immediate response in rise of groundwater level to the infiltration test (Figure 11). The infiltration during the full-scale test is helped by the large amount of precipitation the day and night before, which can be seen by the rise in water level before the time of the infiltration test. The increase of ca. 35 cm in MB07 and 42 cm in MB21 in groundwater level, as seen in Figure 11. The changes in the groundwater lever curves show response to the infiltration test in addition to natural fluctuation from precipitation, as shown in Figure 11.

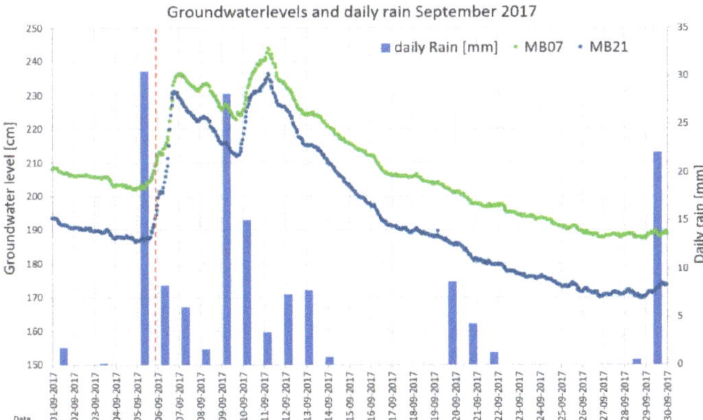

Figure 11. Response in groundwater level (cm) in the monitoring well MB07 (green line) and MB21 (blue line). The red dashed line marks the day of the full-scale infiltration test. Precipitation data from Florida metrological station in the center of Bergen, ca. 1.5 km from Bryggen. Data from Norwegian Metrological Institute @ eKlima.no [33].

Figure 12 shows changes in the groundwater level in boreholes MB 02, MB 06 and MB 14 that are located 75–100 m further downstream from the rain gardens (yellow circles in Figure 2). The direct effect of the infiltration test is not as prominent as in borehole MB 07 and MB 21 (Figure 11), due to the distance from the infiltration point. However, the results still show a clear rise in groundwater level on the day of the full-scale infiltration test in addition to a continued increase with the following days, contributed by precipitation. These boreholes show a delay of ca. 2 days in the response, reflecting the travel time of the groundwater from source to monitoring point. The increase in the groundwater level is 20–25 cm in all three boreholes (Figure 12). Combined with additional precipitation, the infiltration peak has a duration of ca. a week.

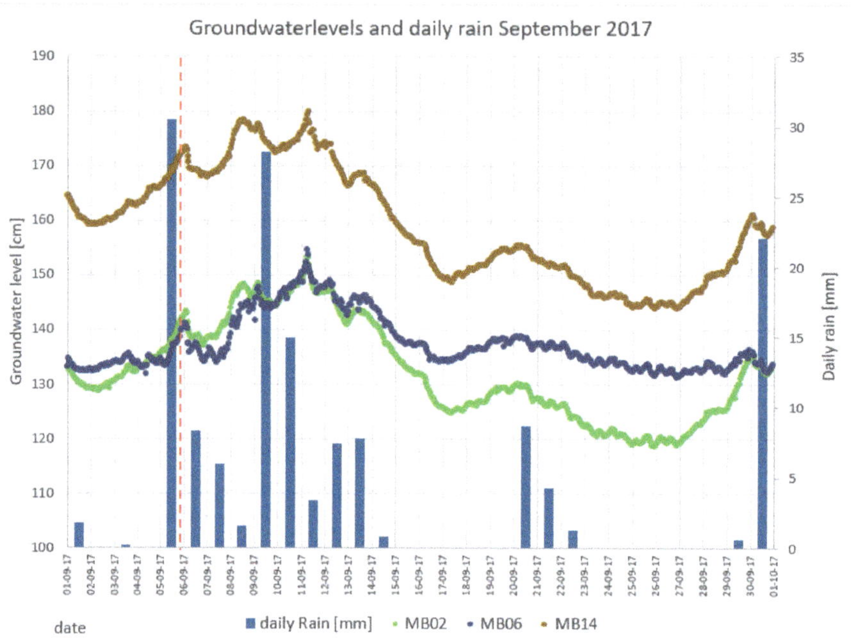

Figure 12. The graph shows variations in the groundwater level (cm) in monitoring wells MB 02 (green line), MB 06 (blue line), and MB 14 (yellow line). The red dashed line marks the day of the full-scale infiltration test. Precipitation data from Norwegian Metrological Institute @ eKlima.no [33].

It could be considered that the rain gardens at Bryggen is constructed with an unnecessarily high infiltration capacity, but if the infiltration capacity were too low, flooding of surface water would occur especially during heavy rainfall of 35–50 mm/day. However, this infiltration system is built for multiple purposes, the main one being to stabilize and increase the groundwater level to protect the cultural layers below Bryggen [16–18]. Other functions of the SuDS are retaining and storing stormwater, filtering pollutants, and increasing the soil moisture [2,9,30]. This study shows good communication between the infiltrated water and the nearby monitoring wells, <30 m distance, with a time delay of ca. 2 days according to the distance from the infiltration point (75–100 m, Figures 2, 11 and 12).

A large infiltration capacity is especially important in locations with a cold climate, as winter will mean a reduced infiltration capacity due to freezing and ice [31,32]. The infiltration system at Bryggen was built for a larger catchment area than presently connected (Figure 2). The infiltration capacity of 1600 mm/h is more than sufficient to handle the current runoff surface water. Therefore, the capacity should be sufficient to expand the catchment area from ca. 8500 m^2 to 31,000 m^2, as shown in the map in Figure 2. The increased infiltration of surface water will contribute to stabilizes the groundwater and prohibit processes driven by lack of groundwater that causes subsidence [56].

Empirical research from this study will improve the groundwater model for Bryggen, and related models [16,36]. In addition, it strengthens the best practices in cultural heritage management that the Bryggen Project has proven to be [21,40]. Bryggen is an example that measures can be taken to infiltrate surface water to restore and stabilize the groundwater, to further delay degradation and subsidence, and as a bonus, prevent flooding.

3.6. Lesson Learned

When comparing small- and large-scale infiltrations tests the tests should be repeated. For small-scale tests, the statistical representative number is 20, as demonstrated by Ahmed et al. [47]. It is a weakness in this study that only four small-scale tests were executed. The large-scale test should also be repeated for better documentation of the infiltration rate and response on groundwater level. Nordic cold climate has different challenges than warmer climate [31,32] and studies that monitor both the stormwater influx as well as groundwater response is needed. As Prudencio and Null [5] point out, SuDS have positive effects as ecosystem services, however, this is not been given any attention in this study at Bryggen.

Especially, a lesson learned from Bryggen is documentation of "as built". There are several gaps of knowledge, there among construction deviations, the compilations of soil used, and sketches of design as built are missing. The lack of documentation make testing and follow ups challenging. For future constructions of SuDS, planning for implementation of monitoring systems, both for stormwater and groundwater, is recommended.

For interactive dissemination and outreach, the tool Climatescan (www.climatescan.org) is applied for engagement with stakeholders where open access results on infiltration of stormwater under extreme climate and hydraulic circumstances are displayed [57]. The Bergen Wharf site with its infiltration systems is continuously updated with pictures, information, research results, and open access publications [58].

4. Conclusions

This study compared the infiltration capacity of the rain gardens at Bryggen with international guidelines [41–45]. It compares small-scale MPD tests to full-scale infiltrations test and further evaluate if the rain garden is preforming as designed. The full-scale infiltration test showed an increase of the groundwater level in several boreholes at Bryggen.

The rain garden and its connected infiltration systems function for the purpose it is built, to infiltrate stormwater into the subsurface to increase the soil moisture as well as groundwater level to protect the cultural layers in the subsurface. The hydraulic conductivity of the SuDS is as designed with an infiltration capacity of 500–1600 mm/h. This study shows that the full-scale infiltration test gives a higher infiltration capacity of the rain gardens, compared to small-scale tests. The effect of the infiltrated volume and the natural precipitation influence the groundwater level, with an immediate response in monitoring wells close to the infiltration system (<30 m) and with a time delay of ca. 2 days in wells 75–100 m away from the infiltration point. The infiltration capacity of the rain garden exceeds the amount of available surface water currently connected to the system. The groundwater level would, in dry periods, benefit from more consistent water input to increase soil moisture and thereby preserve cultural layers below. There is excess capacity, and the connected runoff catchment area can be extended to encompass the total catchment area by 22,500 m^2 or 260 precent (%). There is a need to document the effect of SuDS on the urban water cycle. This study shows how monitoring systems can be implemented in designing and planning, which could help stormwater managers with the scheduling of maintenance requirements for rain gardens with more confidence.

Author Contributions: Conceptualization, G.V. and F.C.B.; methodology, F.C.B.; validation, G.V. and F.C.B.; formal analysis, G.V. and F.C.B.; investigation, G.V.; data curation, G.V.; writing—original draft preparation, G.V.; writing—review and editing, F.C.B. and G.V.; visualization, G.V. All authors have read and agreed to the published version of the manuscript.

Funding: This research is supported through the JPI Water funded INXCES research project "Innovation for eXtreme Climatic EventS" www.inxces.eu. INXCES exchange researchers, methodology and results on an international level, with the aim is to share research results with stakeholders and set up guidelines for design, implementation and maintenance of SuDS to promote sustainable water management systems throughout the world. Within the INXCES project knowledge exchange between disciplines and nations is in focus.

Acknowledgments: Great appreciation to Torstein Dalen at the Water and Wastewater department at Bergen Municipality for help and contributions. Thanks to the Bryggen project for access to borehole data http://prosjektbryggen.no. We thank Bergen Municipality and Hanze University of Applied Sciences Groningen, Deltares and the Geological Survey of Norway for support for this work. Thanks to Riksantikvaren (Directorate for Cultural Heritage) and NIKU (Norwegian Institute for Cultural Heritage) for making data freely available. Thanks to Norwegian Hydrology Council for the NHC2018 conference in Bergen, and this special conference issue. Great appreciation to Malin Andersson at the Geological Survey of Norway (NGU) for constructive feedback on this paper.

Conflicts of Interest: The authors declare no conflict of interest.

References

1. Davis, A.P.; Hunt, W.F.; Traver, R.G.; Clar, M. Bioretention Technology: Overview of Current Practice and Future Needs. *J. Environ. Eng.* **2009**, *135*, 109–117. [CrossRef]
2. Fletcher, T.; Andrieu, H.; Hamel, P. Understanding, management and modelling of urban hydrology and its consequences for receiving waters: A state of the art. *Adv. Water Resour.* **2013**, *51*, 261–279. [CrossRef]
3. Fletcher, T.D.; Shuster, W.D.; Hunt, W.F.; Ashley, R.; Butler, D.; Arthur, S.; Trowsdale, S.; Barraud, S.; Semadeni-Davies, A.; Bertrand-Krajewski, J.-L.; et al. SUDS, LID, BMPs, WSUD and more—The evolution and application of terminology surrounding urban drainage. *Urban Water J.* **2015**, *12*, 525–542. [CrossRef]
4. Majidi, A.N.; Vojinovic, Z.; Alves, A.; Weesakul, S.; Sánchez, A.; Boogaard, F.C.; Kluck, J. Planning Nature-Based Solutions for Urban Flood Reduction and Thermal Comfort Enhancement. *Sustainability* **2019**, *11*, 6361. [CrossRef]
5. Prudencio, L.; Null, S.E. Stormwater management and ecosystem services: A review. *Environ. Res. Lett.* **2018**, *13*, 033002. [CrossRef]
6. Haughton, G.; Hunter, C. *Sustainable Cities*; Jessica Kingsley Publishers: London, UK, 1994; p. 357. ISBN 1134996071, 9781134996070.
7. Bolund, P.; Hunhammar, S. Ecosystem services in urban areas. *Ecol. Econ.* **1999**, *29*, 293–301. [CrossRef]
8. Marsalek, J.; Jiménez-Cisneros, B.E.; Malmquist, P.-A.; Karamouz, M.; Goldenfum, J.; Chocat, B. *Urban Water Cycle Process and Interactions*; UNESCO Publishing and Taylor and Francis Group: Paris, France, 2006; ISBN 0-203-93246-3.
9. Nie, L.; Lindholm, O.; Lindholm, G.; Syversen, E. Impacts of climate change on urban drainage systems—A case study in Fredrikstad, Norway. *Urban Water J.* **2009**, *6*, 323–332. [CrossRef]
10. Ashley, R.; Christensson, A.; De Beer, J.; Walker, L.; Moore, S.; Saul, A. Selling Sustainability in SKINT (SSIS), SKINT Water Series II. 2012. Available online: https://brage.bibsys.no/xmlui/bitstream/handle/11250/176007/Kulturlag_SKINT_Waterseries_2.pdf?sequence=1 (accessed on 1 August 2020).
11. Boogaard, F.C.; Van De Ven, F.; Langeveld, J.G.; Van De Giesen, N. Stormwater Quality Characteristics in (Dutch) Urban Areas and Performance of Settlement Basins. *Challenges* **2014**, *5*, 112–122. [CrossRef]
12. Wakode, H.B.; Baier, K.; Jha, R.; Azzam, R. Impact of urbanization on groundwater recharge and urban water balance for the city of Hyderabad, India. *Int. Soil Water Conserv. Res.* **2018**, *6*, 51–62. [CrossRef]
13. Foster, S.S.D. Impacts of urbanization on groundwater. In *Hydrological Processes and Water Management in Urban Areas*; IAHS-AISH Pub. No. 198; International Association of Hydrological Sciences: Wallingford, UK, 1990.
14. Lerner, D.N. Identifying and quantifying urban recharge: A review. *Hydrogeol. J.* **2002**, *10*, 143–152. [CrossRef]
15. De Beer, J.; Matthiesen, H.; Christensson, A. Quantification and Visualization of In Situ Degradation at the World Heritage Site Bryggen in Bergen, Norway. *Conserv. Manag. Archaeol. Sites* **2012**, *14*, 215–227. [CrossRef]

16. De Beer, J.; Seither, A. Groundwater balance. In *Monitoring, Mitigation, Management: The Groundwater Project—Safeguarding the World Heritage Site of Bryggen in Bergen*; Rytter, J., Schonhowd, I., Eds.; Riksantikvaren: Bergen, Norway, 2015; pp. 106–123.

17. Matthiesen, H.; Hollesen, J.; Gregory, D. Preservation Conditions and Decay Rates. In *Monitoring, Mitigation, Management: The Groundwater Project—Safeguarding the World Heritage Site of Bryggen in Bergen 2015*; Rytter, J., Schonhowd, I., Eds.; Riksantikvaren: Bergen, Norway, 2015; pp. 76–91.

18. Rytter, J.; Schonhowd, I. Operation Groundwater Rescue. In *Monitoring, Mitigation, Management: The Groundwater Project—Safeguarding the World Heritage Site of Bryggen in Bergen 2015*; Rytter, J., Schonhowd, I., Eds.; Riksantikvaren: Bergen, Norway, 2015; pp. 50–59.

19. Christensson, A.; Paszkowski, Z.; Spriggs, J.A.; Verhoef, L. (Eds.) Safeguarding Historic Waterfront Sites. In *Bryggen in Bergen as a Case Study*, 1st ed.; Stiftelsen Bryggen and Polytechnika Szczecinska: Bergen, Norway, 2004.

20. De Beer, J.; Matthiesen, H. Groundwater monitoring and modelling from an archaeological perspective: Possibilities and challenges. In *Geology for Society*; Slagstad, T., Ed.; Geological Survey of Norway Special Publication: Trondheim, Norway, 2008; Volume 11, pp. 67–81.

21. Rytter, J.; Schonhowd, I. (Eds.) *Monitoring, Mitigation, Management: The Groundwater Project—Safeguarding the World Heritage Site of Bryggen in Bergen 2015*; Riksantikvaren: Bergen, Norway, 2015; p. 213. ISBN 978-82-7574-085-2.

22. De Beer, J.; Seither, A.; Vorenhout, M. Effects of a New Hdrological Barrier on the Temperatures in the Organic Ardheaeological Remains at Bryggen in Bergen, Norway. *J. Conserv. Manag. Archaeol. Sites* **2016**, *18*, 99–115. [CrossRef]

23. Nestingen, R.S. The Comparison of Infiltration Devices and Modification of the Philip-Dunne Permeameter for the Assessment of Rain Gardens. Master's Thesis, Department of Civil Engineering, University of Minnesota, Minneapolis, MN, USA, 2007.

24. Asleson, B.C.; Nestingen, R.S.; Gulliver, J.S.; Hozalski, R.M.; Nieber, J.L. Performance Assessment of Rain Gardens. *JAWRA J. Am. Water Resour. Assoc.* **2009**, *45*, 1019–1031. [CrossRef]

25. American Society for Testing and Materials ASTM D3385-09. *Standard Test Method for Infiltration Rate of Soils in Field Using Double-Ring Infiltrometers*; ASTM: West Conshohocken, PA, USA, 2009.

26. Boogaard, F.C.; Lucke, T.; Van De Giesen, N.; Van De Ven, F. Evaluating the Infiltration Performance of Eight Dutch Permeable Pavements Using a New Full-Scale Infiltration Testing Method. *Water* **2014**, *6*, 2070–2083. [CrossRef]

27. Lucke, T.; Boogaard, F.; Van De Ven, F. Evaluation of a new experimental test procedure to more accurately determine the surface infiltration rate of permeable pavement systems. *Urban Plan. Transp. Res.* **2014**, *2*, 22–35. [CrossRef]

28. Boogaard, F.C. Stormwater Characteristics and New Testing Methods for Certain Sustainable Urban Drainage Systems in the Netherlands 2015a. Ph.D. Thesis, Technische Universiteit Delft, Delft, The Netherlands, July 2015; p. 149. [CrossRef]

29. Boogaard, F.C. Stormwater quality and sustainable urban drainage management. In *Monitoring, Mitigation, Management: The Groundwater Project—Safeguarding the World Heritage Site of Bryggen in Bergen 2015b*; Rytter, J., Schonhowd, I., Eds.; Riksantikvaren: Bergen, Norway, 2015; pp. 136–149.

30. Boogaard, F.C.; Lucke, T. Long-Term Infiltration Performance Evaluation of Dutch Permeable Pavements Using the Full-Scale Infiltration Method. *Water* **2019**, *11*, 320. [CrossRef]

31. Muthanna, T.M.; Viklander, M.; Thorolfsson, S.T. Seasonal climatic effects on the hydrology of a rain garden. *Hydrol. Process.* **2008**, *22*, 1640–1649. [CrossRef]

32. Paus, K.H.; Braskerud, B.C. Suggestions for designing and constructing bioretention cells for a nordic climate. VATTEN. *J. Water Manag. Res.* **2014**, *70*, 139–150.

33. MNI Metrological Institute of Norway. Available online: www.eKlima.noorwww.yr.no (accessed on 10 November 2018).

34. Ersland, G.A. The history of Bryggen until C. 1900. In *Monitoring, Mitigation, Management: The Groundwater Project—Safeguarding the World Heritage Site of Bryggen in Bergen 2015*; Rytter, J., Schonhowd, I., Eds.; Riksantikvaren: Bergen, Norway, 2015; pp. 12–23.

35. Jensen, J.A. 2015 Subsidence at the Bryggen site. In *Monitoring, Mitigation, Management: The Groundwater Project—Safeguarding the World Heritage Site of Bryggen in Bergen*; Rytter, J., Schonhowd, I., Eds.; Riksantikvaren: Bergen, Norway, 2015.

36. De Beer, J.; Price, S.J.; Ford, J.R. 3D modelling of geological and anthropogenic deposits at the World Heritage Site of Bryggen in Bergen, Norway. *Quat. Int.* **2012**, *251*, 107–116. [CrossRef]

37. Christensson, A.; Rytter, J.; Schonhowd, I. Management History. In *Monitoring, Mitigation, Management: The Groundwater Project—Safeguarding the World Heritage Site of Bryggen in Bergen 2015*; Rytter, J., Schonhowd, I., Eds.; Riksantikvaren: Bergen, Norway, 2015; pp. 38–47.

38. Schlumberger Water Services, Diver Manual, Delft. November 2014. Available online: https://usermanual.wiki/Datasheet/SchlumbergerDiverCompleteSpecs.428302045.pdf (accessed on 1 October 2018).

39. Boogaard, F.; Wentink, R.; Vorenhout, M.; De Beer, J. Implementation of Sustainable Urban Drainage Systems to Preserve Cultural Heritage—Pilot Motte Montferland. *Conserv. Manag. Archaeol. Sites* **2016**, *18*, 328–341. [CrossRef]

40. De Beer, J.; Boogaard, F. Good practices in cultural heritage management and the use of subsurface knowledge in urban areas. *Procedia Eng.* **2017**, *209*, 34–41. [CrossRef]

41. Woods Ballard, B.; Wilson, S.; Udale-Clarke, H.; Illman, S.; Scott, T.; Ashely, R.; Kellagher, R. *CIRIA—The SuDS Manual*; CIRIA Research Project 2015, (RP)992; Department for Environment Food & Rural Affairs: London, UK, 2015; ISBN 978-0-86017-760-9. Available online: https://www.ciria.org/ItemDetail?iProductCode=C753&Category=BOOK&WebsiteKey=3f18c87a-d62b-4eca-8ef4-9b09309c1c91 (accessed on 1 September 2017).

42. CRC (Cooperative Research Centre for Water Sensitive Cities). *Adoption Guidelines for Stormwater Biofiltration Systems*; CRC: Clayton, Australia, 2015; Available online: https://watersensitivecities.org.au/wp-content/uploads/2016/06/Adoption-Guidelines-for-Stormwater-Biofiltration-Systems-Chapter-1.pdf (accessed on 14 December 2020).

43. Minnesota Pollution Control Agency (MPCA). *Minnesota Stormwater Manual*; Minnesota Pollution Control Agency (MPCA): St. Paul, MN, USA; Melbourne, Australia, 2008. Available online: https://www.pca.state.mn.us/water/minnesotas-stormwater-manual (accessed on 14 December 2020).

44. Prince George's County, Maryland (PGCM). Bioretention Manual. In *Environmental Services Division*; Department of Environmental Resources, Prince George's County: Upper Marlboro, MD, USA, 2007. Available online: https://www.princegeorgescountymd.gov/Government/AgencyIndex/DER/ESG/Bioretention/pdf/Bioretention%20Manual_2009%20Version.pdf (accessed on 14 December 2020).

45. Paus, K.H.; Morgan, J.; Gulliver, J.S.; Leiknes, T.; Hozalski, R.M. Assessment of the Hydraulic and Toxic Metal Removal Capacities of Bioretention Cells after 2 to 8 Years of Service. *Watern Air Soil Pollut.* **2013**, *225*, 1–12. [CrossRef]

46. Vannlaboratoriet Bergen Vann KF. Analysis for Soil Moisture. September 2017. Available online: http://www.bergenvann.com/laboratoriet/om-vannlaboratoriet/ (accessed on 1 October 2017).

47. Ahmed, F.; Gulliver, J.S.; Nieber, J.L. Field infiltration measurements in grassed roadside drainage ditches: Spatial and temporal variability. *J. Hydrol.* **2015**, *530*, 604–611. [CrossRef]

48. Camuffo, D.; Della Valle, A.; Becherini, F. A critical analysis of the definitions of climate and hydrological extreme events. *Quat. Int.* **2020**, *538*, 5–13. [CrossRef]

49. Palhegyi, G.E. Designing storm-water controls to promote sustainable ecosystems: Science and application. *J. Hydrol. Eng.* **2010**, *15*, 504–511. [CrossRef]

50. Bettess, R. *Infiltration Drainage-Manual of Good Practice*; CIRIA R156 1996 CIRIA: London, UK, 1996; ISBN 978-0-86017-457-8.

51. BRE Digest 365. *Soakway Design*; Buildings Research Establishment: Bracknell, UK, 1991; ISBN 0-85125-502-7.

52. DIN 19682-7. *Soil Quality—Field Tests—Part 7: Determination of Infiltration Rate by Double Ring Infiltrometer 2015*; German Institute for Standardization: Berlin, Germany, 2015.

53. Lucke, T.; Beecham, S. Field investigation of clogging in a permeable pavement system. *Build. Res. Inf.* **2011**, *39*, 603–615. [CrossRef]

54. Fassman, E.; Blackbourn, S. Urban Runoff Mitigation by a Permeable Pavement System over Impermeable Soils. *J. Hydrol. Eng.* **2010**, *15*, 475–485. [CrossRef]

55. Pezzaniti, D.; Beecham, S.; Kandasamy, J. Influence of clogging on the effective life of permeable pavements. *Proc. Inst. Civ. Eng. Water Manag.* **2009**, *162*, 211–220. [CrossRef]

56. Venvik, G.; Bang-Kittilsen, A.; Boogaard, F.C. Risk assessment for areas prone to flooding and subsidence: A case study from Bergen, Western Norway. *Hydrol. Res.* **2019**, *51*, 322–338. [CrossRef]
57. Boogaard, F.C.; Kluck, J.; Bosscher, M.; Schoof, G. Flood model Bergen Norway and the need for (sub-)surface INnovations for eXtreme Climatic EventS (INXCES). *Procedia Eng.* **2017**, *209*, 56–60. [CrossRef]
58. Available online: https://climatescan.nl/projects/16/detail (accessed on 1 December 2020).

 *land*

Article

Potentials and Pitfalls of Mapping Nature-Based Solutions with the Online Citizen Science Platform ClimateScan

Britta Restemeyer [1,*] and Floris C. Boogaard [1,2]

1   Hanzehogeschool Groningen, University of Applied Sciences, Zernikeplein 7, P.O. Box 30030 Groningen, The Netherlands; floris@noorderruimte.nl
2   Deltares, Daltonlaan 600, 3584 BK Utrecht Postbus, P.O. Box 85467 Utrecht, The Netherlands
*   Correspondence: b.restemeyer@pl.hanze.nl

Received: 30 November 2020; Accepted: 19 December 2020; Published: 23 December 2020

Abstract: Online knowledge-sharing platforms could potentially contribute to an accelerated climate adaptation by promoting more green and blue spaces in urban areas. The implementation of small-scale nature-based solutions (NBS) such as bio(swales), green roofs, and green walls requires the involvement and enthusiasm of multiple stakeholders. This paper discusses how online citizen science platforms can stimulate stakeholder engagement and promote NBS, which is illustrated with the case of ClimateScan. Three main concerns related to online platforms are addressed: the period of relevance of the platform, the lack of knowledge about the inclusiveness and characteristics of the contributors, and the ability of sustaining a well-functioning community with limited resources. ClimateScan has adopted a "bottom–up" approach in which users have much freedom to create and update content. Within six years, this has resulted in an illustrated map with over 5000 NBS projects around the globe and an average of more than 100 visitors a day. However, points of concern are identified regarding the data quality and the aspect of community-building. Although the numbers of users are rising, only a few users have remained involved. Learning from these remaining top users and their motivations, we draw general lessons and make suggestions for stimulating long-term engagement on online knowledge-sharing platforms.

Keywords: nature-based solutions; online climate adaptation platforms; citizen science; community-building

---

1. Introduction

Online knowledge-sharing platforms could potentially contribute to an accelerated uptake of nature-based solutions (NBS). NBS are considered to be a promising means to address climate change by promoting more green and blue spaces in urban areas. NBS usually have multiple benefits: they increase a city's biodiversity, contribute to more healthy and resilient ecosystems, and provide benefits for health and human well-being [1–3]. However, while advocacy for NBS in the scientific literature is on the rise, implementation in practice is often hampered [4]. Common barriers to implementing NBS are the fear of unknowns (regarding implementation, maintenance, and performance of NBS) and a general lack of understanding the multiple benefits of NBS [4,5]. Moreover, many NBS require the involvement and enthusiasm of a multitude of stakeholders [6]. This is particularly true for small-scale NBS such as bio(swales), green roofs, and green walls, which can be implemented in urban neighbourhoods even if space in a city is scarce. Next to their aesthetical value, they simultaneously help to regulate water flow, prevent floods, and reduce heat stress. However, they often need to be implemented on private land, requiring non-governmental actors such as individual house owners, housing corporations, and property owners of business parks to take action.

To build momentum and foster implementation of NBS, Kabisch et al. (2016) [4] have pointed out that it is important to "learn from action that is already taking place" and share existing approaches and experiences among different countries. Everywhere in the world, stakeholders are experimenting with NBS and climate adaptation. However, many of these experiments are on such a small-scale that they do not reach beyond their regional boundaries.

An online platform could be an ideal tool to increase international knowledge exchange and raise awareness, with the ultimate goal that a good example of an affordable and well-functioning NBS can help actors in other places to move forward in the implementation process. Especially during the current Covid-19 pandemic, we have all come to realize the importance and the opportunities of being connected through the internet. The internet offers the advantage that it is accessible to anyone (with an internet connection) at any time [7,8]. This means that an online platform could potentially reach a multitude of stakeholders: those that already have a say in climate adaptation and water governance, and those who were not involved thus far. Many of these platforms exist, the majority of which have been established with the (financial) support of funding agencies and governments. Examples are Climate-ADAPT, OPPLA, NATURVATION Urban Nature Atlas, BISE, DRMKC, Natural Hazards NBS, NBS Initiative, NWRM, PANORAMA, ThinkNature and weADAPT. However, none of these platforms have a citizen science interactive mapping function.

ClimateScan (www.climatescan.org), the case in this paper, is built on such an interactive mapping function using citizen science. The platform has been set up without any external resources and using only voluntary efforts. It was originally developed by Hanze University of Applied Sciences (HUAS) in 2014 to show the locations of Sustainable Urban Drainage Systems (SUDS) in The Netherlands and publish the research results of functionality tests carried out on these locations. In 2016, the platform has been opened up to the public so that anyone could contribute existing best practices of nature-based solutions and climate adaptation on an interactive map.

This citizen science approach was chosen for two reasons. First, it increases the geographic range of the ClimateScan database. Second, mapping our own examples is in line with other web 2.0 ideas that makes users active knowledge producers instead of passive knowledge consumers [9]. This can increase the "fun factor" and the commitment of users and thereby help to build a more empowered and more sustainable community of practice. Today, the platform has over 800 registered users, which have uploaded more than 5000 projects assigned to seven main themes (Water, People, Nature, Heat, Energy, Urban Agriculture and Air Quality) all around the globe.

In this paper, we discuss how online citizen science platforms such as ClimateScan can stimulate stakeholder engagement and promote NBS. ClimateScan is a unique case because it has adopted an innovative bottom–up approach. This bottom–up approach entails that ClimateScan is independent from the (financial) support of funding agencies and governments, and that there has not been a top–down induced set of rules about what should be uploaded in what way from the start. Accordingly, the platform relies strongly on the self-organizing capacity of the actual users.

One of the authors has been involved with ClimateScan from the outset and throughout. As a result of this, we were able to acquire in-depth and inside knowledge around both the development of the content of the platform and the development of the community behind the platform. We have supplemented these participatory observations with a content analysis of the ClimateScan database and statistical analyses stemming from Google Analytics to analyze what has been uploaded, by whom, and how the content and the community has developed over time. While there have been evaluations of the actual usage and contribution of online decision support tools cf. [9–11], there has not been an evaluation of an online adaptation platform using citizen science. Thereby, we add novel insights regarding the potentials and pitfalls of online citizen science platforms and how they can contribute to climate adaptation and NBS implementation practice.

In the following, we will first present the current state of knowledge regarding the potentials and pitfalls of online adaptation platforms to identify research avenues in the current debate (Section 2). Based on these research avenues, we have designed our empirical strategy for evaluating the

ClimateScan platform. We will explain this methodological approach in Section 3. Section 4 provides more background information regarding the history of ClimateScan, its key design parameters, and how it uses online and offline channels to acquire new users. Section 5 will present the results of our analysis regarding the content and the community of ClimateScan. In Section 6, we will provide concluding remarks on how ClimateScan and similar initiatives could be sustained, with an eye on improving its flaws, without doing harm to its strengths.

2. Online Adaptation Platforms—What We Know So Far

Climate adaptation is eminently a field where Fischer's request to rethink expertise and his plea to build more cooperative relations between experts and lay citizens applies [12,13]. This is simply because climate adaptation requires a rather radical transformation of the built environment, for which agreement and action of many different actors, often non-experts, is needed [6]. With democratizing knowledge about climate adaptation, we mean that knowledge about climate adaptation is made accessible to all actors from the quadruple helix (academia, government, business, and civil society), enabling them to negotiate about goals and how to achieve them [14,15].

A key consideration must be that the adaptation process is often undertaken by people for whom climate change is not a primary concern [16]. Eventually, the success of the world's adaptation efforts will depend on "ordinary" citizens, housing associations, business park owners, and (municipal) civil servants who integrate climate adaptation into the (re-)design of streets, buildings, gardens, parking lots, and other components of our built environment. For them, climate adaptation often has been and still is a voluntary add-on on top of their usual responsibilities. Palutikof et al. (2019) [7] (p. 461) have already pointed out that knowledge adaptation platforms can be a useful and time-saving tool to these typical adaptors, as they can serve as "a comprehensive resource equipping decision-makers with the data, tools, guidance and information needed to adapt to a changing climate". Such platforms are usually online resources, making use of the several benefits the internet offers, which can help to democratize knowledge about climate adaptation.

The first advantage of online platforms is inclusivity: anyone with an internet access can participate, access is free of charge (except for the internet fee itself), and there is no need to be at a specific time at a specific place [8–10]. Second, the internet offers the opportunity to make use of strong visual powers. Videos and photos next to pure text can help in communication [9]. Such visual content is more universal than written text, which can be particularly useful in a field such as climate adaptation where terminology varies strongly between countries and also changes quickly over time (see [17] for an overview of changing terminology in the field of urban stormwater management). Third, web 2.0 developments have made it possible to create interactive platforms, where people can meet and exchange ideas. This increases opportunities to broaden participation and build communities of practice for active knowledge exchange and capacity-building [7,9,10].

However, there are also some common downsides to online adaptation platforms. The downsides of online adaptation platforms are often connected to diverting expectations of users, builders, and funders. A typical problem is that online platforms relate to a specific project and funding horizon; once the project is over and funding stops, information is not updated anymore, and information will ultimately be out of date. Moreover, many platforms need a long time to be designed and built (on average 3-4 years), especially if they aim to function as decision support tools [10]. This can be problematic because the institutional context may have changed by the time the website is up in the air [10]. Particularly in a field such as climate adaptation, regulatory and legislative frameworks are very fluid and can change easily [18]. Users' needs are often not well-understood and hardly evaluated [10]. In one of the few studies focusing on the users' perspective, Hewitson et al. (2017) [11] (p. 16) come to the conclusion that "all climate information websites 'grossly overestimate the ease of use'".

A second pitfall relates to the interactive element of online platforms and the strive to build an online community of practice. Building and maintaining such a community is very resource-intensive, as it requires getting to know the users, engaging with them directly, and regularly tracking how this use is evolving [9]. Commonly, online platforms aim at involving practitioners to make the platform's content more relevant to a wide audience, thereby increasing the frequency of visits and active engagement of users. However, as pointed out above, practitioners often lack the time to actively engage on such a platform because of competing demands. Research has shown that offline interactions next to online activities help in advertising the platform, building user confidence and trust in its content, and facilitating the co-production of knowledge [9,10].

A third pitfall relates to the aspect of inclusivity. The inclusivity of online platforms is based on the idea that everyone has access to the internet. However, access to the internet varies in different geographical contexts. Broadly speaking, only one-third of the people in developing countries are online compared to two-thirds in the developed world, and even in developed countries, internet access is much better in urban areas than in rural areas [9,19–21]. The difference in internet access to climate adaptation platforms between developed and developing countries is particularly striking, bearing in mind that developing countries are likely to suffer more from climate change than developed countries [6,9,22]. Another important point relates to the actual usage and users of climate adaptation platforms. In one of the few studies carried out about the users of what they call "climate knowledge brokering platforms", Hammill et al. (2013) [9] conclude that research-oriented users dominate on these platforms, and that policy-makers, media representatives, and local level actors are often not actively engaging with these platforms. Therefore, Hammill et al. (2013) [9] (p. 90) warn that online adaptation platforms should not become "essentially online spaces established and managed by researchers for researchers in relatively privileged settings".

To conclude, online adaptation platforms are generally held to be a promising tool for knowledge exchange, raising awareness, and building capacity. However, the literature review above also shows that there are three main concerns, and hence research avenues, related to online platforms: (1) Many online platforms are often only "active" for a certain period (i.e., the timeframe of a certain project and related funding), thereby running the risk that the content becomes outdated. (2) Little is known about the actual contributors and users of online platforms and how inclusive these platforms actually are. (3) Creating and maintaining a well-functioning community of practice is a common goal, but it is difficult to achieve. In this paper, we will shed light on these points by turning to the special case of ClimateScan. The bottom–up approach that ClimateScan has adopted offers us the opportunity to analyze what happens if you enable users to create and update content. More specifically, we can analyze what is being uploaded, who contributes to the platform for what reason, and how this could shape an active community of practice.

3. Methodological Approach to Evaluate the ClimateScan Platform

The research avenues identified above have informed a two-fold evaluation of the ClimateScan platform, analyzing (1) the content of the platform and how it has evolved over time, and (2) the users and the community of practice and how this developed over time.

For this analysis, we have applied multiple qualitative methods. First and foremost, we draw on participatory observations from one of the authors, who has initially set up the platform and remained involved throughout. These participatory observations stem from the author's function as a moderator of the website and direct interactions with ClimateScan users. The personal relations with the ClimateScan top users (those who have uploaded 10 or more projects) have helped us to classify which sector they come from (academia, government, business, civil society) and what drives these top users to contribute. That way, we could reflect on the inclusivity of the ClimateScan platform and users' motivations to participate in the ClimateScan community. To ensure reflexivity, the other author, who has not been involved from the beginning, interviewed the participating author on several occasions.

To further substantiate the observations, we have used two additional types of data. The first one is the ClimateScan database itself, which provides information about what has been uploaded, by whom, and when. A content analysis of the database has enabled us to distill an overview of the types of NBS that are uploaded and their geographical distribution. Moreover, by creating user profiles analyzing who has uploaded how many projects over time, we could assess the level of involvement of participants. Assessing the level of involvement of participants was important to draw preliminary conclusions on the aspect of community-building, which were then supplemented by the participatory observations mentioned above.

Last but not least, we have used data from Google Analytics to better understand how many people visit ClimateScan and how this has developed over time. That way, we could assess the overall popularity of the website and which factors have increased its usage. Moreover, with the data from Google Analytics, we could generate an overview of the projects that are visited most on the ClimateScan platform. By analyzing what these projects have in common, we could better understand what makes a project appealing to a wider audience. As a corollary, we gained insights for the further development and improvement of ClimateScan.

4. An Introduction to ClimateScan

ClimateScan's history can be described as an evolution from a research-driven website about Sustainable Urban Drainage Systems to a platform where anyone can register and contribute examples of nature-based solutions on an interactive map. ClimateScan was originally set up in 2014 in the context of a PhD thesis called "Stormwater characteristics and new testing methods for certain sustainable drainage systems in The Netherlands" [23]. In order to make the research transparent and publicly available, the website showed the locations of the NBS that have been studied alongside photos and video footage of a new testing method ("full-scale testing", see [24]) which had been used to study the performance of these specific NBS after they have been in place for a few years. This information was shared with other international researchers as a means to achieve 'international knowledge exchange' (Boogaard, 2015: 119). When the website gradually received more attention and positive feedback from practitioners and other researchers, the recognition grew that ClimateScan could be a powerful tool to showcase also other best practices of climate adaptation and urban resilience. This has led to an opening of the website in 2016, so that anyone could register, download an app, and contribute examples of "blue-green" projects around the globe.

ClimateScan is not the only platform with such an interactive mapping function (other examples in the same field are ClimateADAPT, OPPLA, and NATURVATION). However, ClimateScan is unique in its organization and focus. Organizationally, ClimateScan has been developed in a "learning by doing way". The platform started out without a clear set of rules about what should be uploaded and how it should be described. Instead, suggestions made by users have been used to improve the platform.

To give an example, in the beginning, the entry of a data point could only be assigned to one category (e.g., "Water"). Nowadays, a datapoint can be assigned to several categories (e.g., "Water" and "Heat", which better shows the multiple benefits of NBS. Overall, the focus of ClimateScan is rather small-scale and design-oriented. Many examples on ClimateScan refer to a specific object, e.g., the design of a roundabout with a water storage function. The focus on design implies that ClimateScan does not provide detailed decision support but rather serves as a source of inspiration. ClimateScan showcases many design examples that lend themselves to be easily copied elsewhere. The addition of the function "show related projects" implies that a visitor interested in one example can immediately find similar examples. As such, ClimateScan attempts to be a "solutions broker" [25]. Moreover, because of its origin as a research-driven website, ClimateScan links to location-specific research outcomes wherever possible [26].

Since 2014, several channels—online and offline—have been used to promote and disseminate ClimateScan (see Table 1). ClimateScan makes use of social media, has been linked to other climate adaptation portals, and uses "offline" events called ClimateCafé's to "recruit" new users. ClimateCafés can be characterized as pressure-cooker workshops in which young professionals and practitioners come together for a few days to gather factual data about the vulnerabilities and (potential) solutions of a defined urban or rural area. ClimateScan is a fixed element in each ClimateCafé to scan the study area for existing nature-based solutions. Every ClimateScan workshop results in more datapoints on the map and further stimulates discussions on the type of nature-based solution, its benefits and local characteristics, and governance aspects related to ownership, performance, and maintenance (for an example, see [27]).

Table 1. Promotion and dissemination channels of ClimateScan.

Offline	ClimateCafés	Used in education and (inter)national projects such as WaterCoG, INXCES, IMPETUS, RAAK Infiltrerende Stad and RAAK GroenBlauw oplossingen. for more info visit: https://climatecafe.nl/
Online	**Social media**	
	YouTube	https://www.youtube.com/channel/UCKP6mYu_5ez9gGeL2mLy8XQ
	Twitter	https://twitter.com/Climate_Scan
	Facebook	https://www.facebook.com/climateScanNL/
	Instagram	https://www.instagram.com/climatescan/
	Links on other platforms	
	Delta Programme	https://ruimtelijkeadaptatie.nl/aan-de-slag/docent-student/lesmateriaal/websites/
	Climate Initiative Northern Netherlands	https://climateinitiativenoordnederland.nl/nl/projecten/climatescan-nl/
	Climate Adapt (European portal)	https://climate-adapt.eea.europa.eu/metadata/portals/climate-scan
	Catchment-Based approach (UK portal)	https://catchmentbasedapproach.org/learn/climate-scan/

5. Results from the ClimateScan Evaluation

5.1. Content of ClimateScan: From (Bio)swales to Tornado Trails

A general overview of the interactive map shows that datapoints have been submitted on every continent, which makes ClimateScan a global platform (see Figure 1). However, due to its origin in The Netherlands, most projects have been submitted in The Netherlands and Western Europe. Other hotspots are Australia, South Africa, and the Philippines. While the magnitude of datapoints in South Africa and the Philippines was actively stimulated through international projects and ClimateCafés taking place there, the Australian hotspot can be explained by a more coincidental meeting. Through presentations about ClimateScan on international conferences, contact was established with someone in Australia who has already had a personal database with Australian projects about rainwater harvesting and bioretention, which he then uploaded to ClimateScan. This coincidental meeting has actually resulted in mutual visits and fieldtrips to see Dutch and Australian nature-based solutions. Although anecdotal, it shows that the idea of mapping nature-based solutions can unite people and strengthen knowledge exchange.

Analyzing the types of nature-based solutions that have been submitted (Figure 2) shows that the biggest categories are bioswales, green roofs and walls, and the Australian measures mentioned above. Bioswales are unsurprisingly the biggest category, as this is the type of measure ClimateScan started with (see history) and is a key interest of the founder of the website and his network. A key design principle of ClimateScan is to be as open as possible, which also means that users can create their own categories. The choice for openness over data accuracy has been made consciously, as ClimateScan is more about stimulating people to upload examples instead of creating a perfect database. As mentioned, ClimateScan has a bottom–up approach and offers the opportunity to analyze what happens if you enable users to create and update content, which is illustrated in the next paragraphs.

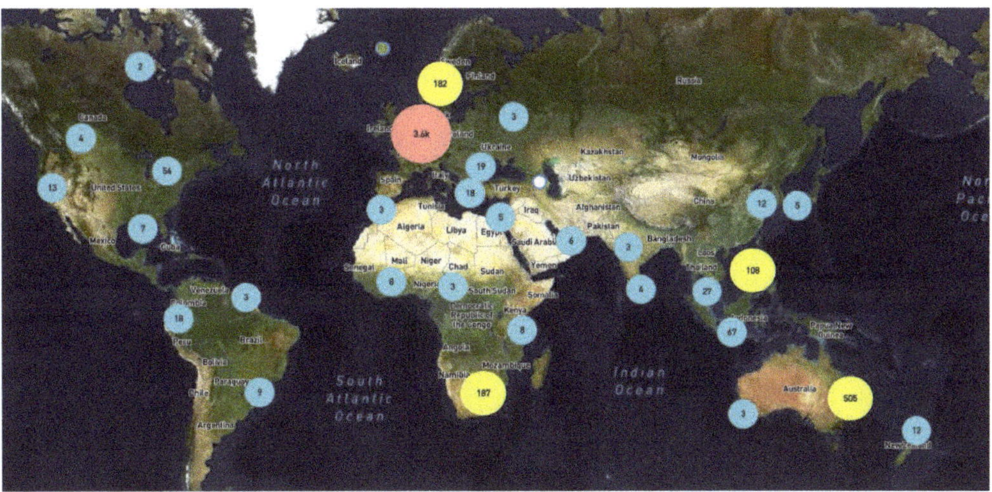

Figure 1. Geographical distribution of ClimateScan datapoints (source: climatescan.org).

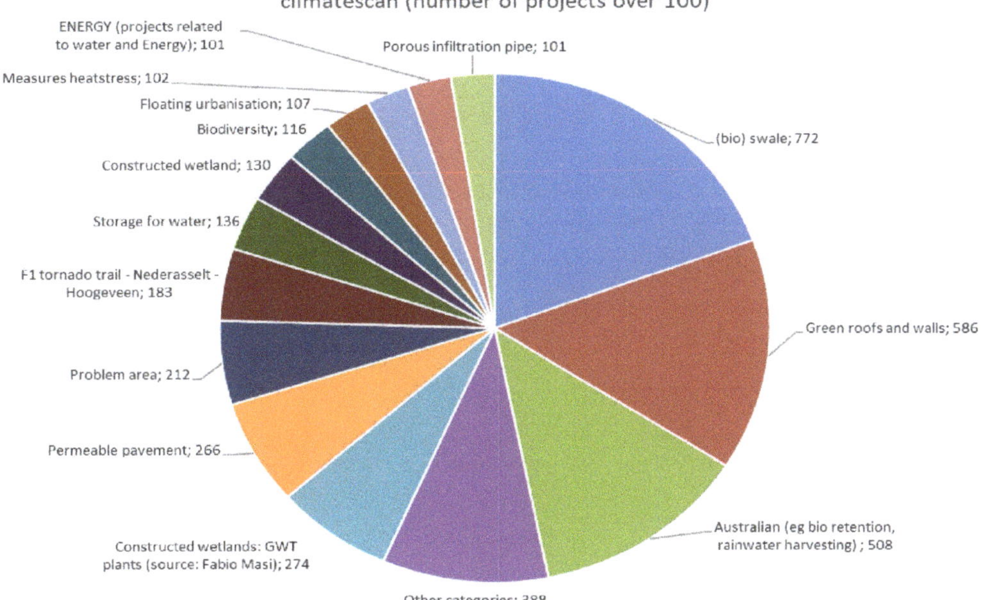

Figure 2. Overview of nature-based solutions uploaded to ClimateScan, to keep the figure readable only categories with more than 100 projects are depicted (based on ClimateScan database).

Moreover, the openness also has the advantage that it leads to surprising content. For example, a more recent example is the category "tornado trails", where users follow the traces of storms and the damage they leave in the landscape in The Netherlands. Looking at the Dutch content in particular also shows that the content uploaded on ClimateScan evolves with the climate adaptation debate. In the beginning, most measures uploaded were related to water storage, as flooding has always been a major concern in The Netherlands. Later, the positive impact of nature-based solutions such as green roofs and green walls on heat stress was emphasized more strongly, as The Netherlands suffered from several heatwaves in a row (visible in the naming and categorization of projects, the request by users to be able to assign more than one category to an example). The new category "tornado trails" shows that storms and wind have gained importance in the debate.

The list with the top 15 most visited projects on the ClimateScan platform (see Table 2) includes some very known examples (e.g., waterquares in Rotterdam), but also some rather unknown examples (e.g., swale in Dalfsen). Projects that are uploaded for a longer period have a higher change to be in the top 15 such as the Swale in Haren that was the second project uploaded to the database (last column of Table 2). Hence, ClimateScan also provides a platform for small-scale initiatives and municipalities that usually do not get so much attention. All the projects have in common that they have good visual content, mostly with photos and video footage. For example, the most visited project, the gully free road in the small town Nieuwleusen, includes a video about how the system functions during a heavy rain event (T100).

To sum up, over the years, ClimateScan had to find a balance between being open and low-threshold to stimulate the upload of examples on the one hand and data maturity on the other. Sometimes, datapoints only include a short text description or only a photo. There is a check on the data quality by the admins, scanning aerial photos and searching the web to validate the datapoint and accompanying information. If possible and in reach, the datapoint is also being visited. In recent years, only few datapoints had to be removed because they included false information or were not meant seriously.

5.2. Users of ClimateScan: Many Registered Users, But Only a Small, Yet Diverse Active Community

Over the years, the attention for ClimateScan has grown, as the timeline with the number of visitors per day shows (see Figure 3). The platform receives on average 100 visits per day, with the highest peak of 1326 (unique) visitors on one day in February 2019 (a day with a university workshop where students were giving an assignment to use ClimateScan as a source for NBS).

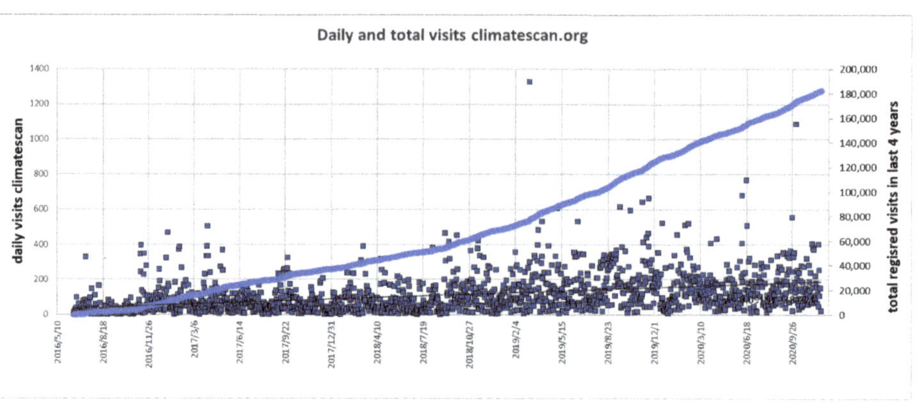

Figure 3. Daily and total visits of ClimateScan (based on data from Google Analytics).

Table 2. Top 15 most visited projects on the ClimateScan platform (based on project site visits retrieved from Google Analytics).

	Name and Explanation of NBS	Illustration [Source: Climatescan.org]	Site Visits	Location	Small Municipality	Good Visuals (Photos and or Video)	Project Number and Hyperlink
1.	Swale combined with a playground function for children in Dalfsen		962	Dalfsen (NL)	yes	yes	935
2.	Gully free road in Nieuwleusen, caught on camera during an intensive rainfall event (T100)		635	Nieuwleusen (NL)	yes	yes	1113
3.	Green infiltration zone/rainwater-garden in Amsterdam, one of the first raingardens in Amsterdam		604	Amsterdam (NL)	no	yes	921
4.	Gully free road, bioswale and permeable pavement in Alkmaar		541	Alkmaar (NL)	yes	yes	9
5.	Watersquare in s'Hertogenbosch, flooded for performance research ('full scale test')		540	Den Bosch (NL)	yes	yes	177
6.	Sustainable water management in UNESCO cultural heritage area Bryggen, including a 'treatment train' infiltration of stormwater with swales, raingardens, sub surface infiltration units and permeable pavement		304	Bergen (Norway)	no	yes	16

Table 2. *Cont.*

	Name and Explanation of NBS	Illustration [Source: Climatescan.org]	Site Visits	Location	Small Municipality	Good Visuals (Photos and or Video)	Project Number and Hyperlink
7.	Swales in Paddepoel Groningen, research conducted on the topsoil quality using XRF		262	Groningen (NL)	no	yes	2528
8.	Waterstoring roundabout in Huizen, multifunctional water storage		258	Huizen (NL)	yes	yes	2050
9.	Swale at school OBS de Wissel, Haren, research conducted on the hydraulic conductivity of this swale		251	Groningen (NL)	no	no	2
10.	Fieldlab permeable pavement in Grubbenvorst, research conducted on the hydraulic conductivity of this permeable pavement		233	Grubbenvorst (NL)	yes	yes	4380
11.	Watersquare Bellamyplein in Rotterdam, one of the first watersquares in The Netherlands		224	Rotterdam (NL)	no	yes	241
12.	Watersquare Benthemplein in Rotterdam, one of the first watersquares in The Netherlands		222	Rotterdam (NL)	no	yes	240

Table 2. *Cont.*

	Name and Explanation of NBS	Illustration [Source: Climatescan.org]	Site Visits	Location	Small Municipality	Good Visuals (Photos and or Video)	Project Number and Hyperlink
13.	Bioswale, nature-friendly, Harkstraat in Amsterdam, research conducted on the hydraulic conductivity of this swale		220	Amsterdam (NL)	no	yes	2224
14.	Restored wetland in urban setting American International School Johannesburg		216	Johannesburg (South-Africa)	no	yes	2689
15.	Swales Ruwenbos in Enschede, oldest swale in The Netherlands (1997), research conducted on the hydraulic conductivity and removal efficiency of these swales		213	Enschede (NL)	no	yes	211

On 24 September 2020, Climate Scan had 806 registered users, and the numbers keep growing. Some of these users can be subtracted because their registration details show clear signs of bots. Of the remaining 766 registered users, only 233 have actively contributed content to the ClimateScan platform (see Table 3). The data show that people register after ClimateScan has been presented on a conference or a meeting. Hence, the story of the platform seems to spark people's interest, but they feel not involved enough to actually contribute.

Table 3. Level of involvement of registered users (based on data from ClimateScan database).

Amount of Projects Submitted	Amount of Users	
0	533	70%
1–4	171	22%
5–<10	37	5%
10–100	21	3%
100 and more	4	1%

While many users can be linked back to a particular event or project, there are also several users that can be considered unexpected users: for example, a user uploading two green roofs in Moscow, although there have never been any links to Russia. Another unexpected user in The Netherlands contributes a rather unconventional solution from his own property: a trampoline with water storage, with a sketch explaining the technical details of how it works (see Figure 4). The Canadian citizen initiative "Depave Paradise" has reached out through social media and subsequently submitted several projects that they have worked on. These projects concerned a transformation of places dominated by asphalt and stones into green spaces, with a "before" and "after" illustration for each project. The online communication with this initiative revealed that they presented themselves on ClimateScan to get in touch with similar initiatives in Europe.

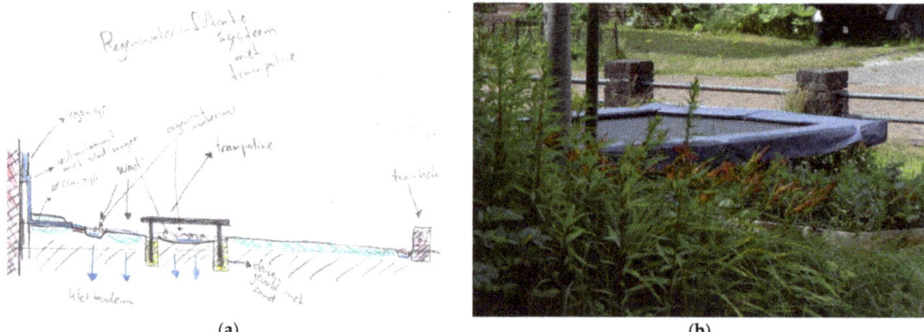

(a) (b)

Figure 4. (**a**) Trampoline with water storage, technical sketch (source: climatescan.org) and (**b**) photo of the actual implementation.

Depave Paradise belongs to the 25 top users that have uploaded 10 or more projects. For the majority of these top users, there has been offline contact first (e.g., through a conference, a project or a meeting), which has stimulated the person to register and contribute to the map. As a result of this offline contact, we could also classify from which sector these top users are coming (see Figure 5).

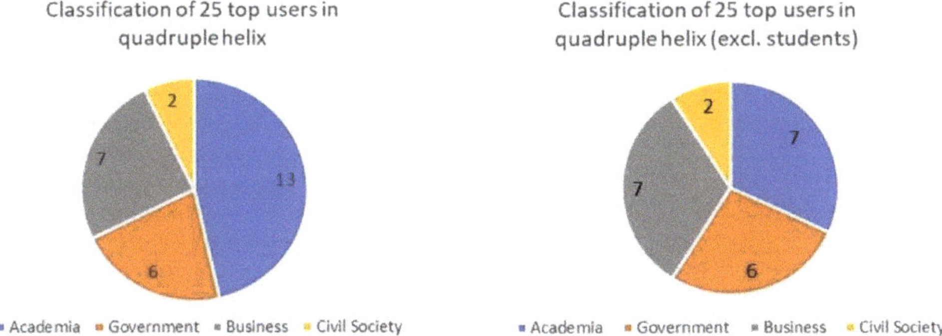

Figure 5. Classification of 25 top users in quadruple helix, with two people having a double function "academia" and "business", and one person having a double function "government" and "business". **Left chart** including students, **right chart** excluding students.

Although most of the users can be related to academia, as Hammill et al. (2013) [9] found in their study, there are also quite a few top users that come from business parties or local governments. The distribution between academia, government and business parties is actually rather even, when students are excluded who usually contribute to ClimateScan as part of an assignment. Only civil society is still rather underrepresented. The low representation of civil society is not surprising, as they have not yet been targeted directly. For business parties, it is an opportunity to present the exact locations of their own innovations in the field (e.g., water harmonica or a particular permeable pavement). Interestingly, only two of the seven top users on ClimateScan coming from business parties use ClimateScan as such; the others also upload other examples simply because they are interested in the topic and experience it as fun to share these examples with a wider audience. The representatives from government mostly come from municipalities and like to share examples of nature-based solutions and climate adaptation from their own city (e.g., Nijmegen, Groningen, and Enschede). While motives such as pride and a certain element of competitiveness certainly play a role, we also know from personal conversations with representatives from municipalities in The Netherlands that ClimateScan gives them the opportunity to create an overview of the nature-based solutions in their municipality. Often, municipalities in The Netherlands do not have such an overview, because measures such as NBS cannot easily be included in the usual sewer asset management software, which is being used to calculate storage capacity and floodings in the urban areas. Therefore, local government officials also use ClimateScan as a free data repository for themselves and in order to draw more attention to these nature-based solutions.

The ClimateScan community that is active and shows a high level of commitment is small yet diverse in their backgrounds. ClimateScan has contributed to intensified contacts between some of the top users. For example, some of the top users have actually made appointments to go "treasure hunting" on the weekends, in which they try to find new and innovative examples of climate adaptation.

6. Discussions and Conclusions

This paper addresses how online knowledge-sharing platforms can stimulate stakeholder engagement and promote NBS by providing in depth insights into the case of ClimateScan—an innovative, bottom–up approach to map best practices of NBS around the world using citizen science.

Based on an analysis of the content, the actual users, and the community behind the ClimateScan platform, we have identified the following potentials and pitfalls of ClimateScan.

A typical concern related to online platforms is that many online platforms are only managed actively for a certain period of time and content becomes quickly outdated [10]. A strength of ClimateScan's citizen science approach is that the database and user numbers only keep growing. Moreover, the content of the platform is "adaptive" and does not easily run the risk of being outdated, as users themselves create and update the content of the website with concrete examples of NBS. Since concrete examples are a manifest of their time, the website can also grow with changing terminology or new developments in the debate (see example "tornado trails"). However, the freedom that is given to users in the way they describe and illustrate examples also leads to varying data maturity and a rather heterogeneous database. Some data points are filled with much information and visual materials, whereas others only contain a dot on the map and a rough description. Although some variation in data maturity will always remain with a citizen science approach, better guidance on the website in the form of video tutorials as well as the training of volunteers would improve the overall data quality. To create a more homogeneous and neater dataset, this training could find inspiration in existing classifications of NBS [1–3]. As visual information seems to be most appealing, the addition of photos and video material could be made a requirement during the data submission process.

Another concern related to online adaptation platforms refers to the actual users and the inclusivity of online adaptation platforms. Online platforms often tend to have an overrepresentation of researchers and the Global North [9]. ClimateScan certainly shows similar characteristics, with a dominance of uploaded projects from The Netherlands and Western Europe due to its origin and active promotion in The Netherlands. However, looking at the top users and their background shows that ClimateScan is appealing not only to researchers but also representatives from business parties and local governments. ClimateScan has managed to give a platform to small-scale initiatives that are otherwise undocumented and receive hardly any attention. To further increase its usage, ClimateScan should maintain and further build on its "niche function". The focus on small-scale measures such as (bio)swales, green roofs, and green walls and their design makes it low key and easy for registered users to upload examples, as the imagery can speak for itself, and no detailed knowledge of the example is necessary.

Last but not least, online adaptation platforms often struggle with creating and maintaining a well-functioning community of practice [9,10]. The openness of the website and active promotion in online and offline fora has resulted in quite a diverse and practice-oriented user group. However, although quite an amount of people have registered over the years, people seem to lose interest after a while. Engaging with the ClimateScan community with workshops and social media updates could stimulate long-term engagement and recruit new users. Here, we can learn from the motivations of top users, which show that mapping can unite people and that mapping together can be a fun activity. Workshops could address different actor groups. Creating teams and using the element of competitiveness (who finds most nature-based solutions in a short amount of time) could be an idea to grow and foster the ClimateScan community in the future.

Author Contributions: Conceptualization, B.R.; methodology, B.R. and F.C.B.; software, B.R. and F.C.B.; validation, B.R. and F.C.B.; formal analysis, B.R. and F.C.B.; investigation, B.R. and F.C.B.; resources, F.C.B.; data curation, F.C.B.; writing—original draft preparation, B.R.; writing—review and editing, B.R. and F.C.B.; visualization, B.R. and F.C.B.; funding acquisition, F.C.B. All authors have read and agreed to the published version of the manuscript.

Funding: ClimateScan was applied as a tool in the international project WaterCoG (WaterCo-Governance) co-funded by the North Sea Region Programme 2014–2021. Ref: http://waterjpi.eu/jointcalls/joint-call-2015/funded-projects-under-the-2015-water-jpi-joint-call and https://northsearegion.eu/watercog/.

Informed Consent Statement: Informed consent was obtained from all subjects involved in the study.

Data Availability Statement: Data available in a publicly accessible repository. The data presented in this study are openly available in climatescan.org.

Acknowledgments: We would like to thank everyone who has contributed to ClimateScan over the years. Without the effort of these volunteers, ClimateScan would not be such a lively platform. Moreover, we would like to thank Koen Salemink for his helpful suggestions during the writing process.

Conflicts of Interest: The authors declare no conflict of interest.

References

1. Babí Almenar, J.; Elliot, T.; Rugani, B.; Philippe, B.; Navarrete Gutierrez, T.; Sonnemann, G.; Geneletti, D. Nexus between nature-based solutions, ecosystem services and urban challenges. *Land Use Policy* **2021**, *100*, 104898. [CrossRef]
2. Raymond, C.M.; Frantzeskaki, N.; Kabisch, N.; Berry, P.; Breil, M.; Nita, M.R.; Geneletti, D.; Calfapietra, C. A framework for assessing and implementing the co-benefits of nature-based solutions in urban areas. *Environ. Sci. Policy* **2017**, *77*, 15–24. [CrossRef]
3. Raymond, C.M.; Berry, P.; Breil, M.; Nita, M.R.; Kabisch, N.; de Bel, M.; Enzi, V.; Frantzeskaki, N.; Geneletti, D.; Cardinaletti, M.; et al. *An Impact Evaluation Framework to Support Planning and Evaluation of Nature-Based Solutions Projects. Report Prepared by the EKLIPSE Expert Working Group on Nature-Based Solutions to Promote Climate Resilience in Urban Areas*; Centre for Ecology & Hydrology: Wallingford, UK, 2017.
4. Kabisch, N.; Frantzeskaki, N.; Pauleit, S.; Naumann, S.; Davis, M.; Artmann, M.; Haase, D.; Knapp, S.; Korn, H.; Bonn, A.; et al. Nature-based solutions to climate change mitigation and adaptation in urban areas: Perspectives on indicators, knowledge gaps, barriers, and opportunities for action. *Ecol. Soc.* **2016**, *21*, 39. [CrossRef]
5. Nesshöver, C.; Assmuth, T.; Irvine, K.N.; Rusch, G.M.; Waylen, K.A.; Delbaere, B.; Haase, D.; Jones-Walters, L.; Keune, H.; Wittmer, H.; et al. The science, policy and practice of nature-based solutions: An interdisciplinary perspective. *Sci. Total Environ.* **2017**, *579*, 1215–1227. [CrossRef] [PubMed]
6. Global Commission on Adaptation. Adapt Now: A Global Call for Leadership on Climate Resilience. 2019. Available online: https://cdn.gca.org/assets/2019-09/GlobalCommission_Report_FINAL.pdf (accessed on 14 December 2020).
7. Palutikof, J.P.; Street, R.B.; Gardiner, E.P. Decision support platforms for climate change adaptation: An overview and introduction. *Clim. Chang.* **2019**, *153*, 459–476. [CrossRef]
8. Uitermark, J. Longing for Wikitopia: The study and politics of self-organisation. *Urban Stud.* **2015**, *52*, 2301–2312. [CrossRef]
9. Hammill, A.; Harvey, B.; Echeverria, D. Knowledge for action: An analysis of the use of online climate knowledge. *Knowl. Manag. Dev. J.* **2013**, *9*, 72–92.
10. Palutikof, J.P.; Street, R.B.; Gardiner, E.P. Looking to the future: Guidelines for decision support as adaptation practice matures. *Clim. Chang.* **2019**, 643–655. [CrossRef]
11. Hewitson, B.; Waagsaether, K.; Wohland, J.; Kloppers, K.; Kara, T. Climate information websites: An evolving landscape. *Wiley Interdiscip. Rev. Clim. Chang.* **2017**, *8*, 1–22. [CrossRef]
12. Fischer, F. *Democracy and Expertise: Reorienting Policy Inquiry*; Oxford University Press: Oxford, UK, 2009.
13. Fischer, F. *Technocracy and the Politics of Expertise*, 1st ed.; SAGE Publications: London, UK, 1990.
14. Carayannis, E.G.; Campbell, D.F.J. "Mode 3" and "Quadruple Helix": Toward a 21st century fractal innovation ecosystem. *Int. J. Technol. Manag.* **2009**, *46*, 201–234. [CrossRef]
15. Colloff, M.J.; Martín-López, B.; Lavorel, S.; Locatelli, B.; Gorddard, R.; Longaretti, P.Y.; Walters, G.; van Kerkhoff, L.; Wyborn, C.; Coreau, A.; et al. An integrative research framework for enabling transformative adaptation. *Environ. Sci. Policy* **2017**, *68*, 87–96. [CrossRef]
16. Porter, J.J.; Dessai, S. Mini-me: Why do climate scientists' misunderstand users and their needs? *Environ. Sci. Policy* **2017**, *77*, 9–14. [CrossRef]
17. Fletcher, T.D.; Shuster, W.; Hunt, W.F.; Ashley, R.; Butler, D.; Arthur, S.; Trowsdale, S.; Barraud, S.; Semadeni-Davies, A.; Viklander, M.; et al. SUDS, LID, BMPs, WSUD and more—The evolution and application of terminology surrounding urban drainage. *Urban Water J.* **2015**, *12*, 525–542. [CrossRef]
18. Tennekes, J.; Driessen, P.P.J.; van Rijswick, H.F.M.W.; van Bree, L. Out of the Comfort Zone: Institutional Context and the Scope for Legitimate Climate Adaptation Policy. *J. Environ. Policy Plan.* **2014**, *16*, 241–259. [CrossRef]
19. West, S.; Pateman, R. How Could Citizen Science Support the Sustainable Development Goals? In *Policy Brief*; Stockholm Environment Institute: Stockholm, Sweden, 2017; p. 8. Available online: https://www.sei-international.org/mediamanager/documents/Publications/SEI-2017-PB-citizen-science-sdgs.pdf (accessed on 14 December 2020).

20. Mariën, I.; Heyman, R.; Salemink, K.; Van Audenhove, L. Digital by Default: Consequences, Casualties and Coping Strategies. In *Social Inequalities, Media and Communication: Theory and Roots*; Servaes, J., Oyedemi, T., Eds.; Lexington Books: Lanham, MD, USA, 2016; pp. 167–188.

21. Salemink, K.; Strijker, D.; Bosworth, G. Rural development in the digital age: A systematic literature review on unequal ICT availability, adoption, and use in rural areas. *J. Rural Stud.* **2017**, *54*, 360–371.

22. IPCC. *Global Warming of 1.5 °C. An IPCC Special Report on the Impacts of Global Warming of 1.5°C above Pre-Industrial Levels and Related Global Greenhouse Gas Emission Pathways, in the Context of Strengthening the Global Response to the Threat of Climate Change*; IPCC: Geneva, Switzerland, 2018. [CrossRef]

23. Boogaard, F. Stormwater Characteristics and New Testing Methods for Certain SUSTAINABLE urban Drainage Systems in The Netherlands. Ph.D. Thesis, Technische Universiteit Delft, Delft, The Netherlands, July 2015. [CrossRef]

24. Boogaard, F.; Lucke, T. Long-term infiltration performance evaluation of Dutch permeable pavements using the full-scale infiltration method. *Water* **2019**, *11*, 320. [CrossRef]

25. Kluck, J.; Boogaard, F. Climate Resilient Urban Retrofit at Street Level. In *Climate Resilient Urban Areas*; De Graaf, R., Ed.; Palgrave Macmillan: London, UK, 2021; pp. 45–66.

26. Tipping, J.; Boogaard, F.; Jaeger, R.; Duffy, A.; Klomp, T.; Manenschijn, M. Climatescan.nl: The development of a web-based map application to encourage knowledge-sharing of climate-proofing and urban resilient projects. In Proceedings of the Amsterdam International Water Week, Amsterdam, The Netherlands, 2–6 November 2015.

27. Boogaard, F.C.; Venvik, G.; de Lima, R.L.P.; Cassanti, A.C.; Roest, A.H.; Zuurman, A. ClimateCafe: An interdisciplinary educational tool for sustainable climate adaptation and lessons learned. *Sustainability* **2020**, *12*, 3694. [CrossRef]

Publisher's Note: MDPI stays neutral with regard to jurisdictional claims in published maps and institutional affiliations.

MDPI

St. Alban-Anlage 66

4052 Basel

Switzerland

Tel. +41 61 683 77 34

Fax +41 61 302 89 18

www.mdpi.com

Land Editorial Office

E-mail: land@mdpi.com

www.mdpi.com/journal/land

www.ingramcontent.com/pod-product-compliance
Lightning Source LLC
LaVergne TN
LVHW070110170726
843515LV00007B/1651